Insuring Against Climate Change

This book provides one of the first systematic in-depth studies on regional catastrophe risk pools. It explores the various goals of these new financial instruments, illustrating how they function on a conceptual, technical and practical level, and reconstructs their political genesis.

With climate-related disasters increasing in frequency and severity, *Insuring Against Climate Change* explores how affected countries, especially those in the Global South, have increasingly turned to innovative index insurance instruments, as demonstrated by the creation of the Caribbean Catastrophic Risk Insurance Facility (CCRIF), the African Risk Capacity (ARC) and the Pacific Catastrophe Risk Assessment and Financing Initiative Facility (PCRAFI Facility). Scherer scrutinizes the formation of this trend, exploring comparatively the goals, characteristics and histories of these tools, and argues that their attractiveness rests more on political than economic benefits and is, in fact, more supply than demand-driven.

Making a significant contribution to current debates on the opportunities and limitations of what are sometimes described as indirect 'climate risk insurance', this book will be of great interest to political scientists with an interest in insurance instruments and climate-related disaster management politics as well as to practitioners working in the insurance, finance and development sectors.

Nikolas Scherer is an advisor at the German Red Cross. Before joining the Red Cross, he was a project manager at Adelphi, a climate and development policy think tank in Berlin, and research associate at the Hertie School of Governance, Germany. His research interests are disaster risk financing, anticipatory humanitarian action, disaster risk management and climate change adaptation. Scherer holds a PhD in governance from the Hertie School of Governance and a Master's degree in international relations from the University of Warwick, UK.

Routledge Advances in Climate Change Research

Culture, Space and Climate Change
Vulnerability and Resilience in European Coastal Areas
Thorsten Heimann

Communication Strategies for Engaging Climate Skeptics
Religion and the Environment
Emma Frances Bloomfield

Regenerative Urban Development, Climate Change and the Common Good
Edited by Beth Schaefer Caniglia, Beatrice Frank, John L. Knott Jr., Kenneth S. Sagendorf, Eugene A. Wilkerson

Local Activism for Global Climate Justice
The Great Lakes Watershed
Edited by Patricia E. Perkins

Political Responsibility for Climate Change
Ethical Institutions and Fact-Sensitive Theory
Theresa Scavenius

Insuring Against Climate Change
The Emergence of Regional Catastrophe Risk Pools
Nikolas Scherer

Climate Justice and Community Renewal
Resistance and Grassroots Solutions
Edited by Brian Tokar and Tamra Gilbertson

Teaching Climate Change in the United States
Edited by Joseph Henderson and Andrea Drewes

For more information about this series, please visit: https://www.routledge.com/Routledge-Advances-in-Climate-Change-Research/book-series/RACCR

Insuring Against Climate Change

The Emergence of Regional Catastrophe Risk Pools

Nikolas Scherer

Routledge
Taylor & Francis Group
LONDON AND NEW YORK

earthscan
from Routledge

First published 2020 by Routledge

2 Park Square, Milton Park, Abingdon, Oxon OX14 4RN
605 Third Avenue, New York, NY 10017

Routledge is an imprint of the Taylor & Francis Group, an informal business

First issued in paperback 2021

Publisher's Note

The publisher has gone to great lengths to ensure the quality of this reprint but points out that some imperfections in the original copies may be apparent.

British Library Cataloguing-in-Publication Data
A catalogue record for this book is available from the British Library

Library of Congress Cataloging-in-Publication Data
Names: Scherer, Nikolas, author.
Title: Insuring against climate change : the emergence of regional catastrophe risk pools / Nikolas Scherer.
Description: Abingdon, Oxon ; New York, NY : Routledge, 2020. | Series: Routledge advances in climate change research | Updated revision of the author's PhD thesis. | Includes bibliographical references and index. | 2019048421 (print) | LCCN 2019048422 (ebook) | ISBN 9780367342470 (hardback) | ISBN 9780429324642 (ebook)
Subjects: LCSH: Climate change insurance. | Disaster insurance. | Climatic changes—Economic aspects—Developing countries. | Climatic changes—Government policy—Developing countries. | Hazard mitigation—Developing countries.
Classification: LCC HG9979 .S34 2020 (print) | LCC HG9979 (ebook) | DDC 368.1/22—dc23
LC record available at https://lccn.loc.gov/2019048421
LC ebook record available at https://lccn.loc.gov/2019048422

ISBN: 978-0-367-34247-0 (hbk)
ISBN: 978-1-03-217357-3 (pbk)
DOI: 10.4324/9780429324642

Typeset in Goudy
by Apex CoVantage, LLC

To my parents, for always encouraging and supporting me
and my grandmother, in remembrance

Contents

Figures

Tables

Abbreviations

ADB	Asian Development Bank
AFD	Agence Française de Développement
ARC	African Risk Capacity
ART	Alternative Risk Transfer
ARV	Africa Risk View
CAADP	Comprehensive African Agriculture Development Programme
CAF	Cancun Adaptation Framework
CARICOM	Caribbean Community
CCRIF	Caribbean Catastrophe Risk Insurance Facility
CDB	Caribbean Development Bank
CDEMA	Caribbean Disaster Emergency Management Agency
CDERA	Caribbean Disaster Emergency Response Agency
CIDA	Canadian International Development Agency
CIMH	Caribbean Institute of Meteorology and Hydrology
CoGS	Certificate of Good Standing
COP	Conference of Parties
COSEFIN	Council of Ministers of Finance of Central America, Panama and the Dominican Republic
CPC	Climate Prediction Center
CSA	Food Security Commission
DFID	UK Department of International Development
EU	European Union
FEMM	Forum Economic Ministers Meeting
FIP	Final Implementation Plan
GDP	Gross Domestic Product
GFDRR	Global Facility for Disaster Risk Reduction and Recovery
GHG	Greenhouse Gas
GRMS	Government Risk Management Scheme (Vanuatu)
IAD	Institutional Analysis and Development
ICRISAT	International Crop Research Institute for the Semiarid Tropics
IPE	International Political Economy
IR	International Relations
IRI	International Research Institute for Climate and Society

KfW	Kreditanstalt für Wiederaufbau (German Development Bank)
MRFC	Malawi Rural Finance Company
MCII	Munich Climate Insurance Initiative
NASFAM	National Smallholder Farmers' Association of Malawi
NDC	National Determined Contribution
NDF	National Disaster Fund
NGO	Non-governmental organization
NOAA	US National Oceanic and Atmospheric Administration
OCHA	United Nations Office for the Coordination of Humanitarian Affairs
OECD	Organization for Economic Co-operation and Development
OECS	Organization of Eastern Caribbean Countries
OIBM	Opportunity International Banking Malawi
Pacific DRFIP	Pacific Disaster Risk Financing and Insurance Program
PIC	Pacific Island Country
PAHO	Pan American Health Organization
PCRAFI	Pacific Catastrophe Risk Assessment and Financing Initiative
PDC	Pacific Disaster Center
PHRD	Japan Policy and Human Resources Development
SEADRIF	South East Asian Disaster Risk Insurance Facility
SOPAC	South Pacific Applied Geoscience Commission
SPC	Segregated Portfolio Company
SPC/SOPAC	Secretariat of the Pacific Community
SPV	Special Purpose Vehicle
UK	United Kingdom
UN	United Nations
UNCTAD	United Nations Conference on Trade and Development
UNECLAC	United Nations Economic Commission for Latin America and the Caribbean
UNFCCC	United Nations Framework Convention on Climate Change
UNISDR	United Nations International Strategy for Disaster Risk Reduction
US	United States
WFP	United Nations World Food Program
WRSI	Water Requirement Satisfaction Index

Acknowledgments

This book presents and updates the findings of my PhD thesis. Writing this book and conducting the underlying research was a tremendous and long journey which I, at times, enjoyed to the fullest and, other times, wished I had not embarked on. Capturing an innovation like catastrophe risk pools and thus a quickly moving research object turned out to be a more much more difficult undertaking than I had imagined when I started this project on a sunny day in Berlin in September 2012. At that time two of the three case studies did not exist yet and I struggled to find empirical material. When I was writing the last lines, I was struggling to review and reframe the book once more in light of the emergence of the newest regional catastrophe risk pool, a potential case number four, and the increasing number of publications on the subject. But every journey has to end. Letting go and learning to live with the imperfections is one of the numerous takeaways of this journey. It remains to others to take this journey into the world of catastrophe risk pools even further. If this book triggers some interest beyond the intellectual silos it has done its job.

Finishing this book was a mentally intense journey that I would not have survived without the support and constructive feedback I was fortunate to receive from colleagues and friends. For their guidance throughout the years I owe special gratitude to Prof. Helmut K. Anheier, Prof. Arne Westad and Prof. Claus Offe. Their critical but pragmatic advice helped me to manage this challenging undertaking and constantly improve this work. They also allowed me to make my mistakes for which I am really very grateful. Likewise, I want to thank Prof. Christopher Browning and Prof. Timothy Sinclair from Warwick University for encouraging me to embark on this journey in the first place. Doing so, they laid important foundations of this book. I am also very thankful to Hillel Schwartz, who kindly took interest in my work and proved to be an inspiring source while putting the research together. I also wish to thank the Hertie School and the LSE IDEAS teams for their great professional support and the Hertie School and Mercator Foundation for the generous financial support. Advice, thoughts and critiques from three anonymous reviewers substantially helped to improve the framing and clarity of this book. For these contributions I owe them many thanks. Moreover, I want to thank Matt Shobbrook and Annabelle Harris for the great editorial assistance in putting this book together and Stephanie Belmer for the copyediting. My greatest appreciation is, however, for those I do not name individually here—the interviewees and practitioners in Johannesburg and Washington DC. They carved out some valuable time in their busy schedules and answered patiently all naive questions and inquiries. Without their input and support this book would not have been possible. I also wish to thank Ferdinand Eibl, Martin Klein, Luis Meja, Sander Werkhoven and Max Linek for the numerous and insightful discussions on various subjects related to this research and beyond. You have been an abundant resource on many occasions. Last but not least, I wish to express my immeasurable gratitude to Katharina Seewald for her unending support and patience when I went through the various emotional stages that, I was told, were obligatory on the way to get a doctorate. I could not have wished for a better friend and partner. You are an inspiration!

Gavdos, August 2019

1 Introduction

The puzzle

Climate change is altering weather extremes around the globe. The years 2014–2018 have not only been the warmest on record (Climate Central, 2019) but have also seen a major increase in the frequency and severity of natural disasters across the globe (Swiss Re Institute, 2018). In particular, the year 2017 stands out: Natural disasters created estimated economic losses of US$ 340 billion—the second highest figure ever (Munich Re, 2018). The hurricane season in the Atlantic basin was thereby one of the busiest in history and characterized by an unusual phenomenon: Three powerful hurricanes—Harvey, Irma and Maria struck the United States (US) and the Caribbean within the space of just a few weeks. Drought conditions in Southern and Eastern Africa related to El Niño and La Niña and extreme rainfall events in China, Bangladesh and India as well as in Western Africa have cost life in the thousands, especially among the poorest segments of society. Late frosts in May 2017 destroyed much of the fruit and wine crops in large parts of Europe. But also 2018 saw major natural catastrophes, leading to overall economic losses of US$ 160 billion and more than 10,000 fatalities. In particular, hurricanes Michael and Florence in the Atlantic basin, typhoons Jebi, Mangkhut and Trami in Asia and wildfires in the US left their mark. What was a more moderate year vis-à-vis 2017 ranks among the ten deadliest and costliest disaster years (Munich Re, 2019). 2019 seem to continue the overall trend of increasing temperature and weather extremes with heat waves across Europe in July, setting all-time temperature records in Belgium, Germany, Luxembourg, the Netherlands and the United Kingdom (UK) (World Meteorological Organization, 2019), a severe drought in Somalia (The Guardian, 2019), the landfall of Tropical Cyclone Idai in Mozambique and hurricane Dorian in the Bahamas. In short: The impacts of climate change are becoming more and more visible.

Accordingly, over the recent years dealing with the consequences of climate change has become a top priority in global policymaking. For years, states have put emphasis on mitigation. Under the umbrella of the United Nations Framework Convention on Climate Change (UNFCCC), the prime global governance goal has been *addressing the causes* of climate change. In recent years, however, the global focus has shifted from mitigation to adaptation and, thus, *to coping with the consequences* of climate change. The Bali Action Plan (2007) was arguably the historical turning point: It identified adaptation as a key building block for a strengthened future response to climate change and claims that adaptation policies should gain 'equal importance' to those of mitigation. In 2010 states adopted the Cancun Adaptation Framework (CAF) with the objective of enhancing global actions on adaption. In 2012, the Doha Conference called the more 'developed states' to support 'developing countries' that are vulnerable to the adverse effects of climate change with more financial and technical resources. In 2014, following two years of deliberations, the 19th session of the Conference of Parties (COP) finally institutionalized goals and policies by establishing the *Warsaw Mechanism for Loss and Damage*. The Warsaw Mechanism for Loss and Damage implemented policies to support 'developing countries' that are particularly vulnerable to the adverse effects of climate change. The Paris Agreement formalized this support. It establishes adaptation as an equal global policy goal next to mitigation (Article 7), highlights the importance of averting, minimizing and addressing loss and damage associated with the adverse effects of climate change (Article 8), calls on 'developing countries' to specify their needs and 'developing countries' to provide regular information on their support (Article 9,10, 11). There is no doubt that adaptation is currently at the forefront in global climate policymaking, in particular in Global North-South relations.

Adaptation can be understood as "measures which enable [persons] to cope with the ill-effects of climate change" (Caney, 2005, p. 752). The unitary label masks, however, different adaptation actions. On a very general level adaptation compromises 扱 *preventive* and *reactive* measures (Duus-Otterström & Jagers, 2011, p. 323). Preventive measures include, for example, the strengthening of a flood defense, the diversification of economic activities and improvements in water irrigation systems, and revolve around mitigating the impact of climate change. Reactive measures, by contrast, center around the access to funding "to rebuild society after already experienced harm" (2011, p. 323), including the compensation for hazard-related loss and damage. They manifest in a number of 'disaster financing' instruments such as contingency credits, catastrophe funds and (index-)insurance instruments (see Table 1.1).

It is in particular insurance instruments which, under the labels '**disaster risk insurance**' and 'climate risk insurance', have gained increasing support from policymakers, NGOs and private insurers alike.[1] The *Compendium of Disaster Risk Transfer Schemes* by the London School of Economics and Political Science reports 110 operating insurance schemes associated with natural hazards in low- and middle-income countries in 2018 (LSE, 2019). In the context of the Paris Agreement and the submission of National Determined Contributions (NDCs), over 38 countries expressed their intention to implement or scale-up insurance schemes (Kreft, Schäfer, Behre, & Matias, 2017) related to different kinds of weather extremes such as floods or droughts. In 2017, at the UN climate conference in Bonn (COP23), the G20—the industrial countries—partnered with the V20—an interest group of the most vulnerable countries to climate change—and representatives from development banks, businesses and NGOs

Table 1.1 Overview of disaster financing instruments

Instrument	*Mechanism*
Budget contingencies	Government sets aside a certain amount of capital of their overall budget and earmarks it for the unforeseen
Budget reallocations	Government reallocates funds from already planned programs, projects or budget lines
Capital market instruments	Government transfers the financial consequences to the capital market actors by issuing catastrophe bonds and other financial market products (e.g., weather derivatives) While government receives a capital infusion in the case of an adverse event, it pays a certain dividend in the case of a non-event
Contingent credit	Government enters into an agreement with a multilateral development bank (e.g., World Bank) which provides low-cost loans for recovery and reconstruction in the case of an adverse event
Credit	Government enters in a loan agreement or issues bonds on the domestic or international market
Donor assistance	Government receives donations and financial grants from donors and international humanitarian programs
(Indemnity/index-based) insurance	Government transfers the financial consequences to a third party (e.g., an insurance company) in exchange for a periodical premium
Reserve fund	Government creates a fund to which they pay in annually
Taxes	Government increases tax rates or introduces a new tax

Source: based on Ghesquiere & Mahul, 2010; OECD, 2012; Suarez & Linnerooth-Bayer, 2011

to launch the *InsuResilience Global Partnership*. The G20 equipped the initiative with US$ 550 million to facilitate the expansion of 'climate risk insurance' for up to 400 million people by 2020. According to Oxfam this is as much as international donors spend each year on disaster risk reduction (Oxfam, 2018). Also, the Sendai Framework for Disaster Risk Reduction 2015–2030, the guiding global framework that steers the efforts of the humanitarian community to build resilience against disasters, calls its stakeholders to promote the development of disaster risk insurance. There is no doubt: Insurance is en vogue.

Introducing index-based insurances

While there are various types of insurance schemes, the most prominent type discussed and put forward within a variety of climate and disaster-related policy initiatives is **'index insurance'**, sometimes also known as 'index-based insurance'.[2]

Simply put, what distinguishes *index insurance* from conventional *indemnity insurance* is that it is not the actual financial loss that triggers a payout but the performance of a model. The performance of this model depends on specific variables (such as the wind-speed, rainfall level, wave height and/or economic or population data) and functions as proxy for the actual loss. If the calculated losses reach a pre-specified threshold, a payout is due, irrespective of whether the model's estimation was accurate or not. The advantage of index insurance is that they allow for quick access to capital since insurers do not need to send a loss adjuster to 'ground zero' for checking and verifying claims. This saves also administrative costs, which, in turn, translates into lower premiums for insurance policyholders. The downside is, however, that the payout may not match the actual losses on the ground. It might turn out to be too low or too high and thus not adequate to compensate for incurred loss and damage. This phenomenon is called 'basis risk'.

Index insurance is typically classified according to the intended beneficiary, that is, the insurance policyholder[3]:

- **Micro index insurance schemes:** Insurance policyholders are individuals (such as farmers or market vendors)
- **Meso index insurance schemes:** Insurance policyholders are societal or financial intermediaries (such as associations, cooperatives, credit unions, banks or NGOs)
- **Macro index insurance schemes:** Insurance policyholders are states

This book is interested in the third type, macro index insurance schemes—that is, in those schemes where sovereign entities themselves resort to index-based insurance coverage. Those are arguably also the schemes that have received the most attention in policy debates surrounding international climate change adaptation.

Introducing regional catastrophe risk pools

Macro-level index insurance schemes are typically provided via newly created 'regional catastrophe risk pools', sometimes also known as 'sovereign catastrophe risk pools'. Regional catastrophe risk pools are effectively supranational, public-private self-insurance **companies**. In exchange for annual premium, they offer states from a specific region index-based insurance coverage to cope with financial consequences of natural hazards such as droughts, tropical storms or earthquakes.

Regional catastrophe risk pools are based on innovative and complex institutional arrangements that **transcend classic public-private boundaries**. While insurance policyholders (= the countries) typically 'own' or control the entities, all **insurance-related services**, including underwriting, are **provided by private actors**. Moreover, to secure cheaper premiums the pools typically retain some of the financial risks of a payout and transfer the financial risks associated with larger payouts—the costs of a large-scale catastrophe—to re-insurers and the global capital markets via the issuance of 'insurance-linked securities' (e.g., weather derivatives, catastrophe bonds or catastrophe swaps). This implies that in these arrangements **financial market investors** act somehow as **'insurer of last resort'**.

Although, relatively new and fairly sophisticated regional catastrophe risk pools have spread around the Southern globe. Their increasing relevance can be demonstrated with reference to the creation of the Caribbean Catastrophic

Table 1.2 Countries covered by regional catastrophe risk pools during the 2018/19 season

CCRIF	*ARC*	*PCRAFI*
Anguilla	Burkina Faso	Cook Islands
Antigua and Barbuda	The Gambia	Marshall Islands
Bahamas	Senegal	Samoa
Barbados		Tonga
Belize		
Bermuda		
British Virgin Islands		
Cayman Islands		
Dominica		
Grenada		
Haiti		
Jamaica		
Montserrat		
Nicaragua		
Panama		
St Vincent and The Grenadines		
St Kitts and Nevis		
St Lucia		
St Maarten		
Trinidad and Tobago		
Turks and Caicos Islands		

Risk Insurance Facility (CCRFI) in 2007, the establishment of the African Risk Capacity (ARC) in 2012, the launch of the Pacific Catastrophe Risk Insurance Pilot as part of the Pacific Catastrophe Risk Assessment and Financing Initiative (PCRAFI) in 2013 and the current technical development of the South East Asia Disaster Risk Insurance Facility (SEADRIF). During the insurance season 2018/19, CCRIF, ARC and PCRAFI provided 28 countries from three regions across the Global South with index-based insurance coverage against natural disasters (see Table 1.2). With the planned inception of SEADRIF in 2020 and the expansion of CCRIF to Central America this number is likely to further increase in the near future.

Insuring against disasters is apparently very attractive. What explains this cross-regional attractiveness? Instrument utility is usually highly context-dependent (Howlett, 2005, p. 40). Were there some international actors such as international financial institutions or global (re-)insurers that pushed countries to come together in regional catastrophe risk pools to buy insurance? Were there some specific events or broader developments such as fiscal crises or institutional developments that incentivized states to purchase insurance coverage? Is there more than insurance that attracts countries to join regional risk catastrophe pools? To scrutinize the formation of this trend, the increasing adoption of regional catastrophe risk pools across the Global South, is the core of this book. To that end the book explores comparatively the goals, characteristics and genesis of CCRIF, ARC and the PCRAFI Pilot and reflects on their commonalities. It does so from a *political* perspective, that is, a perspective that is attentive to the "distribution, exercise and consequences of power" (Hay, 2002, p. 3).

Outline

How to read this book

Analytically speaking, the book conceives CCRIF, ARC and the PCRAFI Pilot as 'policy instruments' and approaches them with four different lenses: a functionalist, a historical, a comparative and a reflexive. Each lens zooms in upon a particular aspect and works on a different level of abstraction:

The **first lens zooms in on the *goals and characteristics* of the regional catastrophe risk pools.** It provides a thick description of the goals of the instruments, their design and the way they work on a technical level. Guiding questions include: What are the goals of the risk pools? How does the insurance arrangement work? What is the reinsurance strategy? Who can take out insurance? Who governs the arrangement?

The **second lens zooms in on the *genesis* of the regional catastrophe risk pools.** It reconstructs the processes through which the instruments have come about. Guiding questions include: When and how did the idea to develop a regional insurance scheme get on governments' agenda? Who was involved in devising the instrument? How was the idea turned into practice? What were the key challenges? How were they overcome? Have the pools been modified? If so, why and how?

The **third lens zooms out to *compare* the various *characteristics* and *histories* of the regional catastrophe risk pools.** The perspective ties together all empirical insights and makes visible commonalities and differences of the various instrument characteristics and instrument histories. Guiding questions include: What are commonalities and differences in terms of goals and functionalities? Are there observable regularities in the genesis of the various instruments (e.g., patterns of events and behavior, institutional changes, 'sticks or carrots' that facilitated or triggered the adoption of the instrument)?

The **fourth lens zooms in on the *commonalities* of the various *characteristics* and *histories* of the regional catastrophe risk pools.** The perspective abstracts away differences and analyzes the commonalities in the characteristics of the instruments and their genesis with regard to their political implications. Guiding questions include: What has made the instruments attractive for states? Why have states from various regions across the Global South come to make use of regional catastrophe risk pools? It, thus, addresses the main research question of this book.

Who should read this book

Social scientists with an interest in insurance instruments and climate-related disaster management politics. Despite the prominence of insurance instruments in global climate policymaking, there exists only limited research in the social sciences on insurance, especially in political science. The limited political science research has mostly focused on the effects of insurance rather than its operation and the political processes that established it as a tool. This book closes this gap. It makes clear how modern catastrophe insurance works in theory and in practice. It strips out technicalities and, in doing so, hopes to make the business of loss more accessible for social science research.

Professionals working in the development, insurance and finance sector. They will probably know more about insurance than this book can ever convey. And, yet, the added value for them is that the book shows that 'disaster risk insurance' and their carriers, regional catastrophe risk pools, differ in terms of their purpose and design. Moreover, it shows how these insurances were implemented across different contexts. In doing so the book highlights the practical challenges encountered by professionals including the ways they were overcome. For professionals the book should provoke some thinking about good practices how to set up insurance schemes, the most sustainable design and also alternatives to insurance.

Scholars with an interest in policy instruments. The book introduces a new family of instruments to the academic literature on policy instruments. It makes visible how a policy instrument entered the global policy arena, how it was translated from one context to another, who the agents were in this process and how they interacted with others to create support. Hence, it speaks especially to scholars interested in policy innovation, policy diffusion and policy learning processes.

Scholars interested in economic history, insurance history and the financialization of capitalism. The book makes visible the newest innovations in the insurance sector, especially the convergence of insurance and capital markets, and illustrates the role of public and societal actors in this market-making exercise. Doing so, it also raises questions about the blurred boundaries between the public and the private sector. Moreover, it shows the difficulties in transferring insurance instruments from one context to another.

How the book is organized

The remainder of **Chapter 1 locates the book within the broader academic literature and explains the approach that underpins this book**. It reviews how insurance in general and index insurances more specifically have been discussed in the social science literature to date and elucidates the analytical framework.

Chapter 2 provides some background knowledge on insurance. Because political scientists have largely ignored insurance as a research subject there exists arguably only a limited understanding of its functioning. To address this shortcoming this chapter presents key concepts of insurance, strips out technicalities and introduces the technical vocabulary used throughout the book. The primer serves to make the book more accessible and to facilitate understanding of the functioning of various regional catastrophe risk pools. The chapter concludes by pointing to recent developments in the insurance sector. The brief history at the very end serves as a bridge between the conceptual and empirical part of this book.

Chapters 3–5 represent the empirical core of this book. The chapters examine CCRIF (Chapter 3), ARC (Chapter 4) and the PCRAFI Pilot (Chapter 5) according to the analytical framework set out in Chapter 1. Each chapter provides a case study and can be read separately. The first part of **each chapter describes the goal and functioning of the regional catastrophe risk pool** and the second part **illustrates its genesis**, paying particular attention to how actors devised and crafted the instrument. The last part provides a concise summary.

Chapter 6 ties the empirical insights together and develops a *political* explanation for the increasing popularity of regional catastrophe risk pools across the Southern globe. The first part summarizes and elucidates commonalities and difference between the instruments and their histories. The second part climbs one step up on the ladder of abstraction and reflects on the commonalities between the various instrument characteristics and histories and develops the core argument of this book.

Chapter 7 reviews core findings and draws out implications for research and policy. It summarizes key insights and relates these back to theoretical discussions in the academic literature. Taking some insights of this book as background, the book concludes with a tentative normative assessment of regional catastrophe risk pools and outlines some suggestions for policymaking on the way forward.

Insurance in the social science literature

Despite its prominence in everyday life, insurance instruments and involved actors (e.g., insurers, re-insurers, underwriters, investment managers) have received comparatively little attention by social scientists in general and political scientists in particular. This is somewhat surprising by virtue of the sheer amount of premiums insurance companies collect worldwide, the scale of investments the companies make and the essential role they play in the political economy. One may claim that insurers are of systemic relevance in enabling investment, securing commerce and lifestyles, if not capitalism as a whole: In 2018, insurance companies collected worldwide US$ 5,193 billion (!) in premiums (including life insurance premiums), according to Swiss Re's *sigma*[4] (Swiss Re Institute, 2019a). That means, the insurance sector alone contributed more than 6.1% to worldwide GDP. Worldwide per capita expenditures for insurance premiums summarized thereby to US$ 663. Taken together, insurance companies are also the world's largest financial investor: According to Insurance Europe, the European insurance and reinsurance federation, in 2014 the global insurance industry held 12% of all financial assets worldwide (Insurance Europe, 2014, p. 9). In 2017, the investment portfolio of European insurer exceeded US$ 10,226 billion. This was equivalent to 63.9% of the GDP of the European Union (EU) (Insurance Europe, 2018).

These numbers indicate that the impact of insurance companies on the political economy can be enormous; a change in operations and investment practices might have tremendous consequences. These aggregated numbers also veil, however, the massive regional differences when it comes to the question of who is insured and who is not: Geographically speaking, global insurance coverage is highly unevenly distributed. According to Swiss Re's *sigma*, in 2018 the 'advanced economies' collected about US$ 4,086 billion in premiums. That made up 79% of all insurance premiums collected worldwide.[5] By contrast, insurance premiums in Africa amounted to just US$ 68 billion and made up only 1.3% of all premiums collected worldwide. The per capita expenditures present a similar picture: The per capita expenditures for insurance ranged from US$ 4,068 in advanced economies to only US$ 1107 in emerging economies, and to just US$ 68 in Africa (see Table 1.3). The picture is even more contrasting on a national scale: The US, for instance, represented by far the biggest market, followed by China. Insurance premiums written in the US summarized to about US$ 1,469 billion; total premiums written in China summarized to about US$ 575 billion (Swiss Re Institute, 2019b, p. 9). This implies not only that US insurance policyholders paid about

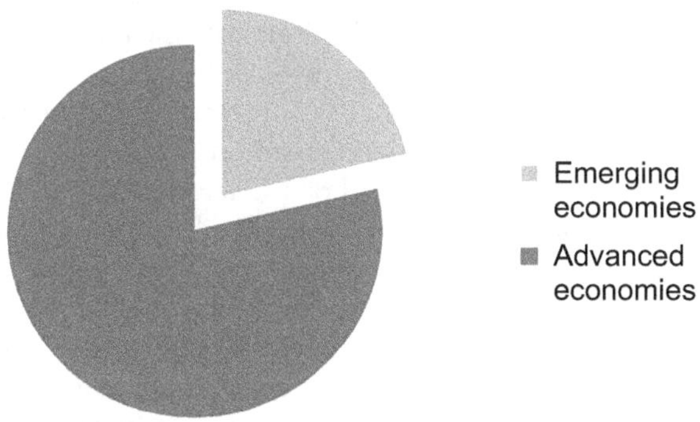

Figure 1.1 Global distribution of insurance premiums in 2018

Source: data obtained from Swiss Re Institute, 2019a

Table 1.3 Total insurance business in 2018

	Premium collected in billion US$	*Insurance density in US$ (premium per capita expenditures)*	*Premiums as a share of GDP*
Advanced economies	US$ 4086	US$ 3737	7,8%
Emerging economies	US$ 1107	US$ 169	3.2%
	Selected regions		
Africa	US$ 68	US$ 54	3.0%
Emerging Asia	US$ 754	US$ 195	3,8%
Latin America and Caribbean	US$ 162	US$ 251	2,8%
US and Canada	US$ 1597	US$ 4377	7.2%

Source: data obtained from Swiss Re Institute, 2019a

28% of global premiums but that the Chinese insurance market, as the second biggest market, makes up less than 40% of the US market.

The illustrative data demonstrates that insurance is largely an OECD phenomenon. In fact, the globe can be divided into an insured and a non-insured world. The 'developed' world is heavily insured whilst the 'developing' world is not. Insurers term this phenomenon 'the protection gap'. The contrasting insurance expenditures correlate with economic development and available market structures (UNCTAD, 1993) but are also related to other non-economic factors such as, for instance, knowledge, culture and the regulatory environment (see the following).

Against the background of these starkly contrasting figures it is irritating that Political Science has so little to say about insurance. Insurance has been approached primarily in economic and thus financial terms (Barrieu & Karoui, 2002; Cummins & Mahul, 2009; Kunreuther & Michel-Kerjan, 2007, 2014; Thoyts, 2010) but also in sociological (Baker & Simon, 2002; Beck, 1986; Ericson & Doyle, 2004; Ewald, 1991; Jarzabkowski, Bednarek, & Spee, 2015; O'Malley, 2007) and increasingly in historical terms (Borscheid, 2012a; Borscheid & Haueter, 2012; Murphy, 2010; Pearson, 2010, 2016). Apart from some notable exceptions in international relations and international political economy (Bougen, 2003; Haufler, 1997; Lobo-Guerrero, 2011, 2012, 2016; Paterson, 2001), political scientists have been largely silent on insurance. The main bulk of research on insurance comes from Economics, Business and Management. The predominance of economic perspectives on the topic leads to an under-appreciation of the political, cultural, moral or ideological forces that have shaped and continue to shape the design, operation, usage or impact of insurance.

The larger purpose of this book is to contribute to offsetting this limitation. In doing so, the book speaks to the growing bodies of insurance research in sociology and history, to which it both owes a debt and also makes a contribution. Since the end of the 1990s sociologists and historians have developed a greater epistemological interest in the political and societal shifts that underpin the operation, development and diffusion of modern insurance and produced some insightful work. While **sociologists started to explore insurance as an *instrument of governance*** with effects beyond the economic realm, **historians commenced to explore insurance as a *historically and culturally specific idea*.** Since this book stands in the intellectual tradition of these sociological and historical ways of thinking, I will first review how these fields have come to approach insurance. My purpose thereby is not to provide a comprehensive review of this literature but to map thematically and chronologically the main lines of sociological and historical thinking that characterize this body of work, including some weaknesses of these accounts.[6] In a second step I will then review in more depth the literature on index-based insurance. As this book approaches index insurance from a *global* perspective, I will put particular emphasis on accounts coming from international relations and international political economy. In this context I will center stage the Foucauldian-inspired takes by Lobo-Guerrero (2011) and Grove (2012). From a political science perspective they offer arguably the most comprehensive accounts of index-based insurances both in theoretical and empirical terms by now.

Insurance in the sociohistorical literature

Perhaps due to the economic nature of insurance instruments, the main bulk of research approaches insurance instruments in financial terms. It is only since the end of the 1990s that sociologists and historians have developed a greater interest in insurance and thus broadened the perspective on insurance (Burgees, 2018). Rather in parallel than in critical engagement with each other sociologists and historians have produced an increasing body of knowledge on the broader context that has shaped the operation, development and diffusion of insurance. What unites sociological and historical account is to locate the rise, operation and development of modern insurance in changing ideological, cultural, societal, regulatory or technological conditions. While sociologists delved thematically more into the cultural and ideational foundations of insurance, especially into the concept of risk (Baker & Simon, 2002; Beck, 1986; Ericson & Doyle, 2004; Ewald, 1991; Lobo-Guerrero, 2011, 2012, 2016; O'Malley, 2007), historians explored the changing political, societal, technological and regulatory practices the underpinned its development (e.g., Bouk, 2015; Clarke, 1999; Horstman, 2001; Murphy, 2010). Embracing a global and comparative perspective, more recent historical studies (Borscheid & Haueter, 2012; James, Borscheid, Gugerli, & Straumann, 2013; Pearson, 2010, 2016) embarked on investigating the inherent difficulties in expanding insurance from one context to another.

Sociological approaches to insurance

Inspired by the Foucauldian-influenced study by Ewald (1986) on insurance and the evolution of the modern social welfare state and the earlier historical work by Zelizer (1979) and Heimer (1985), insurance sociologists started to conceive **insurance as a new 'technology' or 'tool of governance' to manage uncertainty and risk** (Baker & Simon, 2002, p. 13), to explore its rising popularity and to discuss its impact. This 'insurance-as-governance' literature (Baker & Simon, 2002; Ericson & Doyle, 2004; Ewald, 1991; Lobo-Guerrero, 2011, 2012, 2016; O'Malley, 2007) links the rise of insurance in the twentieth century as an instrument to govern populations primarily to cultural changes, notably to the emergence of neoliberalism in the 1980s with its emphasis of efficacy, privatization and deregulation (Grove, 2012; Lobo-Guerrero, 2011; O'Malley, 2007), it problematizes the consequences of addressing societal problems increasingly in terms of risk and through private actors (see in particular Baker & Simon, 2002). For this 'insurance-as-governance' literature insurance exerts a form of social control because it regulates behavior (Heimer, 1985). A key assumption in this literature is that there are no natural risks but only socially constructed risks. A lot of work revolves therefore around the deconstruction of risks, especially the "insurantial imaginary" (Ewald, 1991, p. 198), that is, the political vision behind the development of an insurance scheme. More recent studies in this tradition pushed that approach further and reflected more on how systems of governance are (re-)produced through insurance (e.g., Lobo-Guerrero, 2011, 2012, 2016; Grove, 2012).

In the latest book of his comprehensive trilogy on insurance, Lobo-Guerrero, for instance, posits the thesis that insurance plays a central role in the "transcendence of liberalism" (2016, p. ix), that is, the reproduction of liberalism as dominating system of governance in the Western world. According to Lobo-Guerrero, life insurance contributes to promote and safeguard a political-economic order, which centers on the idea of individuality, the maximization of self-interests and the accumulation of capital. To access insurance products like life insurance requires individuals to adhere to certain behavior. Simply put, individuals need to disclose sensible information on their health to allow insurers to determine their capability to produce economic value throughout their life; otherwise, they are denied of the financial protection provided by life insurance. If individuals are granted life insurance, the financial protection allows them to accumulate more capital as it facilitates the access to credit. This makes it easier for individuals to buy a house (or other expensive goods). At the same time, life insurers use the regular premiums payments as investment capital to generate a profit for the insurance company's shareholder (Lobo-Guerrero, 2016, p. 41). Insurance policyholders and insurers are therefore accomplices in reproducing an order that is not natural but centers around a capitalist economy.

A second sociological approach to thinking about 'insurance-as-governance' has been developed in the context of the 'risk society' thesis popularized by Beck (1986, 1992) and Giddens (1991). Here **insurance is conceived as instrument that represents a new worldview**. Crudely simplified, Beck and Giddens claim that what set modern from traditional societies apart is their preoccupation with and dealing with risks. While societies have always been subject to risks such as natural disasters they have usually been interpreted as driven by external forces (e.g., nature or god) and, therefore, been accepted as fate. Modern, secular societies, by contrast, perceive these risks as being man-made. In response, they have developed a number of practices to assess and manage them. Insurance, in this sense, can be interpreted as a modern risk management tool; that is, a tool to calculate and control these risks. This epochal shift in consciousness and practice led to new discoveries and enabled the expansion of industrial processes and technologies. Yet, this expansion has produced qualitatively new catastrophic threats such as industrial accidents, environmental pollution and global warming. These new catastrophic threats are the unintended consequences of modern human activity. They transcend classes and nations and, in doing so, challenge established risk management practices. Nuclear accidents, global warming and large-scale terrorism are all threats that cannot be longer calculated, quantified and predicted by modern tools. In other words, they represent threats that cannot be covered by insurance. The inability to calculate and manage these threats induces widespread societal doubts and anxieties, undermines trust in scientific expertise and politics and leads ultimately to novel social practices, such as the application of the precautionary principle that seek to avoid risks in the first place. These novel practices alter the course of modernization. In other words, modernity problematizes and turns against itself: It becomes reflexive. In doing so, it ceases to be modern. The limit of insurability marks thereby the beginning of this reflexive modernity and the 'risk society'.

Beck's 'risk society' thesis stimulated a debate on what constitutes the limits of insurability. While some followed Beck's thesis of an epochal shift and explored the applicability of his argument that modernity constrained the limits of actuarial calculation by pointing to changing security politics at the turn of the twenty-first century (e.g., Albert, 2000; Rasmussen, 2004) others were more skeptical (e.g., Aradau, Lobo-Guerrero, & Van Munster, 2008; Bougen, 2003; Collier, 2008; Ericson & Doyle, 2004). They criticized his 'grand theory' approach in terms of its inattention to empirical detail highlighting the various novel ways through which risks have become insurable. Bougen (2003), for instance, pointed to innovations in the finance sectors, notably catastrophe bonds, which combined different elements of the insurance and capital markets that made catastrophe risks insurable. Also, Collier (2008) pointed to innovations in risk modeling and showed that the insurance sectors draw increasingly on new forms of reasoning to assess risk, notably scenario analyses.

A third sociological and very recent way to approach insurance has come from Jarzabkowski et al. (2015). Located in economic sociology and the 'practice turn' (e.g., Schatzki, 1996; Schatzki, Knorr-Cetina, & Savigny, 2001), this ethnographic study **seeks to capture the social practices that constitute the reinsurance market**. The authors provide a thick description of the various, interconnected activities of reinsurance companies—or more specifically of their underwriters—to hedge large-scale catastrophes. Giving credit to the traditional reinsurance practices with their temporally synchronized market cycles and collective pricing and deal making, the book points to changing reinsurance practices, notably the introduction of new models to assess catastrophe risk and the securitization and bundling of catastrophe risks in the form of new financial market products like catastrophe bonds or 'bundled deals'. Drawing parallels to the collapse of the subprime mortgage market, which gave rise to the global financial crisis 2008/09, the authors warn that the introduction of these practices might lead to underestimation of the involved default risk, induce new, potentially harmful dynamics between the capital and reinsurance markets and erode the stability-preserving features of the traditional reinsurance markets such as the temporally synchronized market cycles. The study provides a fresh and very empirical perspective on insurance and some fascinating insights on the making and functioning of the reinsurance market. In theoretical terms this third sociological approach is arguably still very much in its infancy vis-à-vis the other two accounts. In contrast to the other two sociological accounts the book does not refer to a larger cultural shift in (re-)insurance practice or explore a specific concept but highlights the tiny little shifts in everyday practices that change insuring. What unites the approach with the other two is the assumption that risks are ultimately socially produced.

Historical approaches to insurance

At the turn of the twenty-first century and rather in parallel to these sociological approaches, historians moved beyond insurance company histories and commenced to investigate **insurance as a historically and culturally specific idea.** Most of these broader, early historical studies on insurance focused on the development of life insurance in national contexts (e.g., Clarke, 1999; Horstman, 2001; Alborn, 2009; Murphy, 2010; Bouk, 2015), as pointed out by Burgees (2018).

Clarke (1999), for instance, explores the development of life insurance in the Anglo-American context and situated the emergence of betting on one's life within a cultural context that shaped by the idea of discovery and a new world and the desire to make money and diminish risks. Horstman (2001), by contrast, investigates the emergence and evolution of life insurance in the Dutch context. She explores how Dutch physicians become associated with life insurance companies. Her history suggests that the Dutch life insurers used the medical profession to gain the trust of the broader public.

More recent historical insurance studies have moved beyond the focus on life insurance and embarked on comparative studies exploring the difficulties and complexities in expanding insurance from one context to another (e.g., Borscheid & Haueter, 2012; Haueter & Jones, 2016a; James et al., 2013; Pearson, 2010; Pearson & Takau, 2015). This line of research investigates primarily the unevenness of insurance development in the global context.

The edited volume by Pearson (2010), for instance, explores what factors promoted and obstructed the diffusion of insurance in the international context. Focusing on late nineteenth century and the first half of the twentieth century the various essays refer to material innovations such as the substitution of candles by gas and electric lighting (which helped to reduce or eliminate certain hazards), innovations in information technology that reduced risks and facilitated spread (e.g., telegram and telephone), the expansion of trade and cooperation across borders but also regulatory changes and key networks, individuals and transnational bodies that shaped the diffusion of the insurance.

In a similar manner, the collection edited by Borscheid and Haueter (2012) explores the expansion of the insurance industry at the end of the eighteenth century. Illuminating especially economic, political and cultural developments that favored and obstructed the diffusion of a "European invention" (Borscheid, 2012b, p. 1), the volume links the expansion to colonialism and the development of a global political economy. The study points in particular to the role European emigrants and other societal agents (such as actuaries, scientists, traders and entrepreneurs) who from the outset looked beyond national boundaries and worked towards standardization of practices in the internationalization of the modern insurance business. For Borscheid, British entrepreneurs were "the first most effective agents in bringing modern insurance business to the market" (Borscheid, 2012b, p. 5). The collection also touches upon more recent developments in the insurance industry. Referring to Beck's risk society thesis, Borscheid argues that the "very speed of modernization presents insurers with enormous problems" (Borscheid, 2012b, p. 32) that forces them to limit the scope of cover or refuse it outright. This is illustrated by the rapid rise in the number of uninsured economic losses as a consequence of the natural disasters occurring since the 1990s. As a result of this development new innovations are under way in which the state partners with the private industry and acts as a re-insurer of last resort.

By contrast, the edited volumes by James et al. (2013) and by Haueter and Jones (2016a) center stage the reinsurance industry and situate their developments in an international context. While James et al. (2013) provide a historical overview of the history of the reinsurance sector and a more detailed history of Swiss Re, Haueter and Jones (2016a) focus more on how the reinsurance industry changed in dealing with risk over time. Doing so, they explore the changing business model of reinsurance, the development of markets, changing risk management techniques, financial issues and legal as well as political aspects. They conclude that "with few exceptions it has always been an adaptive rather than an innovative industry" (2016b, p. 10).

Tentatively venturing onto the terrain of Sociology of Organizations, the collection "Corporate Forms and Organizational Choice" edited by Pearson and Yoneyama (2015b) represents the first attempt by historians to illuminate in a more systematic way the various organizational forms through which insurance has been offered. It explores the various political, cultural and technological factors that have shaped the organizational form in different historical and national contexts. It considers, among others, what the costs, benefits and relative performances of these forms were in various and changing contexts, traces the learning processes and points to subsequent organizational changes to secure survival and improve performance. The book indicates that there has been a great variety of organizational forms coexisting in parallel throughout history as insurance founders and managers experimented with different corporate forms to adapt to changing circumstances. This insight speaks directly to students of organizational theory and challenges the adequacy of the 'classic' or more static agency theory. The editors conclude that "organizational choice has always been contingent on historical circumstances" (Pearson & Yoneyama, 2015a, p. 26) and underscore that "there is little convincing evidence form insurance history to suggest that some processes of 'natural selection' has ever operated in the evolution of corporate forms" (ibid.).

The contribution of the sociohistorical study of insurance

The great contribution of the growing sociological and historical work on insurance has been to locate the operation, development and more recently also the diffusion of insurance in its broader societal, cultural or political context.[7] By conceiving insurance as a *historically and culturally specific instrument of governance* that has to negotiate its relationship with established policy instruments, regulatory frameworks, cultural practices and societal values, sociologists and historians challenge the decontextualized view of mainstream economics that insurance can and will work in every context. The literature has not only shown that the introduction of insurance is deeply anchored in the (dis-)continuities of modernity and/or Western (neo-)liberalism but also that its broader uptake faces barriers that go beyond economic rationales, risk modeling and pricing and or moral hazard.

For sociologists, insuring is thereby related to major changes in ideas about the world and social practice. Their main goal is to reflect on these shifts, especially their implication, and discuss the empirical validity and/or starting points of these points of change. Also, because of this anchoring in larger political and cultural changes, most sociological accounts on insurance have a strong historical orientation. For historians, insuring tends to be part of a larger tale of modernization. The development and diffusion of (modern) insurance is located in a range of interrelated processes such as industrialization, technological innovations, colonialism, re-regulation and the increase in world trade. A main goal is to dissect the interplay of these factors, in particular how they have facilitate or inhibited insurance. Historians have provided detailed case studies illustrating the inherent complexities in the development and diffusion of various forms of insurance from one context to another. Doing so, they emphasize the historical and cultural peculiarity of insurance.

What unites sociological and historical accounts on insurance is their epistemological interest in the driving forces behind insurance. However—and this is arguably the main shortcoming of this literature—they tend to focus on the macro level, that is on large-scale structural forces (e.g., neoliberalism, industrialization, regulatory changes). As yet, sociologists and historians have been rather reluctant in embracing a meso- or micro-level perspective; a perspective that explores in greater depth *why*, *how* and *to what degree* some actors were able to successfully persuade others to adopt insurance. Such investigations would dissect the processes by which the idea to insure has been inserted, changed or hybridized into another context. They would focus, for instance, on a company's diverging marketization or framing practices (e.g., insurance as investment vs insurance risk management tool) to overcome cultural barriers and attract certain buyers, political practices and considerations (e.g., gain greater share in world trade; stimulate a certain economic sector). While some historians (e.g., Borscheid & Haueter, 2012; James et al., 2013; Pearson & Yoneyama, 2015b) have started to explore these unevenness, much more research needs to be done at the country level on how this unevenness is related to sociological or political questions, such as the performance of traditional risk management practices or existing policy instruments, to increase our understanding of the rise of insurance. I will deepen these points with explicit reference to index insurance in the following section.

Index insurance in the social science literature

Given that 'index insurance' is a relatively new phenomenon academic research on them is limited and highly fragmented. A large part of available studies addresses economists, actuaries and policymakers and belongs to the 'grey literature'. In more general terms research on index insurance can be grouped into two larger accounts: 'problem-solving' and 'critical' accounts. The main bulk of research belongs to the problem-solving accounts.

Problem-solving accounts

The problem-solving accounts seek to address unresolved issues to bring index-based insurance to scale. It is a highly application-orientated line of research. Three larger themes dominate this branch: (1) Barriers to overcome the low take up of index insurance; (2) Impacts of index insurance (3) Rationale and political viability of (macro-index) insurance

1 BARRIERS TO OVERCOME THE LOW TAKE UP OF INDEX INSURANCE

As pointed out previously, index insurance has been met with considerable political enthusiasm among governments, development agencies, businesses and NGOs alike and has been introduced at the micro-, meso- and macro-level in various countries across the Southern globe. Despite the steady increase in index-based insurance schemes in the developing world, the take-up has not matched initial theoretical expectations held by development economists (Carter, de Janvry, Sadoulet, & Sarris, 2017; Cole, Gautam, Sangita, Carina, & Stein, 2012; Jensen & Barrett, 2017; Miranda & Farrin, 2012). The discrepancy between expectation and reality has spurred a discussion about the underlying reasons and potential ways to stimulate the demand for index-based insurance. Identified key barriers involve 'basis risk'—the fact that the payout from an index-based insurance may not match the actual losses on the ground (Bryan, 2010; Clarke, 2016; Clement, Wouter Botzen, Brouwer, & Aerts, 2018; Elabed & Carter, 2015; McIntosh, Povel, & Sadoulet, 2016), cash or liquidity constraints (Giné & Yang, 2009; Vickery et al., 2013), lack of trust in insurers and product (Cai, 2016; Giné, Townsend, & Vickery, 2008), a low degree of financial literacy (Cai, Janvry, & Sadoulet, 2015 for China; Giné & Yang, 2009 for Malawi; Vickery et al., 2013 for India) and the availability of informal risk-sharing mechanism (Mobarak & Rosenzweig, 2013). Carter (2017) provides a recent and excellent summary of this debate for micro-index insurance in the agricultural sector. Depending on the identified causes for the low demand, economists have made various proposals to address them. As largely outlined by Carter (ibid.), they include improvements to reduce basic risk such as relying on proxy data (Clement et al., 2018) or satellite-based measurements (Flatnes, Carter, & Mercovich, 2019) to better estimate losses, improvements in contract design (Carter et al., 2017; Elabed, Bellemare, Carter, & Guirkinger, 2013), new ways of selling insurance such as combining index-insurance with other products like drought-tolerant seeds (Lybbert & Carter, 2014), fertilizer (Carter, Cheng, & Sarris, 2016) or credit (de Nicola & Hill, 2012; Mishra et al., 2018) and financial literacy trainings (Dercon, Hill, Clarke, Outes-Leon, & Taffesse, 2014; Gaurav, Cole, & Jeremy, 2011; Lybbert & Carter, 2014) and 'smart subsidies' (MCII, 2017).

2 IMPACTS OF INDEX INSURANCE

A second, far smaller and more recent line of research investigates the impact of index insurance. The majority of studies focuses on short-term economic effects of index-based insurance in the agricultural sector. Some studies find that index-based insurance have positive impacts on small-scale farmers investment (Cai, Chen, & Zhou, 2015; Cole, Giné, & Vickery, 2017; Elabed & Carter, 2014; Karlan, Osei, Osei-Akoto, & Udry, 2014; Mobarak & Rosenzweig, 2013) and household consumption patterns (Bertram-Huemmer & Kraehnert, 2018; de Nicola & Hill, 2012; Janzen & Carter, 2018; Jensen, Barret, & Mude, 2017). Mobarak and Rosenzweig (2013), for instance, show that insured farmers in India shifted into riskier but higher-yielding rice production. Elabed and Carter (2014) make a similar observation with regard to farmers in Mali. They find that farmers who bought insurance increased the area devoted to higher risk but higher-yielding cotton production. Karlan et al. (2014) on the other hand demonstrate that insured farmers in Ghana increased their expenditures on their farm. Bertram-Huemmer and Kraehnert (2018) demonstrate how the Index-Based Livestock Insurance (IBLI) scheme in Mongolia helped herders avoid selling and slaughtering animals, thus smoothing their productive asset base, and relieving them from credit constraints. Analogously, Jensen et al. (2017) find that IBLI scheme in Kenya helped insured herders to increase expenditures on livestock health services with positive results regarding milk productivity and income. Confirming these results, Janzen and Carter (2018) find that insured households in Kenya's IBLI scheme were much less likely to sell livestock in the case of drought, improving their chances of recovery.

3 RATIONALE AND POLITICAL VIABILITY OF (MACRO-INDEX) INSURANCE

A third and policy-orientated line of research explores the rationale and political viability of (macro-index) insurance. It investigates whether and how a government should use insurance (Cummins & Mahul, 2009; Mahul & Stutley, 2010), collects 'lessons learned' (Surminski & Oramas-Dorta, 2011; Surminski, 2016; World Bank, 2017) and provides guidance for their usage (Clarke, Mahul, Poulter, & Teh, 2017) and implementation (Cremades et al., 2018; Schäfer, Waters, Kreft, & Zissner, 2016). This literature addresses primarily policymakers and development agencies and tends to be descriptive or prescriptive in nature.

While initial studies in this branch of research took stock of available disaster risk financing instruments and explored economic rationales for governments on when to make use of insurance (Ghesquiere & Mahul, 2007, 2010; Suarez & Linnerooth-Bayer, 2011; OECD, 2012), more recent studies have taken a broader view by conceptualizing index-based insurance as part of a larger policy instrument mix to cope with disasters or even as part of overarching disaster risk management strategy (Mahul & Stutley, 2010; World Bank, 2014; Linnerooth-Bayer & Hochrainer-Stigler, 2015; Clarke et al., 2017). An emerging debate in this context is how to link index insurance with other non-financial policy instruments, especially with social protection (Vaananen, Nett, Costella, & de Suarez Mendler, 2019) and risk reduction measures (e.g., ecosystem-based approaches) to smoothen the overall impact of disasters. The theoretical potential of insurance to incentivize disaster risk reduction has been discussed elsewhere (e.g., Warner et al., 2009) but there exist, by now, limited empirical studies demonstrating how index-based insurance has actually incentivized disaster risk reduction. While Clarke and Dercon hope that this might happen mainly through the "development of precise, credible plans" (2016, p. 106) linked to a stand-by finance arrangement, other studies (Boucher & Delpierre, 2014; O'Hare, White, & Connelly, 2016; Müller, Johnson, & Kreuer, 2017) find that index insurance could have the opposite effect and increase vulnerabilities on the long run (see what follows). As such, they might be detrimental to risk reduction measures.

In general this branch of research—which has mainly come from the World Bank and partnered institutions—appears to suggest that macro-index insurance schemes are an appropriate financing instrument for low probability—high impact events in small, highly-indebted developing countries but inappropriate for responding to either high probability, low impact events or for the bulk of reconstruction costs following any catastrophic event (Cummins & Mahul, 2009; Ghesquiere & Mahul, 2010; Mahul & Stutley, 2010; Linnerooth-Bayer, Hochrainer-Stigler, & Mechler, 2012; Linnerooth-Bayer & Hochrainer-Stigler, 2015; Bevan & Adam, 2016). Bevan and Adam (2016), for instance, applying an econometric model to Jamaica to learn about the most appropriate instruments to cover disaster-related expenses conclude that an ex-post tax-financed reconstruction would be more cost-effective than insurance since catastrophe insurers typically charge a high premium that is actuarially not fair. Various frameworks have been proposed to guide the use of (macro-) insurance instruments for policymakers (Clarke et al., 2017; Cummins & Mahul, 2009; Linnerooth-Bayer & Hochrainer-Stigler, 2015; Mahul & Stutley, 2010). The most sophisticated one comes arguably from Clarke et al. (2017). Using insights from actuarial science and financial economics the authors developed an assessment framework for an ex ante evaluation of sovereign disaster risk finance instruments. In general, despite variance in focus and approach, there is a broad agreement for a "risk layering" approach—an approach that differentiates the use of insurance by the severity of the event and related costs. As a rule, developing states should purchase insurance only for large-scale catastrophes.

Critical accounts

The second and far smaller branch of research explores the potential adverse consequences of index-based insurance schemes (Müller et al., 2017; Müller & Kreuer, 2016; O'Hare et al., 2016; Capitanio, Adinolfi, & Santeramo, 2015) and investigates their development in their political economic context (Paterson, 2001; Bougen, 2010; Collier, 2008; Glenk & Fischer, 2010; Grove, 2012; Lobo-Guerrero, 2011; Peterson, 2012; Sturm & Oh, 2010; Keucheyan, 2018). These 'critical' accounts are scattered across disciplines and therefore also lack a common language or methodology. What unites them is not so much a common topic but the emphasis on the 'big picture' and the idea to view the introduction of index-based insurance as a somehow unusual move that must not be taken-for-granted but be questioned. In that sense this line of research resemblances the sociological and historical studies on insurance.

1 CRITICAL REFLECTIONS ON THE IMPACT OF INDEX INSURANCE

The most comprehensive critique on index insurance has come from Müller et al. (2017). In their critical reading of available impact studies on agricultural insurance they point to the bias of available impact studies. The authors claim that policy-oriented reviews "have focused largely on near-term economic effects and practical challenges accompanying the introduction of insurance products in developing countries" (ibid., p. 24). Arguing for a broader framework, Müller et al. provide an inventory of the potential adverse consequences of the introduction of agricultural insurance. Drawing on a range of studies from various disciplines they argue that index-based agricultural insurance may have negative long-term environmental and social effects that may ultimately offset their economic benefits. The change of land use strategies with the increased cultivation of cash crops, for instance, "may reduce the overall resilience of the ecological system by omitting positive effects of intercropping such as improved soil fertility, reduced pest incidence, and increased agrobiodiversity" (ibid., p. 28); the increased pesticide and fertilizer use due to the bundling of insurance product, for instance, may lead to water pollution with negative effects for the entire ecological systems (ibid., p. 29); the introduction of index-based insurance may also weaken informal risk sharing arrangements and the socioeconomic benefits of such networks (e.g., harvest sharing, provision of employment opportunities of households in distress) (ibid., p. 28). Also other studies, also referenced by Müller et al, problematize the adverse social (Boucher & Delpierre, 2014) and environmental (Capitanio et al., 2015; Lubowski et al., 2006) impact of index insurance schemes. O'Hare et al. (2016) and Duus-Otterström and Jagers (2011) instead provide a rather political critique of these instruments. O'Hare et al. (2016) point to contradictions of insurance as a form of resilience and aims to show how insurance regimes serve to embed risky behavior and inhibit change after detrimental events. They conclude that insurance is maladaptive and reinforces exposure and vulnerability. Duus-Otterström and Jagers (2011) review the introduction of climate- or weather-related insurance schemes in the developing world from global climate justice perspective. They find insurance arrangements where those who are least responsible for climate change end up paying the premium unfair. For them, "the only ethically defensible form of insurance is one in which the poor get their premiums subsidised" (ibid., p. 333).

2 CRITICAL REFLECTIONS ON THE DEVELOPMENT OF INDEX INSURANCE

The other critical accounts are located in International Relations (IR) and International Political Economy (IPE) and investigate the development of (index) insurance in their political context and provide critical reflections on the power relations, values and ideologies that underpin this development. The accounts put forward rather structuralist explanations (Bougen, 2003; Grove, 2012; Keucheyan, 2018; Lobo-Guerrero, 2011; Peterson, 2012; Pidgeon & Butler, 2009) as they focus on larger shifts. Only a few contributions (Glenk & Fischer, 2010; Paterson, 2001; Sturm & Oh, 2010) explore the role of and motivations of agents in the development of these schemes in greater detail.

The most comprehensive account within this line of thinking, both in theoretical and empirical terms, is arguably Lobo-Guerrero's volume *Insuring Security—Biopolitics, security and risk* (2011). In his seminal Lobo-Guerrero introduced the phenomenon of index insurance to IR and made it subject to a critical reflection. Grove, a political geographer, builds upon his work (Grove, 2010, 2012) and provides in his later work a detailed case study on the creation of the first regional catastrophe risk pool, the Caribbean Catastrophic Risk Insurance Facility. As this book embarks on this line of research I will now review in more detail Lobo-Guerrero's and Grove's take on index insurance.

Located in a Foucauldian tradition,[8] Lobo-Guerrero approaches insurance instruments as 'biopolitical instruments'. In his first volume (2011), he seeks to expose the influence of statistical techniques on the ways in which people are understood and subsequently governed ('power/knowledge-nexus') and unmasks the 'hidden', (re-)productive function of index insurance from a global IR perspective. A biopolitical instrument can roughly be understood as an interventionist instrument that aims at protecting livelihoods and promoting certain 'lifestyles' of populations. It is based on statistical methods such as counting, classification, connecting and reconnecting data and turns chance or deviations into regularities. Its overall goal is to render the future of populations somehow predictable and thus governable through subsequent assessments and interventions ('biopolitics'). Since biopolitical instruments are designed and used to maximize circulation, flows and movement of people, that is, to control and influence behavior, they possess a productive, transformative capacity. A biopolitical instrument is thus related to a system of numeric distributions, representations, goals and interventionist regulations. By "shifting 'good' elements of the populations from 'bad'" (Peoples & Vaughan-Williams, 2010, p. 67) according to some goals it is not free from but embedded in existing power relations.

Inspired by these Foucault-owned arguments, Lobo-Guerrero traces in an empirically-rich genealogical account the ideological foundations of the concept and practice of insurance back to Europe of the thirteenth century. He highlights the transformative effects by governing through the means of insurance in different security-related policy fields (Lobo-Guerrero, 2011). Lobo-Guerrero basically argues that recent index insurances ultimately *work* to secure and extend global power arrangements, namely the material and ideational power of the West:[9] Governing uncertainty through insurance does not only (un-)intentionally empower the state to control its citizens through surveillance measures and by setting subsequent incentives of right and wrong behavior, but also aligns "'other' non-Western ways of being in the world with a Western financial capitalist rationality of governance" (2011, p. 142). The "Western financial capitalist rationality of governance" (ibid.), or, to put it more simply, the Western form of governance, can roughly be understood as one of the governing through markets. The market-dimension manifests in the idea that positions insurance as *the* tool. The alignment of non-Western forms of governance through a Western form of governance destroys other, non-Western, local cultural understandings and practices to manage contingencies. Moreover, since the insurance instrument is not only of Western origin but also bound by a Western political economy it expands and secures the ideational and material power of the 'Western way of living'. In his subsequent critical reflections on insurance (2012, 2016), especially in the third volume of his trilogy, Lobo-Guerrero pushes that thought further and argues that insurance plays a central role in the "transcendence of liberalism" (2016, p. ix). As highlighted previously, according to Lobo-Guerrero life insurance contributes to promote and safeguard a political economic order, which centers on the idea of individuality, the maximization of self-interests and the accumulation of capital.

Grove (2010, 2012) argues in a similar Foucauldian way, although his main point is a slightly different one: Exposing the discourse—here understood as belief system—that underpins the creation of the Caribbean Catastrophic Risk Insurance Facility to critical scrutiny, he claims that the creation of the regional catastrophe risk pool has been informed by a Western, neoliberal vision of political and economic governance. He interprets CCRIF as a 'biopolitical innovation' that secures and expands a Western form of governance and a neoliberal capitalist mode of accumulation. The adoption of CCRIF reorganizes the public-private relations around "the imperatives of financial speculation" (Grove, 2012, p. 140); it stands for the "financialization of disaster risk management" (ibid.). CCRIF's 'real' goal is not directed towards "reducing a population's existing vulnerabilities" (2012, p. 141) but towards enhancing state security, or, more precisely, its financial capacity to ensure growth, economic development and the circulation of goods and people in the aftermath of natural disaster, and thus to reproduce the capitalist order. The enhancing of states' financial capacity is enabled through the neoliberal logic of capital accumulation, namely the commodification of risks and the trading of these on the capital markets. CCRIF spreads thereby a neoliberal rationality of governance and works to secure the increasing financialization of capitalism by creating a new field of capital accumulation. Akin to Lobo-Guerrero, Grove claims that index insurance ultimately function to secure the present world order that favors the West (Grove, 2010, pp. 556–561).

Both accounts provide some fascinating insights on the relation between culture, political economy and the governance of climate change. Their great contribution has been not only to introduce the topic of index insurances to IR but, in particular, to challenge the taken-for-granted assumptions, the system of knowledge that underpins the development of index insurance schemes and to locate their emergence in the change of modern statehood. Through their deconstructive strategy they have problematized underlying epistemologies informing index insurance schemes and drawn attention to some discomforting potential long-term implications of using index insurance schemes for individuals and populations and the non-Western world. This problematization may be interpreted in normative terms as a respect for marginalized others[10] and alternative practices. And yet, both accounts suffer from several shortcomings:

First, they provide highly structuralist accounts. They downplay the role of local and regional actors, conditions and dynamics in shaping global system logics. More specifically, they 'explain away' the role of local, regional and international actors in promoting and contesting index insurances and rather bypass local conditions and particularities (e.g., limited administrative capacities, financial, technological or legal constraints, increasing number of disasters) which might have played their role in this development.

Second, both accounts suggest a rather unproblematic because unspecified top-down process of 'cultural transmission' whereby an abstract, cultural homogenous 'West' forces an abstract 'non-West' somehow into its 'mode of governance' and 'way of living'. Neither do these categories hold historically nor can these processes of cultural transmission simply be assumed. Both need at least some further specification. What does 'West' precisely mean? Where has it been located 'geographically' and when does it start 'temporally'? Who are the 'carriers' of that imposing cultural 'West'? Why have these Western proponents 'carried over' their Western ideas? Why were these 'Western ideas' at all accepted by these 'non-Westerns'? Also empirically speaking, it is not clear *how* index insurances have emerged as an option among local, non-Western policymakers, *how* and *why* other options were dismissed. Both accounts suffer not only from a rather atomist and antagonist understanding of culture but they ascribe this unspecified West a uniform rationality of power maximization that is, at least from a sociological perspective, highly questionable. Culture is ultimately a negotiation.

Third, the knowledge claims made remain on a high level of abstraction. We do not really learn how this adoption process played out. The global 'risk assemblages', the "shifting networks of insurers, risk management experts, and state and international agencies" (Grove, 2012) that are argued to be responsible for inserting index insurances, remain rather mysterious. Empirically speaking, we learn little regarding why and how these assemblages have come about and why they were at all capable of realizing their goal of power maximization; instead, we learn much about how 'populations at risk' are constructed by the 'risk assemblages' through the practices of collecting and classifying information according to a normative rationality (e.g., in Grove's account a neoliberal one) to make them subject to surveillance and administration thus 'governable'. Yet, there is no policy without actors. Policies do not exist in a social vacuum. Policy ideas must be articulated, advocated and somehow accepted in order to become real. The processes of acceptance are not explicated in the accounts. They might be based on different stimuli and incentives, they range from a perceived 'usefulness' or 'rightfulness', over political 'carrots and sticks' to brute force.

This book seeks to address the shortcomings left by the Foucauldian accounts and provide a more nuanced and agent-centered explanation for why an increasing number of states from various regions across the Global South have come to adopt index-based insurance. To that end it explores comparatively the goals and genesis of the three regional catastrophe risk pools existing to date: CCRIF, ARC and the PCRAFI Pilot.

Studying regional catastrophe risk pools

The theoretical frame

Conceptually speaking, I conceive these regional catastrophe risk pools as 'policy instruments'. Policy instruments can be understood as goal-orientated means, as "tools that policy actors use to attain their goals" (Wurzel, Zito, & Jordan, 2013, p. 12). In Public Policy, the adoption of a policy instrument is often analyzed from a functionalist perspective and as the result of an interplay of various interests, larger institutional developments or ideational discourses (Lascoumes & Le Gales, 2007, p. 4). Following the ideational and critical strands in Public Policy, I claim that such an approach, while having its own strengths, downplays the 'political character' of instruments, especially the "downstream politics" (Jordan, Benson, Wurzel, & Zito, 2011, p. 536) at play when it comes to the selection and implementation of the instrument (Jordan et al., 2011; Lascoumes & Le Gales, 2007).

First, it downplays that **policy instruments are politically not neutral.** This may sound obvious, however it is important to note that the literature on policy instruments in Public Policy *merely*[11] treats policy instruments very much like that (Hood, 2007; Jordan, Benson, Wurzel, & Zito, 2012; Wurzel et al., 2013, p. 12): as neutral, largely exchangeable devices that are equally available. Yet, instruments (re-)structure power relationships. They empower some at the expenses of others. This instrument-as-"concentrate of power"-argument (Capano & Lippi, 2013, p. 14) applies to both the *internal* operation of an instrument and its *external* effect on politics and society at large (Salamon, 2002a). For instance, policy instruments empower some actors by attributing them a functional role in the governing of a problem while denying this role to other actors (ibid.); this enhances their power to influence future public policies. Policy instruments also allocate resources of whatever type (e.g., financial, personnel, competences, etc.) to some groups while denying these to others. Intended or not, some political or societal actors might simply be better off due to the existence of the instrument.[12] It's these distributive consequences that translate typically into political and societal support and opposition for instruments (Peters, 2002, p. 554) and thus also the availability of an instrument.

Second, a functionalist perspective masks that the **adoption of an instrument is often also informed by other criteria than the generic—speak 'technical'—policy goal or economic cost-efficiency and/or effectiveness considerations** (Bressers & Huitema, 1999; Jordan et al., 2011, p. 536). As the more ideational and critical strands in Public Policy and Comparative Politics have demonstrated, the selection of a policy instrument does not happen in a vacuum. "Instrument choice is circumscribed by existing social, political and economic circumstances which may constrain or encourage the use of particular instruments" (Howlett, 1991, p. 4). Sometimes the adoption of an instrument is externally induced as decisions are made under certain conditions. These conditions can be of more material, institutional, ideational, social or situative character (Jordan et al.,

2012, pp. 115–116; Linder & Peters, 1989; Majone, 1976) and might favor or disfavor a certain instrument. The selection of a certain instrument and, thus, the appraisal and legitimation of an instrument as a 'useful', 'rightful' or 'feasible' policy tool might be influenced, for instance, by changes in the recruitment patterns within institutions (Linder & Peters, 1989, p. 50), particular ideological stances (Heclo, 1986; Hall, 1993), legal restrictions or anticipatory electoral cost-benefit calculations (Trebilcock, 1982), "instrument constituencies" (Voß & Simons, 2014) or political 'carrots and sticks' (e.g., financial incentives, political threats) so that it becomes politically opportune for decision-makers to implement the instrument. All these contextual factors associated with the instrument might exercise some form of power on the choice taken by political actors about which instrument to implement.

Following the line of thinking embraced by Lascoumes and Gales (2007) and Jordan et al. (2011), I claim that unraveling the characteristics of an instrument, including its internal governance structure, and the contingent circumstances and associated (de-)politicizing and legitimizing dynamics that surrounded its genesis can reveal less visible and yet crucial aspects about the political reasons why states have come to make use of an instrument. Center staging an instrument and tracking its history may reveal, for instance, that support for an instrument is based not so much on the actual goal but on its distributive consequences. It may reveal that support is based on shakier grounds (e.g., external sponsorship, to pacify critics) than it may look like from the outset. Against the backdrop of these theoretical considerations and potential yield that such an approach may bring, this book takes regional catastrophe risk pools as its key analytical focal point, exploring their characteristics and genesis.

The logic of inquiry

To explain the political attractiveness of regional catastrophe risk pools across the Southern globe this book center stages the regional catastrophe risk pools themselves. To build the argument I will proceed in four steps: I will *explore* (1) the characteristics and (2) the genesis of all three regional catastrophe risk pools existing to date, *compare* (3) their characteristics and genesis to distil commonalties and (4) *reflect* on these commonalties to develop a more general claim why states from various regions across the globe have come to make use of them.

This research strategy can be labeled **"explorative comparison"** (Schmidt, 2012, pp. 100–103) and is couched in a post-positivist epistemology: **Critical realism**[13] (Jackson, 2011, pp. 72–111). An explorative comparison is a highly useful strategy for 'understudied', 'complex' real-word situations (Wright, 2009). It allows for explorations 'off the beaten track', contextualization/localizations of theories, revision/supplementation of applied methods and constant reflections/correction on interpretative claims. In contrast to the most familiar method, the 'structured focused comparison', an explorative comparison is *not* informed by a prefixed analytical framework. Instead, it works with an "emergent research framework" (Wright, 2009). An emergent research framework lists *some* basic topics prior to the empirical investigation but incorporates new topics and questions as understanding evolves throughout the research process. The previous structuring is also what contrasts an emergent research framework from other closely related strategies, namely 'grounded theory'. Grounded theory largely renounces previous structuring; its approach is purely inductive.

The explorative comparison I opted for uses **a small-N, most similar case study design**. A small-N, most-similar case study design is of most value for understudied phenomena because it allows for a detailed examination and for accommodating "causal complexities" (George & Bennett, 2005, p. 19). The exploratory richness of small-N case designs comes, of course, at the cost of parsimony and the difficulty of inferring from the particular to the general (George & Bennett, 2005, p. 23). Generalizations are, strictly speaking, not possible with a small-N case study design. Every case study explanation generates what can be termed "exemplary knowledge" (Thomas, 2011, p. 22); the claims are valid just for the specific case. This impossibility to generalize is, however, in line with comparatively minor ambitions of the post-positivist research tradition with regard to causal inference. Generalizations—the goal to infer from the particular to the universe and generate predictions—is not a goal within the critical realist-informed post-positivist research tradition. Yet, small-N case study designs still allow for *some* generalizations. These generalizations are, however, better understood as *tentative* conclusions on *whether* and, in particular, *how* something mattered to produce a *specific* outcome of interest (George & Bennett, 2005, p. 25)—in this very study, the usage of regional catastrophe risk pools by countries from across the Southern globe.

I will explore comparatively the characteristics and genesis of three regional catastrophe risk pools:

1 **Caribbean Catastrophe Risk Insurance Facility (CCRIF)**
2 **African Risk Capacity (ARC)**
3 **Pacific Catastrophic Risk Insurance Pilot (PCRAFI Pilot)**

Empirically speaking, CCRIF and ARC are regional, public-private insurance entities. The PCRAFI Pilot, by contrast, was a development program[14] in which the World Bank acted as *de facto* insurer. Yet, despite their different organizational and institutional structure, they can all be conceived as *case of* a 'regional catastrophe risk pool'. I define a 'regional catastrophe risk pool' as an "instrument that offers sovereign countries from a specific region index-based insurance coverage in exchange for an annual premium to enable them to achieve a specific goal". CCRFI, ARC and the PCRAFI Pilot are all instruments that offer index-based coverage to the sovereign countries from a specific region in exchange for an annual premium. Their precise instrumental goal is a subject of this inquiry.

As of August 2019 CCRIF, ARC and the PCRAFI (Pilot) **represent the entire population of regional catastrophe risk pools to date.** I examine these instruments case-by-case. Every chapter on CCRIF, ARC and the PCRAFI Pilot (Chapter 3–5) can be read as a separate and largely independent chapter.

I examine the instruments in their chronological order of appearance (CCRIF, ARC, PCRAFI Pilot). I do so to account for potential (causal) linkages between the instrument adoption in various regions such as, for instance, a 'policy transfer'. The main focus of this study is on CCRIF, mainly out of pragmatic reasons: CCRIF was not only the first regional catastrophe risk pools and, thus, may have triggered the adoption of regional catastrophe risk pools across other regions but it represents also the instrument for which most information is available. Hence, the analysis is deliberately focused and ordered in such a way that it starts with the instrument and region that has played a pioneering role.

I used "thick descriptions" (Geertz, 1973, pp. 3–32) of the instruments to delineate their specifics and process tracing (Beach & Pederson, 2013; Checkel, 2008; George & Bennett, 2005; Hall, 2012; Rohlfing, 2013) to reconstruct their genesis. Process tracing has the advantage of allowing for "strong within-case inferences about causal processes whereby outcomes are produced" (Beach & Pederson, 2013, p. 8). The method has a certain affinity with historical analyses in social sciences (Mahoney & Rueschemeyer, 2002) but it differs substantially in the sense that it covers a much shorter time span. What differentiates process tracing from other similar methods such as a thick description or the "congruence method" (George & Bennett, 2005, pp. 181–204) is that process tracing contains a temporal dimension (t=0 → t=1) and is theory-driven, but not necessarily theory-testing (Beach & Pederson, 2013, pp. 9–13). The research relies primarily on primary sources, namely **official documents** (e.g., reports, policy briefs, press statements, presentations) from public and private key actors involved in the creation and operation of regional climate risk insurances (in particular documents issued by CCRIF, ARC and the World Bank) **and original semi-structured interviews** (N=15) with key informants that have been involved in the creation or operation of the instruments.

The book has two overarching interests: It seeks to generate (1) *empirical* insights on the instruments and their history and (2) *analytical* insights on their cross-national attractiveness to explain the cross-regional adoption. The analytical insights are of more general nature, work on a higher level of abstraction and serve to address the main question of this book. To develop this more general claim I will move beyond case-centric analyses and provide a cross-case analysis in Chapter 6. In this chapter will contrast the various instruments and their genesis. The comparison serves to discover if there are commonalities in the goals and characteristics of the instruments and common observable regularities in the history of these instruments (e.g., patterns of events and behavior, institutional changes, etc.). Based on this comparison I will then analyze these commonalities with regard to their *political* implications. Reviewing commonalities I will reflect on the political attractiveness of the instruments. The goal here is **to craft a *minimal sufficient explanation*** for the emergent pattern of why states from various regions of Southern globe have come to make us of regional catastrophe risk pools. With a 'minimal sufficient explanation' I mean an explanation that covers the most relevant aspects that contributed to produce the outcome (= usage of regional climate risk insurances across Global South) but deliberately ignores the necessary case specifics to make the outcome happen[15] (Mackie, 1965). The claim must be understood as an interpretivist abstraction or, as George and Bennett call it, a "contingent explanation[s]" (George & Bennett, 2005, p. 235). Given that this research is couched in a post-positivist methodology, this interpretative claim must *not* be equated with a prediction. It offers, at best, a tentative explanation of what *did* happen (*ex-post* explanation) and is strongly limited to the insurance arrangements investigated.

The analytical framework

To explore regional catastrophe risk pools in greater depth, I developed an eclectic analytical framework (see Table 1.4). It draws upon a range of theoretical notions and concepts originating from different (sub-)disciplines, namely Public

Table 1.4 Analytical framework

Theme	*Topic*	*Analytical purpose*	*Key question*
Instrument characteristics	(1) Goal	Identifies the purpose and scope of the arrangement	What are the generic goals of scheme?
	(2) Operational structure	Identifies the functioning of the arrangement	How does the arrangement work?
	(3) Technical basis	Identifies the technology/ information that underpins the arrangement	What is the technical basis of the arrangement?
	(4) (Re-)insurance structure	Identifies the financial risk management strategy	What is the (re-) insurance strategy?
	(5) Governance structure	Identifies set of entities and their positions within the arrangement	Who steers whom and how?
Instrument genesis	(1) Ideational origin	Identifies key actor(s), events and processes through which the idea to create a regional climate risk insurance has come about and ended on the agenda of governments	When, where, why, how and by whom did the idea to develop a regional pooling mechanism get on the region's governments' agenda?
	(2) Instrument development	Identifies key actors and activities in designing the instrument	Who was involved in devising the policy instrument and what where the main stages of its development?

Theme	*Topic*	*Analytical purpose*	*Key question*
	(3) Instrument implementation	Identifies key actors, activities and challenges in implementing the instrument	How was the idea turned into practice?
	(4) Instrument evolution	Identifies the modifications of the instrument over time and its underpinning reasons	Has the institutional design been modified since its first inception? If so, how and why?

Policy, Comparative Politics and Organization Studies and was developed in critical engagement with the empirical material. The analytical framework served as a research protocol to generate empirical insights on the instrument's goal, operation, origin and implementation. It is structured around two larger themes ('instrument characteristics' and 'instrument genesis') and covers nine topics:

Instrument characteristics

The analytical focal point of the first theme ('instrument characteristics') is the *institutional arrangement* that operates the regional catastrophe risk pools. It seeks to scrutinize what goal states attempt to achieve with the pool and dissects how the institutional arrangement has been composed to achieve that devised goal. The pertinent topics and analytical questions have been developed in critical engagement with the specialized 'policy instrument' literature in Public Policy. I followed thereby an institutionalist approach to study policy instruments. Institutionalist accounts conceive policy instruments as a bundle of rules and practices available to governments (Hood, 2008) and aim to deconstruct their goal(s) and operation (Hood, 1983; Hood & Margetts, 2007; Howlett, 1991, 2011; Lascoumes & Le Gales, 2007; Salamon, 2002b).

For the deconstruction of the instrument characteristics ('Goal', 'Operational structure' and 'Governance structure') I took inspiration from the IAD framework by Ostrom et al. (cf. Ostrom, Cox, & Schlager, 2014, p. 272). The IAD framework is an approach for institutional analyses and helps to organize research that is interested in how governance system seek to solve a certain problem (ibid., p. 270), or more specifically how "people devise institutional arrangements to solve collective action problems and provide shared benefits" (ibid., pp. 270–271). It is of particular use to explore how different elements enable action situations (ibid., p. 276). Starting point for the framework is the assumption that there exists a problem people want to solve through an institutional arrangement—that is, through a structured relationship. The framework then develops a range of questions aimed at uncovering the 'action situation', the structuring rules and practices, in which different actors find themselves in order to 'solve' that problem and realize a common goal (ibid., p. 272). It aims primarily at identifying key rules and practices that structure the relations, that is, the 'action situation' between actors. Although I did not overtly apply the IAD framework, some questions of it informed my framework; it provided on a more general level, the basis for topic no. (2).

The topic 'Technical basis' seeks to uncover the technological innovations upon which the insurance arrangement works. Technical issues are usually not covered by 'classic' generic policy instrument approaches. These approaches are relatively "technology free" (Hood, 2007, p. 129). Yet, as the empirical research revealed the functioning of the instruments depends to a substantial degree on the availability of information technology (e.g., satellites, computers, communication networks, IT software) and the subsequent production and processing of 'big data'—one may argue that the instruments would not be possible without the availability of that technology—I incorporated that topic into the research. In contrast to Hood, I claim in the 'digital age' it is essential to include technical questions into analytical frameworks that explore policy instruments (Hood, 2008).

Instrument genesis

The second theme ('instrument genesis') covers various topics regarding ***how* the instrument has come into being and *how* it has been implemented over time.** Particular attention is paid to the actors that devised and crafted the instruments and, thus, the activities in 'the making' of the instrument. Theoretically speaking, this part of the protocol has its starting point in tracing the *idea* of creating a regional catastrophe risk pool (see 'ideational origin'). As such, it is axiomatically grounded in an *ideational approach* to study politics. Based on the axiom that ideas do not exist independent from actors that "cultivate and enact them" (Voß & Simons, 2014, p. 736) the various topics center stage the activities of those actors that introduced the idea, made it viable and turned it from a concept into a reality.

The first topic ('ideational origin') has the overarching goal to understand where the idea to create a regional insurance arrangement has come from, in particular, ***when, why and how* the idea it took shape and became *relevant* for decision-makers.** The section seeks to identify key actor(s) that came up with the idea to create a regional insurance scheme and reconstructs the context (e.g., events, patterns of behavior) through which this idea has gained viability to end up on the agenda of governments. I seek to scrutinize in particular the extent to which the countries of the region developed the idea to create a regional insurance tool on their own initiative (e.g., through 'policy learning', 'policy transfer', see what follows), respectively the extent to which this initiative has been fueled from the outside (e.g., through 'instrument constituencies', see what follows). The section also explores if alternative instruments had been considered and, if so, scrutinizes on which grounds they were rejected (e.g., technical/political feasibility, cost-benefit considerations).

Analytically this topic resembles the analytical focus of the agenda-setting literature in Public Policy and has been inspired by Kingdon's seminal *Agendas, Alternatives and Public Policies*. The core of this section is to understand *when, why and how* an idea took shape and became *relevant* for decision-makers. This process is usually described as 'agenda setting' (Kingdon, 1995, p. 72). Agenda-setting refers to the way in which a policy proposal gets on the table of a decision-making entity. The process is complex to decipher given the numerous players, policies and processes shaping policymaking. In his seminal *Agendas, Alternatives and Public Policies*, Kingdon puts forward a range of factors for consideration such as the emergence of problems—that is policy failures, crises, disasters, etc. (see also Hall, 1993), the "gradual accumulation of knowledge and perspectives among specialists in a given policy area, and the generation of policy proposal by such specialists" (Kingdon, 1995, p. 17), technological advancements, enduring public discussions on specific issues, existing policies, the "national mood" (Kingdon, 1995, p. 17), feedback-loops from the operation of existing programs, administrative changes, political changes (e.g., elections) etc. Kingdon also suggests to consider that sometimes advocates first generate a policy 'solution' and then look for a problem to which they can attach their proposal (ibid., p. 18).

The second topic '(instrument development') seeks to illuminate the instrument design process. The goal is to **identify key actors and activities in designing the instrument with its specific goal and characteristics**. This section resonates well with what it usually framed as 'policy formulation' and/or 'decision making' in standard Public Policy making accounts (Howlett & Ramesh, 2003, pp. 143–184). The section explores the main stages and activities in devising the instrument. It explores *how* the instrument was designed and *who* was involved in the design process. Doing so, it seeks to scrutinize the extent to which the countries of the region 'own' the instrument, respectively the extent to which the making of the insurance arrangement has been conferred to others (e.g., through collaboration, a mandate etc.).

The third topic ('instrument implementation') covers **the processes through which the decision to set up a regional catastrophe risk pool was translated into action**. As put by Howard et al., "While some decisions have been made on the general shape of a policy, still others are required for it to be set in motion. Funding must be allocated, personnel assigned and rule of procedure developed, among other matters" (Howlett & Ramesh, 2003, p. 185). These are contentious issues as collaboration and coordination efforts can be tremendous. Key questions in this section seek to uncover these efforts and, thus, the division of labor between various actors.

The fourth topic ('instrument modification') **explores *if* and—in case—*why* and *how* the instrument has been modified since its first usage**. Key questions seek to scrutinize the extent to which the instrument has worked as intended and thus the extent to which the instrument has faced (de)legitimation over time. It highlights central instrument modifications (if applicable) and explores the key events and actors that have contributed to these modifications (e.g., in terms of its goals or functioning).

Key concepts

In the attempt to theorize and thus explain observed patterns in the genesis I turned to the following theoretical concepts: (1) 'instrument constituencies' (2) 'policy learning' (3) 'policy transfer':

(1) INSTRUMENT CONSTITUENCIES

'Instrument constituencies' is a recent concept to describe **policy actors that are "strongly oriented towards devising and promoting certain policy instruments"** (Mukherjee & Howlett, 2015, p. 68) regardless of the problem context. They consist of a loose, heterogeneous network of actors hold together by a range of interrelated practices, knowledge and resources. Their activities involve devising policy instruments, delineating alternatives and enacting and assessing/revising policy instruments. Some of the network members might end up existing just "*for* and *by* the instrument" (Voß & Simons, 2014, p. 738).

According to Voß and Simons, 'instrument constituencies' can be distinguished from other familiar concepts (or collective agents) such as "epistemic communities" (Haas, 1992) or "advocacy coalitions" (Keck & Sikkink, 1998). While the notion 'epistemic communities' centers primarily on a group of experts (e.g., scientists, professionals) that is involved in delineating and center staging a specific problem, 'instrument constituencies' are primarily concerned with delineating and promoting a *specific* solution. Epistemic communities are crucial for the agenda-setting process; 'instrument constituencies', by contrast, are crucial in the instrument selection process as they specify (or tailor) the instrument. According to Voß and Simons, 'instrument constituencies' must also be contrasted from 'advocacy coalitions' insofar as 'instrument constituencies' "do not gather around and issue or belief, but rather around a specific solution and related practices" (Voß & Simons, 2014, p. 741). They promote *one* particular instrument across various issue areas. Advocacy coalitions, by contrast, are held together by a shared conviction, but they promote different solutions.

The most important characteristic of 'instrument constituencies' is arguably that they act as "incubator" of new instruments that are then taken up at the political level (Mukherjee & Howlett, 2015, p. 68). Through their loose organizational structure and by bringing together a number of various competences, resources and practices, 'instrument constituencies' are also able to transform an instrument into "a potentially diffusible and adaptable 'tool' in the first place" (Voß & Simons, 2014, p. 736).

(2) POLICY LEARNING

'Policy learning' refers generally to learning from previous experiences as well as to learning from others. It was Peter Hall who first conceptualized the notion of policymaking more systemically as social learning (Hall, 1993). In his famous article, Hall defines social learning **"as a deliberate attempt to adjust the goals**

or techniques of a policy in response to past experience and new information" (Hall, 1993, p. 278). For him "learning is indicated when policy changes as the result of such a process" (ibid.)

In his framework, Hall distinguishes three degrees of policy changes that result from such a learning process: First order, second order and third order changes. *First order* changes refer to minor changes, "the settings of the government's instruments were modified" (Hall, 1993, p. 278); *second order* changes refers to situation where the policy goals remain the same but where new instruments were adopted; *third order* changes, in turn, refer to a radical paradigmatic shift, that is "changes in all three components of a policy: the instrument settings, the instruments themselves, and the hierarchy between goals" (ibid., p. 279).

First and second order changes tend to happen frequently and involve incremental learning through from trial/error experiences and the availability of new information. These changes stand symbolically for the 'normal', risk-averse policymaking process: The gradual adjustments of the existing policies and policy instruments. The result is a policy change in small steps. Paradigmatic third order changes, in turn, tend to realize only very rarely. They happen only in a situation where a puzzling, "not fully comprehensible problem" (ibid., p. 280) in terms of the prevailing way of thinking arises; for example, the accumulation of a number of 'unexplainable' events. Policy failure—the failure of a policy to achieve its goals—and policy experimentations—the development/testing of a new policy—play a key role in the move from one paradigm to another. Faced with these situations of uncertainty, policymakers turn to a range of actors, in particular to 'epistemic communities' (these are specialized experts/professionals) in search for assistance and alternatives. A scientific consensus in the epistemic community of relevance can then inform a policy change. Central to a third order change are thus the ascription of (scientific) authority: "Faced with conflicting opinions from experts, politicians will have to decide whom to regard as authoritative, especially on matters of technical complexity" (ibid., p. 280). The policy community, in turn, "will engage in a contest for authority for the issue at hand" (ibid.).

Indicative for a gradual policy change would be a gradually growing prioritization of a subject matter on the policy agenda and/or experimentations with a policy (e.g., catastrophe coverage for entire population) and subsequent small changes thereof over time (Kingdon, 1995, p. 80). Indicative for non-incremental, discontinuous paradigmatic policy changes would be, for instance, a policy failure following a particular event (e.g., a natural disaster) and a subsequent state of disarray. This messy, uncertain situation typically overlays all other problems on an agenda and may be understood as 'windows of opportunity' for alternative experts/professionals assembled in national or "transnational knowledge networks" (Stone, 2013) to make their voices heard and push for something new. They catch the policymakers' (and also the broader public's) attention. Consequently, these events may also be conceived as 'critical junctures'—turning points that led to policy choices with a long-lasting impact (institutional path-dependencies) by closing off alternatives (Capoccia & Kelemann, 2007).

(3) POLICY TRANSFER

Linked to 'policy learning' is **'policy transfer'** (sometimes also labeled 'policy diffusion'). Simply put, 'policy transfer' refers to learning from (the choices of) others. It was Dolowitz and Marsh who, building on a range of studies in the 1990s including their own landmark study (Dolowitz & Marsh, 1996), conceptualized and analyzed "processes involved in lesson-drawing, policy convergence, policy diffusion and policy divergence" (Dolowitz & Marsh, 2000, p. 5). They developed a synthetizing theoretical framework that draws attention to (1) the *nature* and (2) the *degree* of the 'policy transfer' processes. By a 'policy transfer' they refer to the **"the process by which knowledge about policies, administrative arrangements, institutions and ideas in one political system (past or present) is used in the development of policies, administrative arrangements, institutions and ideas in another political system"** (ibid., p. 5).

Their framework suggests drawing attention to the *nature* of policy transfers, which they characterize either as *voluntary or coercive*. Similar to Hall's 'policymaking as social learning' framework, their research questions point to various exogenous and endogenous sources to uncover a voluntary or coercive policy transfer. The analytical framework revolves around what motivates (domestic) policymakers to transfer policies from other political systems, which actors are involved in this transfer process, which lessons are drawn, how is the policy transfer process related to policy success or policy failure, when does policy transfer occur, among others (Dolowitz & Marsh, 2000, p. 7). Drawing on empirical findings, Dolowitz and Marsh point to a range of conditions that (could) explain the more voluntary or coercive character of a policy transfer. They claim, for example, that "if a government searches hurriedly for a solution to an urgent problem, it is more likely that there will be [voluntary; N.S.] transfer" (ibid., p. 8).

Notes

1 The terms 'disaster risk insurance' and 'climate risk insurance' are neologisms that are used on the policy world. The different popular labels reflect the various communities pushing for these instruments. While the climate policy community refers to them as 'climate risk insurance' (see, for instance, the InsuResilience Secretariat), the disaster management community, including multilateral development banks, prefers to call them 'disaster risk insurance' (see, for instance, the Global Facility for Disaster Risk Reduction).

2 I use the term 'index insurance' and 'index-based insurance' interchangeably.

3 Climate policymakers distinguish here between *direct* and *indirect* climate risk insurance. They use the term 'direct' and 'indirect' to capture the way insurance payouts are distributed to individuals. Consequently, they refer to *micro-level schemes*—schemes which provide direct premium payments to individuals—as *indirect climate risk insurance*; and to *meso-* and *macro-level* schemes—schemes which provide premium payments via public or private entity—as *indirect climate risk insurance*.

4 Swiss Re's annual sigma study contains information on the world's insurance markets. The sigma study is based on the direct premium volume of insurance companies regardless of whether they are privately or state owned. Premiums paid to state social insurers are not included. The web application www.sigma-explorer.com contains the data of Swiss Re's annual sigma report.

5 According to Swiss Re, the designation of the economies mentioned in as 'advanced' or 'emerging' is kept with the conventions of the International Monetary Fund (IMF). Advanced economies include the US, Canada, Western Europe (excluding Turkey), Israel, Oceania, Japan and the other advanced Asian economies (Hong Kong, Singapore, South Korea and Taiwan). All other countries are classified as 'emerging' and correspond to the IMF's 'emerging and developing' economies.
6 It might be noteworthy to emphasize that this is a reading coming from outside of the respective fields. The structure is inspired by the excellent thematic review on the sociohistorical study of life insurance by Burgees (2018).
7 For an excellent thematic review see Burgees (2018). He raises the strengths, weaknesses and relative contributions of sociological and historical accounts on life insurance.
8 The 'Foucauldian tradition' emphasizes why the modern state relies on social knowledge and how such knowledge structures power relations (Boswell, 2009, p. 36). The form of governance that is based on social knowledge, e.g., statistical data, is what Foucault terms 'biopower'.
9 This is *not* to say that index insurance are also intended to work so. It is rather what they unintentionally *do*. It must be understood as a speculative knowledge claim.
10 Hay (2002, p. 218) makes this point about postmodernist accounts more in general.
11 'Merely' implies that there are, of course, notable exceptions (see, for example, Howlett, 2005; Lascoumes & Le Gales, 2007; Linder & Peters, 1989; Peters, 2002; Salamon, 2002b; Wurzel et al., 2013)
12 The goal of redistributive policy instruments is precisely to make some societal groups better off.
13 Key proponents of such a philosophy are, among others, John Searle (2011) and Ian Hacking (1999). Social scientists working in this tradition are, among other, Roy Bhaskar (1975), Margaret Archer (2003), Milja Kurki (2008) and Ulrich Beck (1986).
14 In 2016, the World Bank restructured the development program and a regional facility was created. Unfortunately, due to resources constraints and a lack of publicly available data these developments could not be taken into account.
15 The factors, forces, processes contained in the final causal claim themselves do not produce the outcome in question. The outcome, that is, the causal claim, is therefore best understood as an analytical construct with no real-world existence (see analogously Hedström & Swedberg, 1998).

References

Albert, M. (2000). From defending borders towards managing geographical risks? Security in a globalised world. *Geopolitics*, 5(1), 57–80. http://doi.org/10.1080/14650040008407667

Alborn, T. (2009). *Regulated lives: Life insurance and British society, 1800–1914*. Toronto, Buffalo, and London: Toronto University Press.

Aradau, C., Lobo-Guerrero, L., & Van Munster, R. (2008). Security, technologies of risk, and the political: Guest editors' introduction. *Security Dialogue*, 39(2–3), 147–154. http://doi.org/10.1177/0967010608089159

Archer, M. (2003). *Structure, agency, and the internal conversation*. Cambridge: Cambridge University Press.

Baker, T., & Simon, J. (2002). Embracing risk. In T. Baker & J. Simon (Eds.), *Embracing risk—The changing culture of insurance and responsibility* (pp. 1–26). Chicago and London: University of Chicago Press.

Barrieu, P., & Karoui, N. (2002). Reinsuring climatic risk using optimally designed weather bonds. *The Geneva Papers on Risk and Insurance Theory*, *27*(2), 87–113. http://doi.org/10.1023/A:1021944109402

Beach, D., & Pederson, R. B. (2013). *Process-tracing methods—Foundations and guidelines*. Ann Arbor, MI: University of Michigan Press.

Beck, U. (1986). *Risikogesellschaft auf dem Weg in eine andere Moderne*. Frankfurt am Main: Suhrkamp.

Beck, U. (1992). *The risk society: Towards a new modernity*. London: Sage.

Bertram-Huemmer, V., & Kraehnert, K. (2018). Does index insurance help households recover from disaster? Evidence from IBLI Mongolia. *American Journal of Agricultural Economics*, *100*(1), 145–171. http://doi.org/10.1093/ajae/aax069

Bevan, D., & Adam, C. (2016). *Financing the reconstruction of public capital after a natural disaster* (Policy Research Working Paper 7718). Washington, DC. Retrieved from http://documents.worldbank.org/curated/en/806341468196163691/Financing-the-reconstruction-of-public-capital-after-a-natural-disaster

Bhaskar, R. (1975). *A realist theory of science*. London: Verso.

Borscheid, P. (2012a). Far East and Pacific: Overview. In P. Borscheid & N. V. Haueter (Eds.), *World insurance: The evolution of a global risk* (pp. 415–443). Oxford: Oxford University Press.

Borscheid, P. (2012b). Introduction. In P. Borscheid & N. V. Haueter (Eds.), *World insurance: The evolution of a global risk* (pp. 1–46). Oxford: Oxford University Press.

Borscheid, P., & Haueter, N. V. (Eds.). (2012). *World insurance: The evolution of a global risk*. Oxford: Oxford University Press.

Boswell, C. (2009). *The political use of expert knowledge*. Cambridge: Cambridge University Press.

Boucher, S., & Delpierre, M. (2014). *The impact of index-based insurance on informal risk-sharing* (CEPS/INSTEAD Working Paper). Retrieved from https://pdfs.semanticscholar.org/e6ac/0e462d4fc74808869a6e9b97de5a83cc961e.pdf

Bougen, P. D. (2003). Catastrophe risk. *Economy and Society*, *32*(2), 253–274. http://doi.org/10.1080/0308514032000073428

Bougen, P. D. (2010). The "becoming" insurable of terrorism risk in the US: Imagining systemic risk. *Limn*, (1), 1–2. Retrieved from https://escholarship.org/uc/item/0q574967

Bouk, D. (2015). *How our days became numbered-risk and the rise of the statistical individual*. Chicago and London: Chicago University Press.

Bressers, H. T. A., & Huitema, D. (1999). Economic instruments for environmental protection: Can we trust the 'magic carpet'? *International Political Science Review*, *20*(2), 175–196.

Bryan, G. (2010). *Ambiguity and insurance* (Working Paper). Retrieved from https://pdfs.semanticscholar.org/131f/224c121a377b0d675f43155f0f2245a3041a.pdf

Burgees, A. (2018). *Socio-historical life insurance research, "moral" obstacles and sociology* (Working Paper). Retrieved from www.academia.edu/37026094/Socio-Historical_Life_Insurance_Research_Moral_Obstacles_and_Sociology

Cai, H., Chen, Y., & Zhou, L. (2015). The effect of microinsurance on economic activities: Evidence from a randomized field experiment. *Review of Economics and Statistics*, *97*(2), 287–300. http://doi.org/10.1162/REST_a_00476

Cai, J. (2016). The impact of insurance provision on household production and financial decisions. *American Economic Journal: Economic Policy*, 8(2), 44–88. http://doi.org/10.1257/pol.20130371

Cai, J., de Janvry, A., & Sadoulet, E. (2015). Social networks and the decision to insure. *American Economic Journal: Applied Economics*, 7(2), 81–108. http://doi.org/10.2139/ssrn.2161686

Caney, S. (2005). Cosmopolitan justice, responsibility, and global climate change. *Journal of International Law*, *18*(4), 747–775. http://doi.org/10.1017/S0922156505002992

Capano, G., & Lippi, A. (2013). *Enlightening policy mix complexity: A typology of actors' choices*. Conference Paper, Grenoble. Retrieved from www.icpublicpolicy.org/IMG/pdf/panel_11_s3_capano.pdf

Capitanio, F., Adinolfi, F., & Santeramo, F. G. (2015). Environmental implications of crop insurance subsidies in Southern Italy. *International Journal of Environmental Studies*, *72*(1), 179–190. http://doi.org/10.1080/00207233.2014.947737

Capoccia, G., & Kelemann, D. R. (2007). The study of critical junctures. *World Politics*, 59(349–369). http://doi.org/10.1017/CBO9781107415324.004

Carter, M. R., Cheng, L., & Sarris, A. (2016, January). Where and how index insurance can boost the adoption of improved agricultural technologies. *Journal of Development Economics*, *118*, 59–71. http://doi.org/10.1016/j.jdeveco.2015.08.008

Carter, M. R., de Janvry, A., Sadoulet, E., & Sarris, A. (2017). Index insurance for developing country agriculture: A reassessment. *Annual Review of Resource Economics*, *9*, 421–438. http://doi.org/10.1146/annurev-resource-100516-053352

Checkel, J. T. (2008). Process-tracing. In A. Klotz & D. Prakash (Eds.), *Qualitative methods in international relations: A pluralist guide* (pp. 114–127). Basingstoke: Palgrave Macmillan.

Clarke, D. (2016). A theory of rational demand for index insurance. *American Economic Journal: Microeconomics*, 8(1), 283–306. http://doi.org/10.1257/mic.20140103

Clarke, D., & Dercon, S. (2016). *Dull disasters? How planning ahead will make a difference*. Oxford: Oxford University Press.

Clarke, D., Mahul, O., Poulter, R., & Teh, T. L. (2017). Evaluating sovereign disaster risk finance strategies: A framework. *The Geneva Papers*, *42*, 565–584. http://doi.org/10.1057/s41288-017-0064-1

Clarke, G. (1999). *Betting on lives: The culture of life insurance in England 1695–1755*. Manchester: Manchester University Press.

Clement, K. Y., Wouter Botzen, W. J., Brouwer, R., & Aerts, J. C. J. H. (2018, June). A global review of the impact of basis risk on the functioning of and demand for index insurance. *International Journal of Disaster Risk Reduction*, *28*, 845–853. http://doi.org/10.1016/j.ijdrr.2018.01.001

Climate Central. (2019). *The 10 hottest global years on record*. Retrieved September 8, 2019, from https://climatecentral.org/gallery/graphics/the-10-hottest-global-years-on-record

Cole, S., Gautam, B., Sangita, V., Carina, W., & Stein, D. (2012). *The effectiveness of index-based microinsurance in helping smallholders manage weather-related risks* (Working Paper). London. Retrieved from http://r4d.dfid.gov.uk/pdf/outputs/systematicreviews/

Cole, S., Giné, X., & Vickery, J. (2017). How does risk management influence production decisions? Evidence from a field experiment. *The Review of Financial Studies*, *30*(6), 1935–1970. http://doi.org/10.1093/rfs/hhw080

Collier, S. (2008). Enacting catastrophe: Preparedness, insurance, budgetary rationalization. *Economy and Society*, *37*(2), 224–250. http://doi.org/10.1080/03085140801933280

Cremades, R., Surminski, S., Máñez Costa, M., Hudson, P., Shrivastava, P., & Gascoigne, J. (2018). Using the adaptive cycle in climate-risk insurance to design resilient futures. *Nature Climate Change*, 8(1), 4–7. http://doi.org/10.1038/s41558-017-0044-2

Cummins, D., & Mahul, O. (2009). *Catastrophe risk financing in developing countries—Principles for public intervention*. Washington, DC: World Bank. Retrieved from http://documents.worldbank.org/curated/en/377421468136797977/Catastrophe-risk-financing-in-developing-countries-principles-for-public-intervention

de Nicola, F., & Hill, R. V. (2012). *Interplay among credit, weather insurance and savings for farmers in Ethiopia*. Retrieved from www.researchgate.net/publication/268293441_Interplay_among_credit_insurance_and_savings_for_farmers_in_developing_countries

Dercon, S., Hill, R. V., Clarke, D., Outes-Leon, I., & Taffesse, A. S. (2014). Offering rainfall insurance to informal insurance groups: Evidence from a field experiment in Ethiopia. *Journal of Development Economics*, *106*, 132–143. http://doi.org/10.1016/j.jdeveco.2013.09.006

Dolowitz, D. P., & Marsh, D. (1996). Who learns what from whom: A review of the policy transfer literature. *Political Studies*, *44*(2), 343–357. http://doi.org/10.1111/j.1467-9248.1996.tb00334.x

Dolowitz, D. P., & Marsh, D. (2000). Learning from abroad: The role of policy transfer in contemporary policy-making. *Governance*, *13*(1), 5–23. http://doi.org/10.1111/0952-1895.00121

Duus-Otterström, G., & Jagers, S. C. (2011). Why (most) climate insurance schemes are a bad idea. *Environmental Politics*, *20*(3), 322–339. http://doi.org/10.1080/09644016.2011.573354

Elabed, G., Bellemare, M. F., Carter, M. R., & Guirkinger, C. (2013). Managing basis risk with multiscale index insurance. *Agricultural Economics*, *44*(4–5), 419–431. http://doi.org/10.1111/agec.12025

Elabed, G., & Carter, M. (2014). *Ex-ante impacts of agricultural insurance: Evidence from a field experiment in Mali* (Working Paper). Retrieved from https://arefiles.ucdavis.edu/uploads/filer_public/2014/04/25/elabed-impact_evaluation_0422_vdraft2_1.pdf

Elabed, G., & Carter, M. (2015). Basis risk and compound-risk aversion: Evidence from a WFP experiment in Mali. *Journal of Economic Behavior and Organization*. Retrieved from https://econpapers.repec.org/RePEc:ags:aaea13:150353

Ericson, R., & Doyle, A. (2004). Catastrophe risk, insurance and terrorism. *Economy and Society*, *33*(2), 135–173. http://doi.org/10.1080/0308514042000225716

Ewald, F. (1986). *L'Etat providence*. Paris: Grasset.

Ewald, F. (1991). Insurance and risk. In G. Burchell, C. Gordon, & P. Miller (Eds.), *The Foucault effect: Studies in governmentality* (pp. 197–210). Chicago: University of Chicago Press.

Flatnes, J. E., Carter, M., & Mercovich, R. (2019). *Improving the quality of index insurance with a satellite-based conditional audit contract* (Working Paper). Retrieved from 6e566e2fb062e53014ba10c03c7684239660a828

Gaurav, S., Cole, S., & Jeremy, T. (2011). *Marketing complex financial products in emerging markets: Evidence from rainfall insurance in India* (Working Paper). Retrieved from www.people.hbs.edu/scole/webfiles/journals/07-gct-2011-06-23.pdf

Geertz, C. (1973). Thick description: Toward an interpretive theory of culture. In *The interpretation of culture: Selected essays*. New York: Basic Books.

George, A., & Bennett, A. (2005). *Case studies and theory development in the social sciences*. Cambridge, MA: MIT. http://doi.org/10.1017/s0022381607080231

Ghesquiere, F., & Mahul, O. (2007). *Sovereign natural disaster insurance for developing countries: A paradigm shift in catastrophe risk financing* (Policy Research Working Paper No. 4345). Washington, DC. Retrieved from http://documents.worldbank.org/curated/en/538391468150588169/Sovereign-natural-disaster-insurance-for-developing-countries-a-paradigm-shift-in-catastrophe-risk-financing

Ghesquiere, F., & Mahul, O. (2010). *Financial protection of the state against natural disasters—A primer*. Washington, DC: World Bank. Retrieved from http://elibrary.worldbank.org/content/workingpaper/10.1596/1813-9450-5429

Giddens, A. (1991). *Modernity and self identity*. London: Sage.

Giné, X., Townsend, R., & Vickery, J. (2008). Patterns of rainfall insurance participation in rural India. *The World Bank Economic Review*, *22*(3), 539–566. http://doi.org/10.1093/wber/lhn015

Giné, X., & Yang, D. (2009). Insurance, credit and technology adoption: Field experiment evidence from Malawi. *Journal of Development Economics*, 89(1), 1–11. http://doi.org/10.1016/j.jdeveco.2008.09.007

Glenk, K., & Fischer, A. (2010). Insurance, prevention or just wait and see? Public preferences for water management strategies in the context of climate change. *Ecological Economics*, 69(11), 2279–2291. http://dx.doi.org/10.1016/j.ecolecon.2010.06.022

Grove, K. (2010). Insuring "our common future"? Dangerous climate change and the biopolitics of environmental security geopolitics. *Geopolitics*, *15*(3), 536–563. http://doi.org/10.1080/14650040903501070

Grove, K. (2012). Preempting the next disaster: Catastrophe insurance and the financialization of disaster management. *Security Dialogue*, *43*(2), 139–155. http://doi.org/10.1177/0967010612438434

The Guardian. (2019). *Two million people at risk of starvation as drought returns to Somalia*. Retrieved September 1, 2019, from www.theguardian.com/global-development/2019/jun/06/two-million-people-at-risk-of-starvation-as-drought-returns-to-somalia

Haas, P. M. (1992). Epistemic communities and international policy coordination. *International Organization*, 46(1), 1–35.

Hacking, I. (1999). *The social construction of what?* Cambridge, MA and London: Harvard University Press.

Hall, P. (1993). Policy paradigms, social learning, and the state: The case of economic policymaking in Britain. *Comparative Politics*, *25*(3), 275–296. http://doi.org/10.2307/422246

Hall, P. (2012). Tracing the progress of process tracing. *European Political Science*, *12*(1), 20–30. http://doi.org/10.1057/eps.2012.6

Haueter, N. V., & Jones, G. (Eds.). (2016a). *Managing risk in reinsurance: From city fires to global warming*. Oxford: Oxford University Press.

Haueter, N. V., & Jones, G. (2016b). Risk and reinsurance. In N. V. Haueter & G. Jones (Eds.), *Managing risk in reinsurance: From city fires to global warming* (pp. 1–65). Oxford: Oxford University Press.

Haufler, V. (1997). *Dangerous commerce: Insurance and the management of international risk*. Ithaca, NY: Cornell University Press.

Hay, C. (2002). *Political analysis—A critical introduction*. Basingstoke: Palgrave Macmillan.

Heclo, H. (1986). Reaganism and the search for a public philosophy. In J. L. Palmer (Ed.), *Perspectives on the Reagan years* (pp. 87–124). Washington, DC: Urban Institute Press.

Hedström, P., & Swedberg, R. (1998). *Social mechanisms: An analytical approach to social theory*. Cambridge: Cambridge University Press.

Heimer, C. (1985). *Reactive risk and rational action: Managing moral hazards in insurance contracts*. Berkeley: University of California Press.

Hood, C. (1983). *The tools of government*. London: Palgrave Macmillan.

Hood, C. (2007). Intellectual obsolescence and intellectual makeovers: Reflections on the tools of government after two decades. *Governance*, *20*(1), 127–144. http://doi.org/10.1111/j.1468-0491.2007.00347.x

Hood, C. (2008). The tools of government in the information age. In *The Oxford handbook of public policy* (pp. 469–481). Oxford: Oxford University Press.

Hood, C., & Margetts, H. Z. (2007). *The tools of government in the digital age*. Basingstoke: Palgrave Macmillan.

Horstman, K. (2001). *Public bodies, private lives: The historical construction of life insurance, health risks, and citizenship in the Netherlands*. Rotterdam: Erasmus Publishing.

Howlett, M. (1991). Policy instruments, policy styles, and policy implementation: National approaches to theories of instrument choice. *Policy Studies Journal*, *19*(2), 1–21. http://doi.org/10.1111/j.1541-0072.1991.tb01878.x

Howlett, M. (2005). What is a policy instrument? Policy tools, policy mixes, and policy-implementation styles. In P. Eliadis, M. M. Hill, & M. Howlett (Eds.), *Designing government: From instruments to governance* (pp. 31–50). London and Montreal: McGill-Queen's University Press.

Howlett, M. (2011). *Designing public policies: Principles and instruments*. London: Routledge.

Howlett, M., & Ramesh, M. (2003). *Studying public policy-policy cycles and policy subsystems*. Oxford: Oxford University Press.

Insurance Europe. (2014). *European insurance in figures*. Retrieved from www.insuranceeurope.eu/uploads/Modules/Publications/european-insurance-in-figures-2.pdf

Insurance Europe. (2018). *Insurance data*. Retrieved June 1, 2019, from www.insuranceeurope.eu/insurancedata

Jackson, P. T. (2011). *The conduct of inquiry in international relations: Philosophy of science and its implications for the study of world politics*. London and New York: Routledge.

James, H., Borscheid, P., Gugerli, D., & Straumann, T. (2013). *The value of risk*. Oxford: Oxford University Press.

Janzen, S. A., & Carter, M. R. (2018). After the drought: The impact of microinsurance on consumption smoothing and asset protection: Evidence from a drought in Kenya. *American Journal of Agricultural Economics*, *101*(3), 659–671. http://doi.org/10.1093/ajae/aay061

Jarzabkowski, P., Bednarek, R., & Spee, P. (2015). *Making a market for acts of God—The practice of risk-trading in the global reinsurance industry*. Oxford: Oxford University Press.

Jensen, N., & Barrett, C. (2017). Agricultural index insurance for development. *Applied Economic Perspectives and Policy*, 39(2), 199–219. http://doi.org/10.1093/aepp/ppw022

Jensen, N., Barret, C., & Mude, A. (2017). Cash transfers and index insurance: A comparative impact analysis from Northern Kenya. *Journal of Development Economics*, *129*(1), 14–28. http://doi.org/10.2139/ssrn.2547660

Jordan, A., Benson, D., Wurzel, R., & Zito, A. (2011). Policy instruments in practice. In J. Dryzek, R. Norgaard, & D. Schlossberg (Eds.), *Oxford handbook of climate change and society* (pp. 536–549). Oxford: Oxford University Press.

Jordan, A., Benson, D., Wurzel, R., & Zito, A. (2012). Environmental policy: Governing by multiple policy instruments? In J. Richardson (Ed.), *Constructing a policy-making state? Policy dynamics in the EU* (pp. 104–124). Oxford: Oxford University Press.

Karlan, D., Osei, R., Osei-Akoto, I., & Udry, C. (2014). Agricultural decisions after relaxing credit and risk constraints. *The Quarterly Journal of Economics*, *129*(2), 597–652. http://doi.org/10.1093/qje/qju002

Keck, M., & Sikkink, K. (1998). *Activists beyond borders: Advocacy networks in international politics*. Ithaca, NY: Cornell University Press.

Keucheyan, R. (2018). Insuring climate change: New risks and the financialization of nature. *Development and Change*, 49(2), 484–501. http://doi.org/10.1111/dech.12367

Kingdon, J. W. (1995). *Agendas, alternatives and public policies*. New York: Longman.

Kreft, S., Schäfer, L., Behre, E., & Matias, D. (2017). *Climate risk insurances for resilience: Assessing countries' implementation plans* (Report No. 23). Bonn: Munich Climate Insurance Initiative. Retrieved from http://collections.unu.edu/view/UNU:6321

Kunreuther, H., & Michel-Kerjan, E. (2007). *Climate change, insurability of large-scale disasters and the emerging liability challenge* (NBR Working Paper Series 12321). Cambridge, MA.

Kunreuther, H., & Michel-Kerjan, E. (2014). Economics of natural catastrophe risk. In *Handbook of the economics of risk and uncertainty* (Vol. 1, pp. 651–699). Oxford: Elsevier.

Kurki, M. (2008). *Causation in international relations reclaiming causal analysis*. Cambridge Studies in International Relations. Cambridge: Cambridge University Press. http://doi.org/10.1017/CBO9780511491481

Lascoumes, P., & Le Gales, P. (2007). Introduction: Understanding public policy through its instruments? From the nature of instruments to the sociology of public policy instrumentation. *Governance*, *20*(1), 1–21. http://doi.org/10.1111/j.1468-0491.2007.00342.x

Linder, S., & Peters, G. (1989). Instruments of government: Perceptions and contexts. *Journal of Public Policy*, 9(1), 35–58. http://doi.org/10.1017/S0143814X00007960

Linnerooth-Bayer, J., & Hochrainer-Stigler, S. (2015). Financial instruments for disaster risk management and climate change adaptation. *Climatic Change*, *133*(1), 85–100. http://doi.org/10.1007/s10584-013-1035-6

Linnerooth-Bayer, J., Hochrainer-Stigler, S., & Mechler, R. (2012). *Mechanisms for financing the costs of disasters*. Retrieved from https://assets.publishing.service.gov.uk/government/uploads/system/uploads/attachment_data/file/287474/12-1308-mechanisms-financing-costs-of-disasters.pdf

Lobo-Guerrero, L. (2011). *Insuring security: Biopolitics, security and risk*. Abingdon and New York: Routledge.

Lobo-Guerrero, L. (2012). *Insuring war: Sovereignty, security and risk*. Abingdon and New York: Routledge.

Lobo-Guerrero, L. (2016). *Insuring life*. Abingdon and New York: Routledge. http://doi.org/10.4324/9781315880228

LSE. (2019). *Compendium of disaster risk transfer schemes*. Retrieved June 4, 2019, from www.lse.ac.uk/GranthamInstitute/wp-content/uploads/2019/05/Diaster_risk_transfer_data-.xlsx

Lubowski, R. N., Bucholtz, S., Claassen, R., Roberts, M. J., Cooper, J. C., Gueorguieva, A., . . . Cooper, J. C. (2006). *Environmental effects of agricultural land-use change—The role of economics and policy* (Economic Research Report No. 25). Retrieved from https://naldc.nal.usda.gov/download/18042/PDF

Lybbert, T., & Carter, M. (2014). Bundling drought tolerance and index insurance to reduce rural household vulnerability to drought. In A. Balisacan, U. Chakravorty, & M. Ravago (Eds.), *Risk, resources, and development: Foundations of public policy* (pp. 410–414). Amsterdam: Elsevier.

Mackie, J. L. (1965). Causes and conditions. *American Philosophical Quarterly*, *2*(2), 245–264.

Mahoney, J., & Rueschemeyer, D. (2002). *Comparative historical analysis in the social sciences*. Cambridge: Cambridge University Press.

Mahul, O., & Stutley, C. (2010). *Government support to agricultural insurance: Challenges and options for developing countries*. Washington, DC. Retrieved from http://documents.worldbank.org/curated/en/698091468163160913/Government-support-to-agricultural-insurance-challenges-and-options-for-developing-countries

Majone, G. (1976). Choice among policy instruments for pollution control. *Policy Analysis*, *2*(4), 589–613.

MCII. (2017). *Weathering climate change through climate risk transfer solutions*. Bonn. Retrieved from www.climate-insurance.org/fileadmin/user_upload/20170316_MCII_G20_Position_2017.pdf

McIntosh, C., Povel, F., & Sadoulet, E. (2016). *Utility, risk, and demand for incomplete insurance: Lab experiments with Guatemalan cooperatives*. http://are.berkeley.edu/~sadoulet/papers/InsuranceGame.pdf

Miranda, M., & Farrin, K. (2012). Index insurance for developing countries. *Applied Economic Perspectives and Policy*, *34*(3), 391–427. http://doi.org/10.1093/aepp/pps031

Mishra, K., Gallenstein, R., Sam, A. G., Miranda, M. J., Toledo, P., & Mulangu, F. (2018, July 28–August 2). *You are approved! Insured loans improve credit access and technology adoption of Ghanaian farmers*. Conference Paper, International Association of Agricultural Economists, Conference. Vancouver. Retrieved from https://ideas.repec.org/p/ags/iaae18/277089.html

Mobarak, A. M., & Rosenzweig, M. R. (2013). Informal risk sharing, index insurance and risk taking. *American Economic Review*, *103*(3), 375–380. http://doi.org/10.1257/aer.103.3.375

Mukherjee, I., & Howlett, M. (2015). Who is a stream? Epistemic communities, instrument constituencies and advocacy coalitions in public policy-making. *Politics and Governance*, *3*(2), 65–75. http://doi.org/10.17645/pag.v3i2.290

Müller, B., Johnson, L., & Kreuer, D. (2017, June). Maladaptive outcomes of climate insurance in agriculture. *Global Environmental Change*, *46*, 23–33. http://doi.org/10.1016/j.gloenvcha.2017.06.010

Müller, B., & Kreuer, D. (2016). Ecologists should care about insurance, too. *Trends in Ecology and Evolution*, *31*, 1–2. http://doi.org/0.1016/j.tree.2015.10.006

Munich Re. (2018). *A stormy year: Natural catastrophes 2017*. München. Retrieved June 8, 2018, from www.munichre.com/topics-online/en/2018/topics-geo/topics-geo-2017.

Munich Re. (2019). *Extreme storms, wildfires and droughts cause heavy nat cat losses in 2018*. Retrieved January 8, 2019, from www.munichre.com/en/media-relations/publications/press-releases/2019/2019-01-08-press-release/index.html

Murphy, S. (2010). *Investing in life: Insurance in antebellum America*. Washington, DC: John Hopkins University Press.

OECD. (2012). *Disaster risk assessment and risk financing—A G20/OECD methodological framework*. Retrieved from www.oecd.org/gov/risk/g20oecdframeworkfordisasterriskmanagement.htm

O'Hare, P., White, I., & Connelly, A. (2016). Insurance as maladaptation: Resilience and the "business as usual" paradox. *Environment and Planning C: Government and Policy*, *34*(6), 1175–1193. http://doi.org/10.1177/0263774X15602022

O'Malley, P. (2007). *Risk, uncertainty and government* (Vol. 39). London: The GlassHouse Press. http://doi.org/10.4324/9781843146025

Ostrom, E., Cox, M., & Schlager, E. (2014). An assessment of the institutional analysis and development framework and introduction of the social-ecological systems framework. In *Theories of the policy process* (pp. 267–306). Boulder, CO: Westview Press.

Oxfam. (2018). *Facing risk: Options and challenges in ensuring that climate/disaster risk finance and insurance deliver for poor people*. Oxford. Retrieved from www.oxfam.org/en/research/facing-risk-options-and-challenges-ensuring-climatedisaster-risk-finance-and-insurance

Paterson, M. (2001). Risky business: Insurance companies in global warming politics. *Global Environmental Politics*, *1*(4), 18–42. http://doi.org/10.1162/152638001317146354

Pearson, R. (Ed.). (2010). *The development of international insurance*. London: Pickering & Chatto.

Pearson, R. (2016). The evolution of the industry structure. In N. V. Haueter & G. Jones (Eds.), *Managing risk in reinsurance: From city fires to global warming* (pp. 70–91). Oxford: Oxford University Press. http://doi.org/10.1093/acprof:oso/9780198754916.003.0003

Pearson, R., & Takau, Y. (2015). *Corporate forms and organizational choice in organizational forms in insurance*. Oxford: Oxford University Press. http://doi.org/10.1093/acprof

Pearson, R., & Yoneyama, T. (2015a). An overview of the history and theory. In *Corporate forms and organizational choice in international insurance* (pp. 1–31). Oxford: Oxford University Press. http://doi.org/10.1093/acprof

Pearson, R., & Yoneyama, T. (Eds.). (2015b). *Corporate forms and organizational choice in international insurance*. Oxford: Oxford University Press. http://doi.org/10.1093/acprof:oso/9780198739005.003.0001

Peoples, C., & Vaughan-Williams, N. (2010). *Critical security studies: An introduction*. London: Routledge.

Peters, G. (2002). The politics of tool choice. In L. M. Salamon (Ed.), *The tools of government—A guide to the new governance* (pp. 552–564). Oxford: Oxford University Press.

Peterson, N. D. (2012). Developing climate adaptation: The intersection of climate research and development programmes in index insurance. *Development and Change*, *43*(2), 557–584. http://doi.org/10.1111/j.1467-7660.2012.01767.x

Pidgeon, N., & Butler, C. (2009). Risk analysis and climate change. *Environmental Politics*, *18*(5), 670–688. http://doi.org/10.1080/09644010903156976

Rasmussen, V. M. (2004). "It sounds like a riddle": Security studies, the war on terror and risk. *Millennium—Journal of International Studies*, *33*(2), 381–395. http://doi.org/10.1177/03058298040330020601

Rohlfing, I. (2013). Varieties of process tracing and ways to answer the why-question. *European Political Science*, *12*(1), 31–39. http://doi.org/10.1057/eps.2012.7

Salamon, L. M. (2002a). The new governance and the tools of public action. In *The tools of government—A guide to the new governance* (Vol. 28, pp. 1–47). Oxford: Oxford University Press.

Salamon, L. M. (2002b). *The tools of government—A guide to the new governance*. Oxford: Oxford University Press.

Schäfer, L., Waters, E., Kreft, S., & Zissner, M. (2016). *Making climate risk insurances work for the most vulnerable* (UNU-EHS Publication Series No. Policy Report No. 1). Bonn: Munich Climate Insurance Initiative. Retrieved from www.climate-insurance.org/fileadmin/mcii/documents/MCII_PolicyReport2016_Making_CRI_Work_for_the_Most_Vulnerable_7GuidingPrinciples.pdf

Schatzki, T. R. (1996). *Social practices: A Wittgensteinian approach to human activity and the social*. Cambridge: Cambridge University Press.

Schatzki, T. R., Knorr-Cetina, K., & von Savigny, E. (Eds.). (2001). *The practice turn in contemporary theory*. New York: Routledge.

Schmidt, R. (2012). *Soziologie der Praktiken*. Berlin: Suhrkamp Verlag.

Searle, J. (2011). *Making the social world: The structure of human civilization*. Oxford: Oxford University Press.

Stone, D. (2013). *Knowledge actors and transnational governance: The public-private Nexus in the global age*. Basingstoke: Palgrave Macmillan. http://doi.org/10.1057/9781137022912

Sturm, T., & Oh, E. (2010). Natural disasters as the end of the insurance industry? Scalar competitive strategies, alternative risk transfers, and the economic crisis. *Geoforum*, *41*(1), 154–163. http://doi.org/10.1016/j.geoforum.2009.09.010

Suarez, P., & Linnerooth-Bayer, J. (2011). *Insurance-related instruments for disaster risk reduction*. Background Paper for the 2011 Global Assessment Report on Disaster Risk Reduction. Submitted to the International Strategy for Disaster Reduction (UNISDR). Geneva. Retrieved from www.preventionweb.net/english/hyogo/gar/2011/en/bgdocs/Suarez_&_Linnerooth-Bayer_2011.pdf

Surminski, S. (2016). *Submission to the call for "best practices, challenges and lessons learned from existing financial instruments at all levels that address the risk of loss and damage associated with the adverse effects of climate change" by the UNFCCC's executive council*. London. Retrieved from www.lse.ac.uk/GranthamInstitute/publication/submission-to-the-executive-committee-of-the-warsaw-international-mechanism-for-loss-and-damage/

Surminski, S., & Oramas-Dorta, D. (2011). *Building effective and sustainable risk transfer initiatives in low- and middle-income economies: What can we learn from existing insurance schemes?* London. Retrieved from http://www2.lse.ac.uk/GranthamInstitute/Home.aspx

Swiss Re Institute. (2018). *Closing the protection gap—Disaster risk financing: Smart solutions for the public sector*. Retrieved from www.swissre.com/Library/closing-the-protection-gap-disaster-risk-financing.html

Swiss Re Institute. (2019a). *Sigma explorer*. Retrieved September 1, 2019, from www.sigma-explorer.com/

Swiss Re Institute. (2019b). *World insurance: The great pivot east continues*. Sigma. Retrieved from www.swissre.com/dam/jcr:b8010432-3697-4a97-ad8b-6cb6c0aece33/sigma3_2019_en.pdf

Thomas, G. (2011). The case: Generalisation, theory and phronesis in case study. *Oxford Review of Education*, *37*(1), 21–35. http://doi.org/10.1080/03054985.2010.521622

Thoyts, R. (2010). *Insurance theory and practice*. London and New York: Routledge.

Trebilcock, M. (1982). The choice of governing instrument. *International Review for Law and Economics*, *2*, 29–46. http://doi.org/10.1016/0144-8188(82)90012-6

UNCTAD. (1993). *Insurance in developing countries: An assessment and review of developments (1989–1992)*. Geneva: UNCTAD.

Vaananen, E., Nett, K., Costella, C., & de Suarez Mendler, J. (2019). *Linking climate risk insurance with shock-responsive social protection contribute to climate resilience*. Retrieved from www.insuresilience.org/wp-content/uploads/2019/03/insuresilience_policybrief_1-2019_190312_web.pdf

Vickery, J., Cole, S., Gine, X., Tobacman, J., Topalova, P., & Townsend, R. (2013). Barriers to household risk management: Evidence from India. *American Economic Journal: Applied Economics*, 5(1), 104–135. http://doi.org/10.1257/app.5.1.104

Voß, J. P., & Simons, A. (2014). Instrument constituencies and the supply side of policy innovation: The social life of emissions trading. *Environmental Politics*, *23*(5), 735–754. http://doi.org/10.1080/09644016.2014.923625

Warner, K., Ranger, N., Surminski, S., Arnold, M., Linnerooth-Bayer, J., Michel-Kerjan, E., . . . Herweijer, C. (2009). *Adaptation to climate change: Linking disaster risk reduction and insurance*. Retrieved from www.climate-insurance.org/fileadmin/mcii/documents/MCII_UNISDR_2009_Adaptation_CC_Linking_DRR_and_Insurance.pdf

World Bank. (2014). *Financial protection against natural disasters: An operational framework for disaster risk financing and insurance*. Washington, DC. Retrieved from http://documents.worldbank.org/curated/en/2014/01/24156733/financial-protection-against-natural-disasters-products-comprehensive-strategies-operational-framework-disaster-risk-financing-insurance

World Bank. (2017). *Sovereign climate and disaster risk pooling: World Bank technical contribution to the G20*. Washington, DC. Retrieved from http://documents.worldbank.org/curated/en/837001502870999632/World-Bank-technical-contribution-to-the-G20

World Meteorological Organization. (2019). *July matched and maybe broke the record for the hottest month since analysis began*. Retrieved August 1, 2019, from https://public.wmo.int/en/media/news/july-matched-and-maybe-broke-record-hottest-month-analysis-began

Wright, H. (2009, December). Using an "emergent design" to study adult education. *Educate*, 62–73.

Wurzel, R., Zito, A., & Jordan, A. (2013). *Environmental governance in Europe*. Cheltenham and Northampton, MA: Edward Elgar Publishing. http://doi.org/10.4337/9781849804721

Zelizer, V. (1979). *Morals and markets: The development of life insurance in the United States*. New York: Columbia University Press.

2 Understanding insurance

A primer

Despite being an element of everyday life—at least in the OECD world—insurance is not self-explanatory. Quite in contrast, the functioning of insurance is a rather complex issue and deserves some scrutiny to allow for a critical engagement. Throughout the book I will use a good deal of very technical vocabulary such as, for instance, 'risk pooling', 'hazard', 'indemnity-trigger', 'attachment point' or 'risk appetite'. To make the book more accessible and facilitate understanding of the functioning of the various regional catastrophe risk pools, this chapter provides a brief introduction to insurance and its recent history. The primer can, of course, be jumped by readers who have already a decent knowledge of insurance and its recent developments.

What is insurance?

Simply put, insurance is a contract for indemnity (Bolt, 2012, p. 20; Thoyts, 2010, p. 4). It is a (annual) financial contract between two parties in which one party (the insurer) agrees to compensate the financial losses of the other party (the insurance policyholder) suffered from an *uncertain* but *to a certain degree predictable event*[1] in exchange for a premium. Insurance is thus essentially about **financial compensation**: compensation against damage or loss. In other words, by buying insurance one buys **the right to some level of compensation** for the financial consequences for an **uncertain but to certain degree predictable event**.

The uncertain but to a certain degree predictable event is typically labeled '**risk**'. It describes **a potential exposure to a financial loss**. What constitutes a risk is not as easy as it may at first sound. In a broad sense it may be any 'event that has the potential to cause a financial loss'. The **event that has the potential to cause a loss** is usually termed '**peril**'. It describes "the precise means by which the risk could come about" (Thoyts, 2010, p. 5). A peril may be, for instance, a theft, a car accident or a hurricane. The peril is typically influenced by a '**hazard**', the 'something' that **"affects the probability of a risk occurring"** (ibid.). Hazards are, for instance, unlocked car doors, defective breaks or damaged property.

The important point is, however, that not every risk can be insured. For a risk to be insurable a number of criteria must be fulfilled. On general level, the occurrence of the event must be uncertain and to some degree predictable. Insurance is about possibilities, not certainties. It is crucial that insurance policyholders may not be able to "deliberately bring about the loss" (ibid., p. 17 my emphasis). In addition, the costs must be quantifiable, that is, translatable in financial replacement costs. The event that triggers a loss must be clearly definable and thus sufficiently particular. Moreover, it is crucial that the size of the potential losses is manageable for the insurance company.

These are only five characteristics and they are far from clear-cut: For instance, how much worth for instance is one's life? When does a potential loss become unmanageable for an insurance company? It is noteworthy that there exists no common catalogue in the insurance literature what precisely constitutes an insurable risk and what does not. Some authors list more criteria than others (e.g. Haubner, 2007; Nguyen, 2007; Thoyts, 2010; Wolgast & Theis, 2013). A reason for this variety is that what constitutes an insurable risk and what does not has been changing throughout insurance history (Carter & Falush, 2009; James, Borscheid, Gugerli, & Straumann, 2013). The expansion of the boundary of insurability is, in fact, also somehow the empirical topic of this book.

Insuring as risk-transfer mechanism

From a financial-sociological perspective, insuring may be better understood as a financial **"risk-transfer mechanism"** (Thoyts, 2010, p. 4): **one party shifts the *financial* consequences of a potential event to another party in exchange for a premium.** Yet, in contrast to the common perception, this other party is not the insurance company—although this is, admittedly, the legal relationship. A more accurate view is that the insurance policyholder shifts the financial consequences of a potential event, that is, its indemnification, to a '**common pool**'. The practice of transferring the risk of suffering a financial loss from an individual to a collective pool is termed '**risk pooling**'.

The common pool comprises a group of individuals (or entities such as firms or even states as in the case of regional catastrophe risk pools) agreeing to set aside a smaller amount of money, termed 'premium', to put it into a collective fund (see Figure 2.1). This collective fund then compensates them in case a pre-agreed event occurs. Thus, in contrast to insurers' frequent advertisement 'you will not be paying for the mistakes of others', this is *de facto* precisely what insurance policyholders are doing when they sign an insurance contract (ibid., p. 11). In this sense, insurance may also be understood as a solidarity mechanism: a solidarity mechanism where pool members compensate each other by paying a premium into a collective fund.

This solidarity principle is, however, limited insofar as the degree of solidarity depends on the approach taken to define the premium level each participant pays. In private insurance, the premium level is typically tied to the pool member's exposure to the potential financial loss, respectively the financial consequences it wishes to transfer (see the following, *On insurance premiums*). In public insurance schemes, the premium level depends typically not only on the financial exposure but also on a pool member's ability to pay.

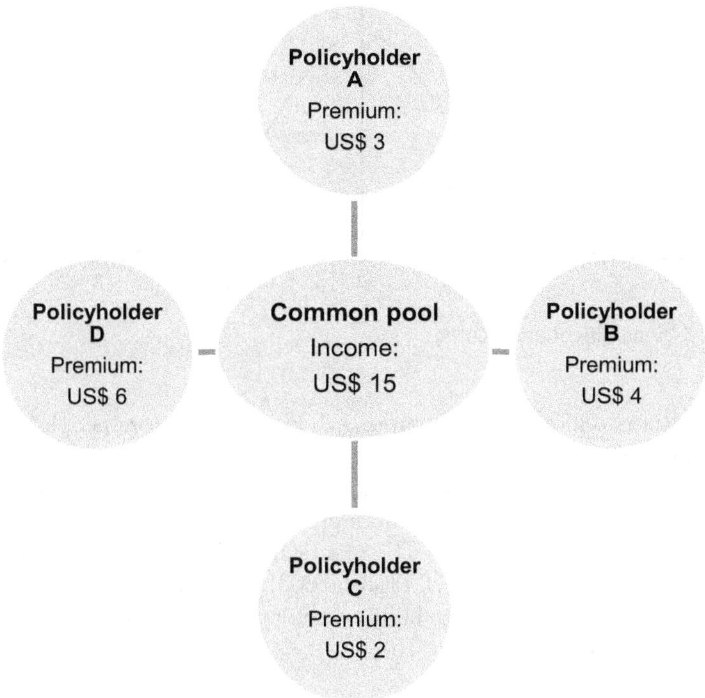

Figure 2.1 The concept of risk pooling
Source: modified from Thoyts, 2010, p. 302

Those who organize the risk transfer and take care of the **common pool** are termed '**insurance companies**'. Conceptually speaking, insurance companies can be understood as financial intermediaries or financial service providers (Feldmann, 2002, p. 187; Thoyts, 2010, p. 10) of the common pool (see the following, *The different roles of insurer*). The term 'common pool' refers thereby "to the fact that the risks borne by the pool are all of a similar nature, even if they are not identical" (Thoyts, 2010, p. 11). In other words all insurance policyholders face the same peril—e.g., a theft, a car accident, a hurricane—yet with different probabilities and different financial consequences. This **heterogeneity of risk** is **crucial** to the **functioning of the insurance** concept. If all pool members would face the identical risk, insurance would not work.

Insurance works best the greater the pool. The reason is that the majority of a sufficiently large number of insurance policyholders will typically not experience a major financial loss. This might sound counterintuitive, yet, this statistical phenomenon is best illustrated by the bell-shaped normal distribution curve (see Figure 2.2): where there is person with risk +3, there is also a person with risk -3.

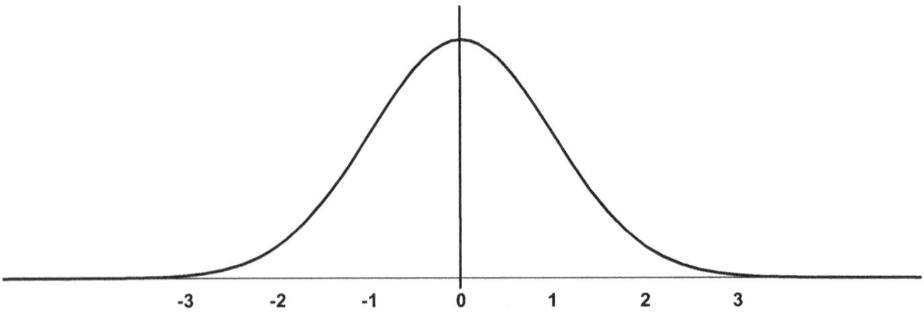

Figure 2.2 Normal distribution curve

Yet, the majority will concentrate around 0. Hence, the more people in a pool, the more likely the losses will even out.

In probability theory this phenomenon is described as the **law of the large numbers:** The greater the pool the more predictable the losses. It is the second axiom that makes insurance work. The practical implication is that insurers will always seek to enlarge the pool (cf. Pfister, 2003, p. 5). Against this background, it becomes clear that insurance is built on the principle where "the many compensate the few" (Thoyts, 2010, p. 16).

The benefits of insuring

Why do individuals, groups thereof or larger entities buy at all insurance? The main benefit of insuring is first of all of financial nature.[2] For some events the costs could be immense and overwhelm the financial capacity of an individual; in the worst case, they could lead to bankruptcy and wipe out a certain lifestyle. In contrast to the costs of such a personal financial disaster, the costs of regular premium payments are small and, by nature, predictable. Insurance spreads the costs of such a disaster over a larger time span. This cost predictability also facilitates financial planning and leads, psychologically speaking, to a 'peace of mind'. Another merit is that insurance provides readily accessible funds that otherwise would have to be retained or at least somehow be mobilized (e.g., in the form of a credit) if one wants to restore the previous status. Insurance provides "the means to repair lives and livelihoods under adverse circumstances" (Lobo-Guerrero, 2011, p. 7). In other words, it limits the measures that must be undertaken to stay resilient and keep one in business. That means, insurance fulfills primarily a **conservative function**: It provides **financial security**. Yet, it offers financial security that then also translates into socioeconomic security.

Insurance has also an **empowering function**: it frees up money—the money that would have been retained by risk-averse actors—for other purposes such as investment but also trivial activities such as consumption. Many banks, for instance, would only give credit for an entrepreneur if s/he has a liability insurance for errors and failures in place (Pfister, 2003, p. 4). This implies that insurance enables or even **promotes business activities, investments and certain lifestyles.** In a more general sense, insurance allows its holders to use financial resources more productively (if spending money for consumption can be labeled productively).

The insurance premium

As mentioned earlier, the regular amount of money paid to insurers is termed '**premium**'. What makes up a premium? A premium contains typically four components: (1) pure risk premium (2) administrative costs (3) a contingency load and (4) a profit (Thoyts, 2010, p. 14, see Figure 2.3).

The pure risk premium is the notional sum of money the insurance company needs to settle claims. To set the risk premium, insurers estimate both the frequency and severity of the losses an event is likely to generate and create an annual average-loss. The information underpinning such actuarial calculations is typically based on experiences, that is, available historical loss data, and supplemented with other publicly available data (Parsons, 2015, p. 1239). The pure risk premium is thereby adjusted over time to reflect changes in risk and prices. The general rule is, the greater the expected loss, the greater the premium. Usually, the pure risk premium makes up about 65–75% of the total premium. 'Administrative costs' refer to the costs of running the pool such as underwriting, settling claims, investing premiums, etc. (Thoyts, 2010, p. 14). The 'contingency load' represents a sum to cover miscalculations. The 'profit' should be self-explaining: Insurance is the business of loss and insurance companies are not altruists.

Pricing insurance premiums

Economic insurance theory suggests that a pool should not be based on '**flat rate premiums**' but on '**fair premiums**'. A fair premium is understood as a premium that reflects an individual's probability to suffer a loss and the financial severity involved in suffering this loss; this requires, of course, knowledge thereof. As

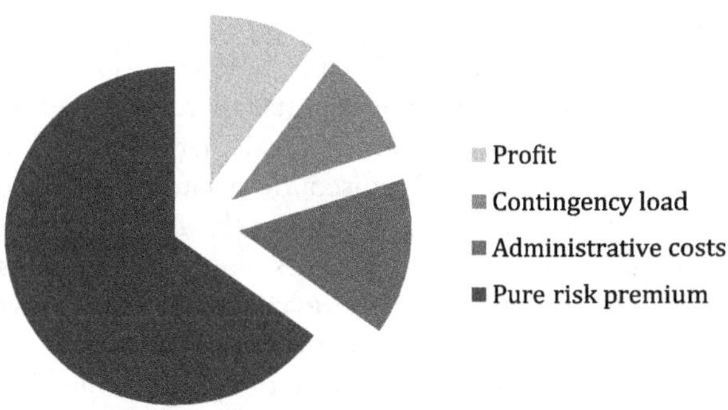

Figure 2.3 Traditional model of an insurance premium

Source: adapted from Thoyts, 2010, p. 14

insurers usually lack data of the individual being interested in purchasing insurance, they calculate the premium basically "at arm's length" (ibid., p. 12). The calculation involves typically components such as a hazard analysis, a vulnerably analysis and a financial replacement costs analysis. Simply put, the calculation is based on available actuarial data and some statistical, that is, stochastic operations to understand the chances and distribution of hazardous events (Collier, 2008, p. 227) and to estimate the subsequent of losses for the insurance policyholder. This **practice of valuation** is termed '**underwriting**'. As one can easily imagine, the quality thereof depends substantially on the data available and the quality thereof. Private insurance pools are usually based on fair premiums to prevent cross-subsidization and make insurance equally attractive to all (at least in theory). Flat rate premiums are the exception and are typically for pools where risk is nearly identical and a differentiation in premiums is not cost-effective (Thoyts, 2010, p. 12).

The different roles of insurers

As mentioned earlier, insurance is not only a financial contract and a risk transfer mechanism but also the business of loss. So what are the roles of insurance companies in an insurance arrangement? Insurers perform different tasks. In general terms insurers can be conceived as financial service providers. Insurers are **financial service providers** insofar as they control the membership of the pool, calculate the premiums, control the claim payments, guarantee the solvency of the pool, try to improve the risk insured to their favor and conclude reinsurance arrangements for the case an insured loss exceeds the pool's paying capacity (ibid., pp. 12–13). As already mentioned, the costs of these services add to the premium in the form of administrative costs. A closer view reveals, however, that insurers are also 'risk managers' and investment houses. The latter functions are crucial to make insurance work in practice.

The role of insurer as 'risk managers'

It is important to point out that insurers generally seek to modify their proper calculation to their favor. That means they seek to prevent or reduce the potential for financial vulnerability (cf. Renn, 2008, p. 170). They do so, for example, by "imposing some minimum security requirements before offering the cover . . . by physically inspecting the risk and requiring particular improvements" (Thoyts, 2010, p. 13) or by offering a financial discount (that is, sending price signals) if some measures are undertaken or by requiring the insured to pay for the loss up to a certain amount.[3] In this sense insurers may also be understood as 'risk managers'. **Activities to alter behavior to prevent or at least to lower the occurrence of an event** can be labeled **'risk management practices'** (Pfister, 2003, p. 20). These sort of steering or policing activities are informed by the insurers' fear of '**moral hazard**', the problem of '**adverse selection**' and the '**precautionary principle**'.

Moral hazard describes the threat that the insured changes its behavior after s/he is insured. The insured might, for instance, be less concerned about the consequences of his behavior and engage in a more risky behavior and thereby provoke a loss. Such behavior, this "lack of moral character" (Moss, 2002, p. 37), can increase the likelihood or magnitude of a loss and thus of a payout.

Adverse selection describes an imbalance in the common pool due to an information problem. It follows from the fact that the insured typically knows more about his own level of risk than insurers (ibid., p. 36). People might falsify or hide critical information from insurers; since they pay lower premiums if insurers put them in a lower risk class they have a strong incentive to do so. The problem arising from this information asymmetry is that insurers cannot calculate the risk profile and the price tag correctly. Such behavior has the potential to distort the financial viability of common pool and even render it dysfunctional due to the fact that insurers might end in a situation where claim settlements exceed available financial resources.

The precautionary principle, in turn, derives from insurers' insecurity towards their own model. As pointed out earlier, insurers calculate the risk and associated losses with the help of actuaries. The calculations are based on a range of assumptions and depend substantially on data availability and quality. Actuarial calculations might be incorrect, ignoring or over-/downplaying some factors. This risk is particular prominent when it comes to larger natural catastrophes. Larger natural catastrophes such as earthquakes do not occur very often. This leads to a scarcity of historical loss data and thus to poor statistics. The usefulness of historical loss data from catastrophes is limited insofar as the number of properties, the value thereof, construction material costs, building codes as well as repairing costs has most likely changed since the last recorded event. This makes extrapolating from the past to the future extremely difficult (Poliquin & Lalonde, 2012, p. 196). To deal with these 'known unknowns' as well as the 'unknown unknowns' insurers run sophisticated catastrophe models ('vendor models') based on scenarios that estimate event probability and losses (Jarzabkowski, Bednarek, & Spee, 2015). These models are built on assumptions and thus are prone to mistakes. With risk management practices, insurers seek to mitigate potential errors and enhance their financial robustness against an uncertain event.

The role of (re-)insurers as 'investment houses'

(Re-)insurers are also massive **"investment houses"** (Bougen, 2003, p. 258 my emphasis). (Re-)insurers do not keep the premium payments in cash, as this would imply extremely high opportunity costs. Instead, they **use the temporal space between the premium payment and the claim settlement by investing the premiums** in assets to generate further revenues and a profit. The investments allow (re-)insurers to lower premiums and/or generate sufficient reserves to settle claims.

In economic language this practice "guarantees the solvency of the pool" (Thoyts, 2010, p. 13). Guaranteeing the solvency of the pool is particularly an issue where the number of pool members is comparatively low (ibid., p. 13) and/or in the case of a greater financial loss, associated, for example, with a natural disaster. In this case (re-)insurers need to raise massive amounts of capital in a relatively short time span to settle claims; doing so might not be as easy. This condition has implication for the investment practices of (re-)insurers. It requires that the capital (re-)insurers invest is highly liquid (Pfister, 2003, pp. 24, 32). Consequently, (re-)insurers usually tend to invest in assets which can be sold fairly quickly (e.g., stocks, bonds).

What is reinsurance?

While insurers should have pooled enough financial resources to pay compensations, a big loss event—such as a major hurricane—or a series of many smaller events in a short time period could lead to claims that exceed the capital reserves available. It could lead to a 'liquidity crisis' or even to the insolvency of the pool. To prevent this from happening, insurers usually put some reinsurance arrangement in place.

In a reinsurance arrangement an **insurer transfers the sourcing of funds to settle claims for larger losses to a third party**, namely a re-insurer, in exchange for a premium ('catastrophe load'). As re-insurers pay the losses that typically exceed the financial capacities of insurers, they are frequently labeled as the 'insurers of insurer'. On a basic level a reinsurance arrangement parallels the characteristics and functioning of an insurance arrangement. Conceptually speaking, it reflects a risk-transfer mechanism just on a higher level: now its insurers that pool risks. In exchange for a premium, re-insurers assume the costs of less predictable events—typically man-made or natural catastrophes—from insurer via an indemnity contract (Thoyts, 2010, p. 165).

Insured event: catastrophic losses

Importantly, re-insurers assume losses only from catastrophic events. A **'catastrophic event'** is an event of **'low frequency and high severity'**—that is, an event with high financial impact; an event, for instance, that affects all pool members in parallel or larger number of members such as a *major* hurricane or earthquake. Catastrophic events can be contrasted with the 'normal' events insurer focus on, namely high frequency—low severity events (Pfister, 2003, p. 6). Statistically speaking, these events are outliers, or **"black swans"** as Taleb (2007) famously put it; they can be imagined as being distributed along the 'fat tails' on the bell-curve shaped normal distribution curve. The infrequent realization of catastrophic events—and thus the lack of empirical data—is also what makes them so hard to predict (Jarzabkowski et al., 2015). Probability models are typically based on past events; the models extrapolate from the past to the future. Yet, a large-scale catastrophe may realize only 1:10,000 years and the subsequent, actual (financial) impact thereof is simply not known. As already mentioned earlier, the (non-)occurrences of these 'outliers' and the financial costs of such an event involve a great degree of uncertainty and thus unpredictability. To come up with probabilities, re-insurers work at the edge of knowledge, using complex models that combine historical loss data with scenarios about possible future developments.

The functioning of a reinsurance arrangement

Reinsurance works similar to the logic presented previously: **In reinsurance arrangements insurers transfer a part of their financial costs with other insurers into a higher-level pool** administered by a reinsurance company in exchange for a premium. In return, re-insurers will keep enough capital reserves to indemnify insurer.

Usually insurers do not transfer all the costs to re-insurers but only a portion of it (Jarzabkowski et al., 2015, p. 10), namely those losses associated with catastrophes. Mostly, insurers retain a certain amount of losses and buy compensation for events that exceed this level. Such arrangements are termed '**excess-of-loss arrangements**' and were developed in the mid-twentieth century (Gugerli, 2013).

Excess-of-loss arrangements are characterized by an '**attachment point**' and an '**exhaustion point**'. An attachment point reflects the specified dollar amount of a loss (e.g., US$ 100 million) until the primary insurers believe they can handle the claims in-house by using their own financial resources. For costs beyond that level, insurers buy the right to compensation from re-insurers (Jarzabkowski et al., 2015, pp. 9, 201). That means, if the losses for insurers breach that level they become a loss for re-insurers.[4] Yet, it is also important to note that re-insurers do not resume unlimited losses. In excess-of-loss arrangements re-insurers compensate only up to a certain limit ('exhaustion point'). For costs beyond this limit re-insurers will not provide further coverage. In other words, compensations to insurers are capped after an exhaustion point. Insurers are typically then again liable for the rest.

Excessive-of-loss arrangements are usually structured into various layers (see Figure 2.4). The practice of dividing losses in various layers is termed '**risk layering**'. In exchange for a premium various re-insurers resume a certain risk layer (or loss level) from the primary insurer. They compensate them, so to say. To take the illustrative example by Jarzabkowski et al. (2015, p. 9): Layer 1, for example, would involve compensations from re-insurers to the insurer for losses falling between US$ 100–150 million, layer 2 compensations for losses in the range of US$ 150–300 million, layer 3 compensations US$ 300–600 million, with US$ 600 million being the defined limit. Re-insurers would not provide any coverage for losses beyond that level. These various layers would also come with different premium costs for the insurer. The principle is the higher the coverage—that is, the higher the protection the primary insurer seeks—the higher the premium it has to pay. This practice of risk layering is central to make reinsurance coverage possible in the first place, as these layers are traded as separate financial objects, as I will explain in the next section.

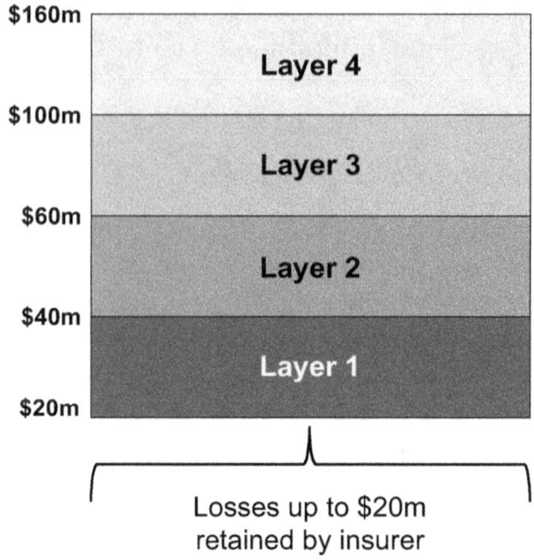

Figure 2.4 Structure of an excess-of-loss arrangement

Source: (Jarzabkowski et al., 2015, p. 67) © Jarzabkowski, Bednarek, Spee reproduced with permission of the licensor through PLSclear

The practice of collective risk spreading and consensus pricing

Typically reinsurance coverage is not provided by a single re-insurer. The claims re-insurers need to cover can easily reach several billion US dollars. The terrorist attacks on the World Trade Center on 9/11, for instance, cost the reinsurance industry about US$ 35,5 billion (Jarzabkowski et al., 2015, p. 4). As emphasized by Jarzabkowski et al., the costs of such large-scale catastrophes threaten the financial survival of even the largest reinsurance companies (ibid., p. 4).

To make sure re-insurers are still in business when such a sudden and big loss event happens, they neither take all layers nor a single complete layer. Instead, they typically take just a share of a layer or shares of various layers. According to Thoyts, it is not an exception to have about 30 re-insurers taking one share (Thoyts, 2010, p. 167). This '**collective risk bearing**' or 'risk spreading' across a range of re-insurers **ensures that the loss is distributed among a number of re-insurers**. It secures the survival of re-insurers in the face of the great uncertainty and value of a potential loss when it comes to catastrophic event. Re-insurers "have a better chance of survival if they take a share in a deal, and all have to pay some, but not all, of any subsequent losses" (Jarzabkowski et al., 2015, p. 11). From a legal perspective, it might be noteworthy that the shares are all separate reinsurance contracts. That means, no re-insurer is liable in case another fails to meet its paying obligation; the credit risk is with the insurer (Thoyts, 2010, p. 167).

Insurers sell the shares within one layer typically for the same price. This principle is termed '**consensus price**'. This consensus price is **negotiated in a complex process** involving insurers, re-insurers and brokers (see Figure 2.5). Brokers have a key role in there. They are intermediaries. They help insurers "to structure a

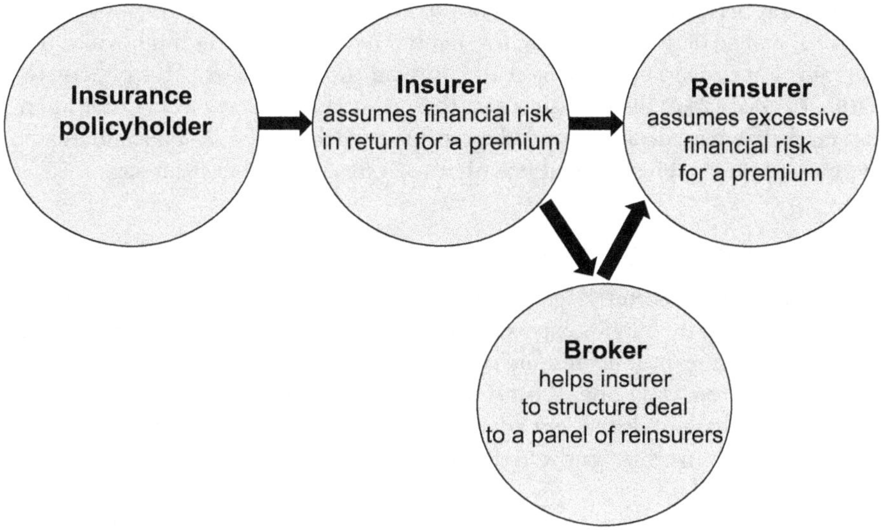

Figure 2.5 Overview of reinsurance market actors and their relationships

Source: modified from Jarzabkowski et al., 2015, p. 5

deal for transferring their risk, and then help them find re-insurers who want to supply capital to such deals" (Jarzabkowski et al., 2015, p. 4).

The specificities of the reinsurance market

Competition within reinsurance market

The principle of a consensus price does not mean that the reinsurance market is not competitive. It is just competitive in a different way, as emphasized by Jarzabkowski et al.: re-insurers compete for shaping the consensus price, different risk types—that is, different perils (e.g., flood, hurricane, tsunami etc.)—different deals (entailing lower/higher risk layers) and larger amounts of shares of the same deals to maximize their profit (Jarzabkowski et al., 2015, p. 25).

The level of risk re-insurers want to resume is termed '**risk appetite**'. The risk appetite of re-insurers is typically fluid and depends, among other things, on the insurance deals re-insurers already hold, their strategy to diversify risks and the overall market conditions—primarily the availability of capital (ibid., p. 18). The competition for shares is, however, also shaped by the (long-term) business relationships between re-insurers and insurers. Insurers usually grant those re-insurers which provided them with reinsurance coverage over a longer time period or who have recently paid for losses a greater share. According to Jarzabkowksi et al., this reciprocity may be read as a form of payback by insurers (ibid., p. 40). It serves to stabilize "long-term trading relationships in a context where the risk of loss, and thus of its appropriate price, is so uncertain" (ibid., p. 57).

Reinsurance deals are traded 'over-the-counter'—that means, **in direct negotiation** between the parties and not in an institutionalized forum—and only on pre-specified dates. There exist five main renewal dates. This implies also that the reinsurance market exists basically only on those five days. This punctuated nature of the market distinguishes also the reinsurance market from other financial markets where deals are traded continuously (ibid., p. 37). This punctuated nature serves to stabilize capital flows between insurers and re-insurers.

The global nature of the reinsurance market

The **reinsurance market is global in nature** (ibid., pp. 11, 34). That means, re-insurers typically bundle excessive risks from different insurers of the globe; hence, the risks re-insurers resume originate from different local insurance markets and not from only one. Re-insurers are able to do so as these catastrophic events are globally independent from each other when it comes to their cause (e.g., after all, hurricanes in the Atlantic basin occur independently from earthquakes in Japan).

Yet, the global nature of the market implies that a massive catastrophic event in, say, the Caribbean, can have severe effects on the balance sheet of European re-insurers and, thus, affect the prices and coverage re-insurers grant local insurers in, say, Japan. Re-insurers will not only need to allocate massive amounts of capital to settle insurance claims but re-stock their capital reserves. The typical way they do is by tightening coverage and raising premiums. Accordingly, large catastrophes "spread throughout the market through global rate rises, thus making these events coverable" (ibid., p. 36). Big catastrophes—and thus the demand for a large amount of capital—lead to **'hard markets'** (high prices), the non-occurrence thereof—and thus the over-supply of capital—to **'soft markets'** (low-prices) (ibid., p. 33). A big event could therefore be a game changer and turn the market from a soft to hard one.

Geographically speaking, re-insurers are mainly located in London, Bermuda and Continental Europe. Munich Re and Swiss Re are the biggest players in the field. Together they hold about 33% of the global reinsurance market by premium ceded in 2012 (ibid., p. 12). Lloyd's of London, albeit not a company but a syndicate that assembles a number of corporations and private individuals to its own specialized market, which underwrites catastrophe risks, is another big corporate body that makes up 6,6% of the global reinsurance markets ceded by premium in 2012 (ibid.).

What is Alternative Risk Transfer (ART)?

The **provision of (re-)insurance coverage via financial market products** is termed 'alternative risk financing' or **'alternative risk transfer'** (ART). The respective **financial market products** that serve to fund these insurance losses are termed **'insurance-linked securities'** (ILS).

Crudely simplified, in ART, (re-)insurers transfer the costs of large, unexpected loss associated with catastrophes to financial market investors. Financial market investors—typically hedge funds, pension funds or specialized investment

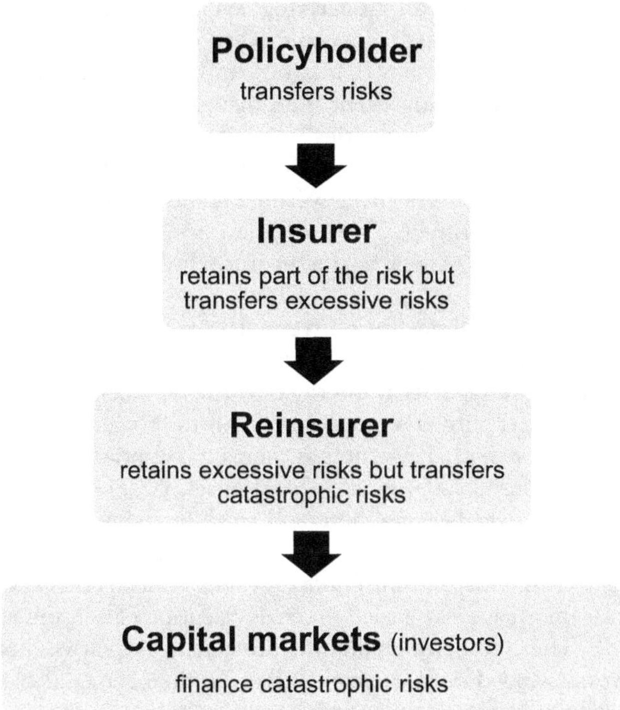

Figure 2.6 Overview of an Alternative Risk Transfer mechanism

funds—assume the costs (associated with large-scale catastrophes) in exchange for an extraordinary financial benefit, for instance, above-average interest payments.

In other words, financial market actors add an additional financial layer to the 'traditional' reinsurance arrangement. It allows (re-)insurers to expand their financial capacity and thus coverage (see Figure 2.6). In arrangement compromising both insurer and re-insurers, it is not the re-insurers but the capital markets that function as "financiers 'of last resort'" (Bougen, 2003, p. 259).

The conceptual building blocks of ART

ART works similar to insurance but is more complicated. ART is enabled a through a fairly complex but in the finance world a very popular process termed '**securitization**'. Simply put, a securitization **transforms financial risks** (such as the excessive insurance losses associated with a large-scale catastrophes) **into commodities** that can be traded on the financial markets (Bougen, 2003, p. 254). The commodities resulting from these transformations are labeled 'securities'.

Securities are financial contracts that entail payment obligations; more precisely speaking, they are '**derivative contracts**' and come in different forms such as 'futures', 'swaps' and 'options'. In derivatives, contracts **payment obligations *derive* from the realization of an 'underlying' event**, an index or price of a commodity (e.g., a certain price level of a commodity, exchange rate, credit default, profits and losses of firms) that may or may not realize in the future; in insurance-linked securities this **underlying event** is an ***insurance loss***. Therefore they are also termed '***insurance*-linked securities**'.

The economic value of derivatives lies in "the rights to dividends from a corporation, to interest payments from a government and so on" (MacKenzie, 2007, p. 357) specified in the contract. The realization or non-realization of a certain event triggers a specified payment to the securities holder (that is, the investor) or the issuer.

Historically seen, derivatives were invented in the 1970s in the context of the demise of Bretton Woods system of stable exchange rates and the associated financial market re-regulation informed by the neoliberal turn in economic thought. The (original) purpose of derivatives was to protect buyers against dramatic, spontaneous price changes such as jumps in oil prices or currency volatilities (Carter & Falush, 2009, p. 125; MacKenzie, 2007). However, over the course of time derivatives became more and more instruments of opportunistic risk-taking and speculation (if that distinction exists at all) (Bolt, 2012, p. 20). Derivatives became infamous in the context of the financial crisis 2008/09, when unpaid credits by mortgage owners led to the devaluation of housing-related securities in the US. The credit crunch and the mistrust in derivatives held by banks and other investors ended in a near standstill of the financial market system with a subsequent fallout on the real economy.

Insurance-linked securities are differently structured in terms of what triggers a payment (e.g., real loss, modeled loss, performance of an index, etc.) and how the payment obligations are contractually fixed (e.g., continuous cash flows, payments in case a pre-defined event occurs). Depending on their precise form they are termed '**catastrophe bonds**' (cat bonds), '**catastrophe swaps**' (cat swaps), '**catastrophe options**' (cat options), sometimes also '**weather derivatives**'.

While insurance-linked securities all differ in their precise structure it is, once again, of critical importance to understand that they are, by definition, **financial derivatives and *not* insurance contracts.** That means that they are also *not* bound by the principles of insurance law. Instead, they underlie specific regulations concerning liability-backed securitizations. These regulations are captured in a different body of law. As introduced previously, an *insurance* contract is a *contract for indemnity*. An insurance contract "requires the insured to have an indemnifiable interest in the object or activity being protected" (Bolt, 2012, p. 22). This is not necessarily the case with insurance-linked securities. Securitization is the trick that removes of the barrier of an indemnifiable interest. It transforms a financial risk into a commodity.

Financial securitization in practice

How does a (re-)insurer transform financial risks associated with catastrophes into securities? The (re-)insurer 'sponsors'—that is, sets up—a '**Special Purpose**

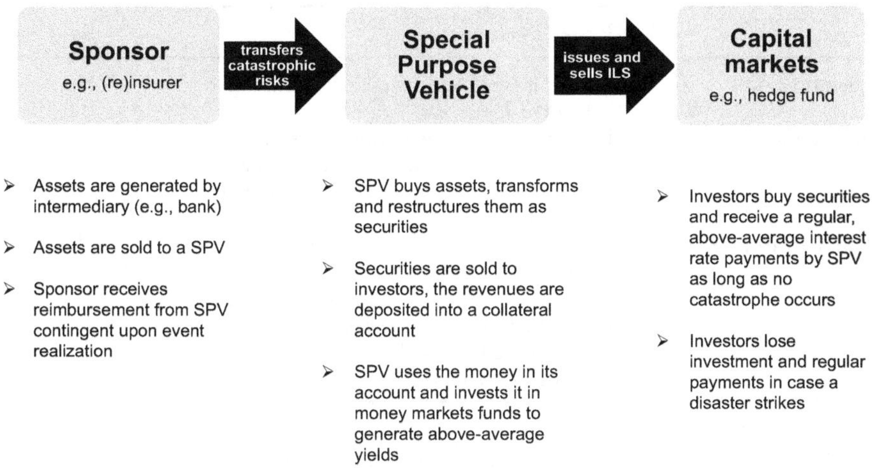

Figure 2.7 Key actors and building blocks of a securitization

Vehicle' (SPV) to which it temporarily transfers its financial obligation to pay insurance-related claims exceeding a certain level of loss.[5] The SPV **restructures this payment obligation** in different shares (e.g., smaller/larger, riskier/non-riskier, short-term/long-term deals) and then **sells them to interested investors** in the form of securities[6] (e.g., catastrophe bond, catastrophe swap) (see Figure 2.7).

Through the sale of these insurance-linked securities to investors the SPV receives revenues. The SPV uses these revenues by investing them into money markets with the goal to make a profit.[7] Part of this profit is then transferred to investors in the form of regular payments ('coupon payments'). The **coupon payments** investors receive reflect usually **above-average dividends** and vary according to type and amount of shares. Between 1997 and 2015 the annual average coupon payments for insurance-linked securities ranged from 5–12% (Artemis, 2016a). It is important to note, however, that investors receive regular **cash payments only in the absence of a catastrophe**. If a catastrophe occurs and generate losses for the sponsor, investors lose all (or at least some) of their initial payments to the SPV. Investors receive no further interest rates from the SPV (Carter & Falush, 2009, p. 125). The SPV transfers the investment to the sponsor—that is, to the (re-)insurer. If no catastrophe occurs, investors get their money back upon maturity of the bond.

On triggers

What triggers the payment from financial market investors to the (re-)insurer via a SPV varies from contract to contract. It can be triggered by the actual loss, by a modeled loss, but also by the magnitude of a pre-defined event. The overarching logic is akin to excess-of-loss arrangement: if a certain threshold is breached, a certain payout is due and the investors need to transfer its investment to the

Table 2.1 ILS-trigger types

Trigger	*Payout mechanism*	*Basis risk*
Indemnity	Payout is tied to the final calculation of insured-losses.	None
Modeled-loss	Payout is tied to the performance of a model that estimate losses based on different static and variable components (e.g., indices).	Low
(Parametric) index	Payout is tied to the performance of an (parametric) index and linked to a certain amount of money. The index is based on reported intensity measurements (e.g., wind-speed, ground-shaking, wave-height in a certain area).	High

(re-)insurer. In broad terms three different trigger mechanisms can be distinguished (see Table 2.1): **(1) an indemnity-trigger (2) a modeled-loss trigger** and **(3) a (parametric) index trigger**. They all differ regarding **'basis risk'**. Basis risk describes **a situation where the estimated losses do *not* match the real losses.**

Currently, the large majority of insurance-linked securities are based on the performance of a model and (parametric) indices (Artemis, 2016b). That means, in most cases (re-)insurers do not dispatch a loss adjuster to 'ground zero' to estimate losses and establish claims; instead, they use the performance of a model or indices as an indicator to establish claims. Using a modeled-loss or parametric index-approach reduces the transaction costs for (re-)insurers, accelerates the process of loss adjustment and reduces fraud as the process is more transparent. Investors can review the performance of the model/indices (Poliquin & Lalonde, 2012, p. 211). The downside is with the insured. The model/indices might simply not be accurate.

On insurance-linked securities

ART enables (re-)insurers to tap new resources to fund financial risks associated with the coverage of 'mega catastrophes' and thus helps to reduce their inherent insolvency risk associated with the coverage of 'mega catastrophes'. Through this complex mechanism and products (re-)insurers gain "access to alternative markets populated by investors like hedge funds who are comfortable in taking unusual risks in exchange for a high expected return" (Bouteille & Coogan-Pushner, 2013, p. 255). In other words, insurers can access new and thus broaden their capital reserves; they can obtain protection from a pool *separate* from traditional re-insurers. This helps insurers to reduce their dependence on just one market, respectively their dependency on re-insurers. By expanding their capital base, insurers can offer protection for events they would possibly not cover otherwise. In other words, it allows them to provide coverage for events that normally would be deemed as insurable.

What attracts investors such as hedge funds and pension funds to insurance-linked securities and to engagement in what could be interpreted as 'speculations on the non-occurrence of catastrophes' is that **insurance-linked securities** represent an **asset class whose value is—at first sight—uncorrelated to their investment in 'traditional' asset classes** (e.g., equities, corporate bonds) and thus market risks and dynamics like an economic recession (Bougen, 2003, p. 256; Pfister, 2003, pp. 10–11): Catastrophes occur independent of stock market booms and crashes. The value of insurance-linked securities does not fall when 'traditional' classes fall. That means, investors are able to obtain high returns in times of financial or economic distress (Constantin, 2011, p. 79). Even during the fallout of the financial crisis 2008/09, the average yield of insurance-linked securities revolved around 8,25% (Artemis, 2019). The fact that insurance-linked securities are not depended on market dynamics makes them particularly attractive products for investment diversification (Swiss Re Capital Markets, 2012, p. 10). Moreover, insurance-linked securities are also particular attractive for gamblers on the secondary market seeking a 'quick and dirty' profit based on a speculation on changing weather conditions—not to speak of risk modelers themselves who can cash-in their expertise. As Nakada and Smith from Risk Management Solutions Inc., one of the world-leading catastrophe modelers, advertise: "'Live cat' analytics allow portfolio managers to understand the potential impact of unfolding catastrophic events, and potentially take action to change their risk profile in response" (Nakada & Smith, 2012, p. 193).

If catastrophe bonds are really uncorrelated to market dynamics is, however, questionable. (Re-)insurer's risk of insolvency is not entirely eliminated. Quite, in contrast: the SPV faces its own investment risks which, in the end, must be assumed by the sponsor, that is, the (re-)insurer. The reason is that in asset-backed securitization insurance-linked liabilities are not transferred. As Bouteille et al. claim: "If the investments fall in value, the SPV has no other sources of funds" (Bouteille & Coogan-Pushner, 2013, p. 255). Most likely the SPV will have some financial arrangements in place to prevent an investment fallout (such as *return swap*) and, yet, financial hedges are no absolute guarantee for financial stability as the financial crisis 2008/09 has vividly illustrated (Bouteille & Coogan-Pushner, 2013, p. 255). This implies, in contrast to conventional wisdom, insurance-linked securities and associated regular payments to investors are in practice not totally uncorrelated from major market development and dynamics. The correlation depends to a great degree on the investment portfolio of the SPV.

ART in historical context

ART is not only a complex but also a fairly recent way to provide insurance coverage. It is only in recent years that the insurance sector has started to employ these mechanisms. In this section I will briefly outline how this has come about historically. This section serves thereby as bridge between the theoretical and the empirical part of this book. (Regional) catastrophe risk pools are, in fact, only the latest chapter of this historical trajectory of shifting boundaries between insurance and finance.

Neoliberal changes

Despite some limited experiments in the health insurance sector in the early 1970s (Sandor & Bouriaux, 2012), it was only in the beginnings of the 1990s that insurers and re-insurers alike have started to embrace financial market practices to finance large-scale catastrophes (Bougen, 2003, p. 254; Sandor & Bouriaux, 2012, p. 28). **ART** is a rather new way to shift risk and source capital. Put in a broader historical context, it could be seen as the **outcome of "crossbreeding of insurance and capital markets"** (Bougen, 2003, p. 255), a process that started in the 1980s. This 'crossbreeding of insurance and capital markets' can be subsumed under a broader phenomenon which in Global Political Economy is commonly described as *financialization*—the increasing relevance of the finance sector in the political economy (Mader, 2015, p. 27).

The 'crossbreeding of insurance and capital markets' was **largely the result of the neoliberal liberalization agenda** which took root first in the US and the UK in the 1970s and then within the EU with the implementation of the European Single Market in 1992. On a step-by-step basis national governments broke down impediments to trade and financial services, giving up the protection of domestic markets while at the same time aiding national export industries and capital providers to enter new markets (Gugerli, 2013, p. 215). Ever since national lawmakers on both sides of the Atlantic have focused on increasing world trade by coordinating differing markets through accepting or unifying different regulatory standards; the TTIP negotiations—buried under Trump administration—were only the latest cornerstone of that larger development.

The **deregulation of the European insurance industry**—the regulatory space where most key re-insurers are located—was completed in 1994 (Borscheid, 2013, p. 115). From 1994 onwards "the principle of mutual recognition of regulatory provisions was applied to banks and insurance companies" (Gugerli, 2013, p. 216). With the introduction of a Single European License, the Home Country Control principle and the Solvency Requirements the European insurance market was re-regulated (Sterzynski, 2003). As a consequence, insurers and financial service providers could offer their products in other jurisdictions freely, that is without an additional permission of the host country. As a consequence, they went on to open up local branches and offices elsewhere. The changing conditions—in particular with regard to market access—transformed the way insurance industry and financial service providers were operating in a general way (Gugerli, 2013, p. 217). **The (neo-)liberalization of financial services and investment practices led increasingly to takeovers of activities** with banks acting as insurance salespeople and local insurer starting to operate internationally but also to cross-national acquisition and mergers, in particular in the 1990s (Gugerli, 2013, p. 221). The (neo-)liberalization also enabled institutional investors such as pension funds to invest in other markets than domestic ones and thus to flood formerly segregated markets with capital, fueling the creation of new businesses (Borscheid, 2013, pp. 112–115; Gugerli, 2013, p. 219). Put differently, insurers and banks sought to exploit these (neo-)liberal policies. They were further enhanced through technological advancement as well as lower information and transaction costs (Borscheid, 2013, p. 121).

While the neoliberal regulatory changes provided the context for ATR, the major turning point for insurers to embrace these complex mechanisms and products can be attributed to Hurricane Andrew in 1992 and its major impact on the balance of (re-)insurers (James, 2013, p. 16).

The impact of Hurricane Andrew

Hurricane Andrew brought major devastation to Florida and the Gulf of Mexico and created about US$ 15–17 billion in insured losses. These US$ 15–17 billion may look small from today's perspective —Hurricane Katarina in 2005 was responsible for insured losses between US$ 40–60 billion—and yet, at that time losses of that financial magnitude were unprecedented and deemed unthinkable (McDonnell, 2002, p. 81). **Hurricane Andrew** reflected **a 'black swan'**, an outlier. The event revealed not only the huge exposure of the (re-)industry, it also demonstrated the industry that the coverage it provided went, in fact, far beyond its actual financial capacities: Following Hurricane Andrew, 11 insurance companies went insolvent as they had assumed too much risk or were not re-insured enough (James, 2013, p. 16). The industry did not seem to be capable of dealing with catastrophes of that magnitude. The event was a **game changer**. In the follow-up of Hurricane Andrew **(re-)insurers** turned their attention inwards and revised their practices.

First, they **tightened terms and limited their exposure** through so-called industry loss warranties (McDonnell, 2002)—a second trigger point that considered the total losses of the aggregated insurance industry. According to this new trigger, re-insurers would only pay out if the losses incurred to the insurers exceeded a certain threshold (Bolt, 2012, p. 22).

Second, the (re-)insurance industry **started to employ new, more sophisticated statistical risk modeling practices** to calculate the likelihood and the financial impact of an event (James, 2013, p. 16; Jarzabkowski et al., 2015, p. 68). The application of these new computer-based *vendor models* drew on fine-grained data from seismology, meteorology, engineering and actuarial science and incorporated information such as the geophysical characteristics of a specific area, past destructions or hypothetical scenarios—e.g., potential passages and speeds of a hurricane if the climate changes. The models were originally developed during the late 1980s and not by (re-)insurers themselves but by new and highly specialized risk or catastrophe modeling agencies, namely AIR (founded in 1987), RMS (founded in 1988) and, later, EQECAT (founded in 1994). Yet, despite being available from the end of the 1980s onwards, vendor models lived a life at the margins. They became the new 'gold standard' for catastrophe modeling only after Hurricane Andrew (Jarzabkowski et al., 2015, p. 70).

Third, the **(re-)insurance industry considered new approaches to gain liquidity and diversify its access to capital**. It began to experiment with financial market instruments and practices. The turn to the financial markets must also be seen in the broader context of the hard market cycle characterizing the (re-)insurance industry since 1987. Between 1987 and 1990 a range of major catastrophes, including a number of North European storms (1987), the destruction of the Piper Alpha oil platform and Hurricane Gilbert (1988), the San Francisco earthquake, Hurricane Hugo, the grounding of the Exxon Valdez tanker (all 1989) and further North European storms (1990) impacted the insurance industry heavily, limiting capital availability (Bain, 1999, p. 1). Hurricane Andrew contributed to further harden the market, as (re-)insurers had to pay huge claims (Gugerli, 2013, p. 212). Accordingly, (re-)insurers experimented with a range of differently structured insurance-linked securities to tap new financing sources. In September 1992 the first catastrophe swap was created, in 1994 the first catastrophe bond and in 1995 the first catastrophe option (Bolt, 2012, pp. 31–35; James et al., 2013, p. 221). In these early years, these were all limited over-the-counter deals, meaning they were non-standardized products and not traded in an institutional setting. However, these initial over-the-counter deals gave quickly way to an institutionalized trading forum. In 1996 a screen-based Catastrophe Risk Exchange (CATEX) was set up in New York to facilitate a swapping of catastrophe risks between insurer and re-insurers (Carter & Falush, 2009, p. 125). While there exist a range of different insurance-linked securities, cat bonds emerged as the most dominant form of insurance-linked securities over time. The market took off when rating agencies started to rate cat bonds in 1997 (Bolt, 2012, p. 34). According to Bolt, cat bonds turned out to be particular attractive because they "replicate a reinsurance contract and are generally accountable as such. They are also 'flexible' instruments in that they can fit all bond sponsors' needs. Cat bonds can have payoffs tied to either an indemnity trigger or an index trigger" (Bolt, 2012, p. 34).

Fourth, **Hurricane Andrew brought about new and highly specialized reinsurance companies.** As Hurricane Andrew wiped out 11 insurers, a number of new, specialized insurance companies in catastrophe were founded. Due to the (neo-)liberalization of the insurance industry, insurer could (re-)locate their offices and businesses to Bermuda, an offshore jurisdiction known for its loose tax and regulatory standards, more easily, circumventing the regulatory standards of national insurance supervision authorities and taxes (Gugerli, 2013, pp. 219–220). As Gugerli points out: "Bermuda became the world's fourth-largest reinsurance market. In 1993 alone, USD 4 billion flooded into the creation of eight new catastrophe re-insurers in Bermuda" (ibid., p. 220). It is, however, noteworthy that Hurricane Andrew only reinforced a trend that had its origin already in the 1960s and 1970s when Bermuda emerged as primer jurisdiction for captives—basically self-insurer of multinationals (Gugerli, 2013, p. 214).

The impact of Hurricane Katrina, Rita and Wilma: the 'sidecar moment'

ART became another boost towards the end of 2005 in the context of a busy hurricane season in the Atlantic basin. Some (re-)insurance professionals describe the year as the "Sidecar Moment" (Perez, 2012, p. 87). Sidecar is another technical name for SPVs. The 2005 hurricane season was the costliest insurance loss year in history. Insured losses due to Hurricane Katrina, Rita and Wilma summarized to about US$ 123 billion (ibid., p. 90). Similar to Hurricane Andrew, the events led to a shortage in capital and higher premium levels. Moreover, two main catastrophe risk modelers (AIR and RMS) revised their catastrophe models. Also, rating agencies reviewed their rating practices and demanded higher capital requirements from insurers to get the same rating. Consequently, insurers were in need for additional capital. In this context of this hard market a range of highly specialized catastrophe insurance companies were formed. This, however, would not have been possible without the immense investment interest to participate in future underwriting profit, as highlighted by Peréz. The investor demand led to the establishment of more than 20 SPVs between 2005 and 2006, a moment that reinsurance professionals label as "side car moment" (ibid., p. 90).

Alternative risk transfer today

With the invention of derivatives in general and insurance-linked securities more specifically insurers could tap new capital resources to finance catastrophic risks. **ART shows how (re-)insurers have managed to move the frontier of what can be insured once more further out**, that is, to new territories. If one considers the history of (re-)insurance as a struggle to expand the boundary of insurability (James, 2013, p. 3), the turn to use capital market instrument is only the most recent chapter in that process.

Yet, with the invention of insurance-linked securities there emerged also a new opportunity for financial market investors, namely "the ability to separate and repackage various insurance risks irrespective of form" (Bolt, 2012, p. 21). Institutional investors, such as investment banks (notably Goldman Sachs) and newly formed capital arms of insurance brokers (notably Swiss Re Capital Markets) have tapped into that possibility (ibid., p. 21). Accordingly, **some professionals view ART also as a competitor** to traditional (re-)insurance arrangements (as already the branding as 'alternative' indicates), **others**, in turn, see it more as a **complement** (Sandor & Bouriaux, 2012). (Re-)insurers fear that financial market participants, by offering insurance-linked securities (will) drive drown (re-)insurance premium levels and thus erode the business model for 'traditional' (re-)insurer. In contrast to the limited capacity of (re-)insurers, financial market participants have, collectively, an abundant amount of capital available. How that develops remains to be seen. By now, it is only clear that the market for insurance-linked securities keeps growing independent of the issuer. According to a market analysis by Artemis, an online platform dedicated to ART, the market size increased from US$ 785 million in 1997 to US$ 37,9 billion in 2019 (Artemis, 2019). This implies that the relation between insurers, re-insurers and financial market investors continues to be undergoing a transformation.

Notes

1 Uncertainty refers here to the fact that we do not know whether the event will happen. In contrast to uncertainty, risk entails some degree of measurement; that means it entails the idea that one can at least estimate the probability of the occurrence of the realization of such an uncertain event. This measurement is based on statistics from previous events (Knight, 1921, p. 233). Based on stochastic operations, insurers try to calculate the probability of the occurrence of such an event in the future.

2 It may also be of psychological or macroeconomic nature. Risk adverse people, for instance, may not want to invest if there exists no insurance in case this undertaking fails.

3 Some international historians and cultural sociologists see these promoting activities as a rather new cultural trend in insurance and have termed this development 'risk embracing'. Risk embracing refers basically to insurers' activities to use policyholders' exposure to risk as an incentive to modify their behavior and thus to reduce insurer' claim payments (Baker & Simon, 2002; Lobo-Guerrero, 2011). According to Baker and Simon, this is based on the "idea that some amount of risk is good for people and too much protection is harmful" (Baker & Simon, 2002, p. 10). The issue here is that it is not based on eliminating the risk but about transforming people into responsible risk-managers.

4 A simplified, everyday illustration of such an excess-of-loss deal would be the self-retention when renting a car: If one rents a car, usually one has to pay the costs for damages up to a certain level. Beyond that level, the car insurer will take over the costs.

5 The sponsor literally 'sponsors' the vehicle: That means it takes over the costs associated with the setting up and maintaining of the vehicle. The relationship between the SPV and the sponsor is structured through a reinsurance arrangement (Perez, 2012, p. 87).

6 In other words, the SPV does more or less the same work as insurers do when they transfer and sell risks to re-insurers, as explained previously.

7 In money markets, participants trade highly liquid products of short maturity (one year or less—mostly, however, of three months or less).

References

Artemis. (2016a). *Cat bonds and ILS average coupon by year*. Retrieved March 13, 2016, from www.artemis.bm/deal_directory/cat_bonds_ils_expected_loss_coupon.html

Artemis. (2016b). *Catastrophe bonds and ILS risk capital outstanding by trigger type*. Retrieved April 14, 2016, from www.artemis.bm/deal_directory/cat_bonds_ils_by_trigger.html

Artemis. (2019). *Catastrophe bonds & ILS issued and outstanding by year*. Retrieved June 15, 2019, from www.artemis.bm/dashboard/catastrophe-bonds-ils-issued-and-outstanding-by-year/

Bain, A. (1999). *Insurance spirals and the Lloyd's market* (Working Paper). Glasgow. Retrieved from www.gla.ac.uk/media/media_219037_en.pdf

Baker, T., & Simon, J. (2002). Embracing risk. In T. Baker & J. Simon (Eds.), *Embracing risk—The changing culture of insurance and responsibility* (pp. 1–26). Chicago and London: University of Chicago Press.

Bolt, T. (2012). A history of direct insurance-linked investments. In L. Morton (Ed.), *Alternative (re-)insurance strategies* (pp. 17–26). London: Risk Books.

Borscheid, P. (2013). Global insurance networks. In H. James, P. Borscheid, D. Gugerli, & T. Straumann (Eds.), *The value of risk* (pp. 23–146). Oxford: Oxford University Press.

Bougen, P. D. (2003). Catastrophe risk. *Economy and Society*, *32*(2), 253–274. http://doi.org/10.1080/0308514032000073428

Bouteille, S., & Coogan-Pushner, D. (2013). *The handbook for credit risk management-originating, assessing, and managing credit exposures*. Hoboken: Wiley.

Carter, R. L., & Falush, P. (2009). *The British insurance industry since 1900*. Basingstoke: Palgrave Macmillan.

Collier, S. (2008). Enacting catastrophe: Preparedness, insurance, budgetary rationalization. *Economy and Society*, *37*(2), 224–250. http://doi.org/10.1080/03085140801933280

Constantin, L-G. (2011). Portfolio diversifiaction through structured catastrophe bonds amidst the financial crisis. *Buletinul*, *LXIII*(3), 75–84.

Feldmann, R. J. (2002). Government insurance. In L. M. Salamon (Ed.), *The tools of government—A guide to the new governance* (pp. 186–216). Oxford: Oxford University Press.

Gugerli, D. (2013). Cooperation and competition: Organization and risks in the reinsurance business 1860–2010. In *The value of risk* (pp. 147–237). Oxford: Oxford University Press.

Haubner, D. (2007). *Grenzen der Versicherbarkeit: Möglichkeiten des Versicherungsschutzes gegen Katastrophen*. München and Oxford. Retrieved from http://fzk.rewi.hu-berlin.de/doc/sammelband/Grenzen_der_Versicherbarkeit.pdf

James, H. (2013). Introduction: The insuring instinct. In H. James, P. Borscheid, D. Gugerli, & T. Straumann (Eds.), *The value of risk* (pp. 1–22). Oxford: Oxford University Press.

James, H., Borscheid, P., Gugerli, D., & Straumann, T. (2013). *The value of risk*. Oxford: Oxford University Press.

Jarzabkowski, P., Bednarek, R., & Spee, P. (2015). *Making a market for acts of god—The practice of risk-trading in the global reinsurance industry: Making a market for acts of God*. Oxford: Oxford University Press.

Knight, F. (1921). *Risk, uncertainty and profit*. Boston: Houghton Mifflin.

Lobo-Guerrero, L. (2011). *Insuring security: Biopolitics, security and risk*. Abingdon and New York: Routledge.

MacKenzie, D. (2007). The material production of virtuality: Innovation, cultural geography and facticity in derivatives markets. *Economy and Society*, *36*(3), 355–376. http://doi.org/10.1080/03085140701428332

Mader, P. (2015). *The political economy of microfinance*. New York: Palgrave Macmillan. http://doi.org/10.1057/9781137364210

McDonnell, E. (2002). Industry loss warranties. In M. Lane (Ed.), *Alternative risk strategies* (pp. 81–97). London: Risk Books.

Moss, D. (2002). *When all else fails*. Cambridge, MA: Harvard University Press.

Nakada, P., & Smith, D. (2012). Advances in ILS portfolio: Management analytics. In L. Morton (Ed.), *Alternative (re)insurance strategies* (pp. 177–194). London: Risk Books.

Nguyen, T. (2007). *Grenzen der Versicherbarkeit von Katastrophenrisiken: Erweiterungsmöglichkeiten durch Rückversicherungen, Katastrophenanleihen und Versicherungsderivate*. Hagen: Deutscher Universitätsverlag.

Parsons, C. (2015). Insuring the unknown. *Human & Experimental Toxicology*, *34*(12), 1238–1244. http://doi.org/10.1177/0960327115601765

Perez, A. (2012). Sidecars. In L. Morton (Ed.), *Alternative (re-)insurance strategies* (pp. 85–102). London: Risk Books.

Pfister, G. (2003). *Zur Versicherungsfähigkeit von Katastrophenrisiken*. Stuttgart: Akademie für Technikfolgenabschätzung in Baden-Württemberg. Retrieved from http://elib.uni-stuttgart.de/opus/volltexte/2004/1726/pdf/ab232.pdf

Poliquin, B., & Lalonde, D. (2012). Role of catastrophe risk modelling in ILS. In L. Morton (Ed.), *Alternative (re-)insurance strategies* (pp. 195–218). London: Risk Books.

Renn, O. (2008). *Risk governance*. London and Sterling, VA: Earthscan.

Sandor, R., & Bouriaux, S. (2012). Good derivatives: Why securitisation of insurance risks matters and more. In L. Morton (Ed.), *Alternative (re-)insurance strategies* (pp. 27–44). London: Risk Books.

Sterzynski, M. (2003). The European single insurance market: Overview and impact of the liberalization and deregulation processes. *Belgian Actuarial Bulletin*, *3*(1), 42–49.

Swiss Re Capital Markets. (2012, September). *What are Insurance Linked Securities (ILS) and why should they be considered?* (pp. 1–17). Presentation to the CANE Fall Meeting. Retrieved from www.casact.org/community/affiliates/CANE/0912/Cat-Bond.pdf

Taleb, N. (2007). *Black swan—The impact of the highly improbable*. London: Allen Lane.

Thoyts, R. (2010). *Insurance theory and practice*. London and New York: Routledge.

Wolgast, M., & Theis, A. (2013, January). Risiko, Unsicherheit, Versicherbarkeit. *Wirtschaftswissenschaftliches Studium; WiSt—Zeitschrift Für Studium Und Forschung*, *1*, 29–34. Retrieved from www.gdv.de/wp-content/uploads/2014/04/Rolle-der-Versicherer-beim-gesellschaftlichen-Risikomanagement-WiST01-2013.pdf

3 The Caribbean Catastrophe Risk Insurance Facility (CCRIF)

I will now investigate the first case: The Caribbean Catastrophe Risk Insurance Facility (CCRIF). Following the analytical framework developed in Chapter 1, this chapter is divided into three parts.

The first part of this chapter (Part I) center stages the goal and functioning of CCRIF. It explores the problems CCRIF seeks to solve and illuminates how the insurance arrangement works on a practical level to fulfill its objectives. To guide the reader through these complexities, it first provides a short overview of CCRIF forestalling key takeaway points. It then elaborates on these points by zooming onto the goals, the operational structure, technical basis, the reinsurance structure, the financial strategy and the governance structure of CCRIF.

The second part of this chapter (Part II) explores how CCRIF came about. It explores how the *idea* to set up CCRIF came about and examines the making of the arrangement. Doing so, it illuminates the various processes and activities leading up to the creation of CCRIF and the adoption of its insurance policies by the Caribbean states and analyzes modifications of CCRIF and its policies over time.

The third part of this chapter (Part III) provides a concluding summary.

Part I: understanding CCRIF

The Caribbean Catastrophe Insurance Facility (CCRIF) is a **regional, state-owned insurance company** ('captive'/'segregated portfolio company') that offers 19 countries **from the Caribbean and two Central American countries insurance coverage** against the financial consequences of catastrophic events in exchange for an annual premium.

States receive rapidly limited funds following the occurrence of a ***major* hurricane** (modeled to happen 1:15 years) or **earthquake** (modeled to happen 1:20 years) or **excessive rainfall**[1] (modeled to happen 1:5 years). The level of the payout is thereby modeled on real-time hazard and archived exposure data and capped at a certain, pre-defined level. Insurance payouts to governments flow into national treasuries and are not tied to specific expenditures. Governments are free to decide where they want to allocate a payout.

CCRIF is a self-insurance vehicle with a club character: To be eligible for the purchase of insurance coverage, a country must first of all become a member of CCRIF. To become a member, countries must apply and a pay a one-time membership fee. CCRIF is open to all Caribbean Community (CARICOM) and Council of Ministers of Finance of Central America, Panama and the Dominican Republic (COSEFIN) countries. It was created under the technical leadership of the World Bank in 2007, following the massive destruction left by Hurricane Ivan on Grenada in 2004.

The **primary generic goal** of CCIRF is **to insure the liquidity of governments** by providing them with *limited funds*—that is, with a small amount of capital—following the occurrence of a *major* disaster. The insurance has been structured akin to a private business interruption insurance—an insurance that provides cash to bridge revenue shortfalls. It secures governments a limited budget for 'unforeseen' (disaster-related) tax revenue shortfalls. The **secondary generic goals are to mitigate the long-term impact of catastrophes on public finances by empowering countries to accelerating the launch of rebuilding measures and to supply countries with tools and knowledge for disaster management.**

From an organization studies perspective, CCRIF is a 'virtual organization'—meaning **a letterbox company—registered in the Cayman Islands**. It is formally owned by the Caribbean countries (including Nicaragua) and **operated as public-private partnership**: While the public stakeholders—the countries—legally own the facility all services are, in fact, outsourced and thus delivered by private service providers. The 'virtual existence' is enabled through a number of service provision agreements with private subcontractors. This implies that CCRIF exists 'in reality' just online (www.ccrif.org). There exist neither physical buildings made of 'brick-and-mortar' nor staff *per se*. CCRIF becomes 'real' only on the website and in the locations where the private service providers are at home. Representatives from the **Caribbean Development Bank (CDB) and the CARICOM** act as the entity's supervisor. That means, while the member states formally own the facility it is the CDB and the CARCICOM that***, in practice,* govern CCRIF** on the countries' behalf.

Historically seen, **CCRIF was created in 2007** under the technical leadership of the World Bank (see Part II). The World Bank, the Caribbean Development Bank and a number of donors provided the start-up capital and financed a part of the premiums for the first insurance policy years. CCRIF was the first multi-country insurance company in the world, providing governments with parametric catastrophe insurance policies against hurricanes and earthquakes backed by re-insurers and capital market investors. Since the insurance policy year 2014/15, CCRIF has expanded its 'product line' by offering insurance coverage not only to countries but also to Caribbean financial institutions (such as banks) and low-income groups (such as small farmers and day laborer). During the course of 2015, CCRIF expanded its geographic reach by opening membership to the Central American COSEFIN countries.

As of September 2019, 21 countries are members of CCRIF: Anguilla, Antigua and Barbuda, Bahamas, Barbados, Belize, Bermuda, British Virgin Islands, Cayman Islands, Dominica, Grenada, Haiti, Jamaica, Montserrat, Nicaragua, St Kitts and Nevis, St Maarten, Saint Lucia, St Vincent and the Grenadines, Trinidad and Tobago and Turks and Caicos Islands. Nicaragua joined CCRIF in May 2015 as the first Central American country (CCRIF SPC, 2015b, p. 7). The Bahamas, the British Virgin Islands, Montserrat and St Maarten joined CCRIF only very recently, during the 2018/2019 season.

Between 2007 and December 2018 CCRIF made 38 payouts to 13 countries totaling about US$ 138 million. CCRIF appears to work and figures prominently as a model of reference to other regions (see other case studies).

Generic goals of CCRIF

CCRIF has been designed to serve three principal objectives: (1) to provide short-term liquidity after a major catastrophe (2) to mitigate long-term negative macroeconomic impact of disasters (3) to supply tools and knowledge for disaster risk management. Why these goals? What is the reasoning behind these goals? And how have they come about?

1 To provide short-term liquidity after a major catastrophe

The primary goal 'ensuring short-term liquidity' relates back to the experiences made with Hurricane Ivan in Grenada and the lessons that have been learned by regional policymakers. The insurance seeks to prevent the situation of a 'liquidity gap' as it happened to the island in 2004, a gap between available financial needs and means.

Hurricane Ivan was a category-5 hurricane on the Saffir-Simpson Hurricane Wind Scale and passed the Caribbean in September 2004. It had a destructive impact on eight Caribbean countries and was particular devastating for Jamaica and Grenada. It caused estimated economic losses to Jamaica of about US$ 549 million or about 8% of the country's GDP (Grove, 2012, p. 143) and approximately US$ 800 million to Grenada or about twice of the country's GDP

(Ghesquiere, Mahul, Forni, & Gartley, 2006, p. 2). Hurricane Ivan destroyed an estimated 89% of Granada's housing stock and damaged more than 80% of its public and commercial buildings (Ghesquiere & Mahul, 2012). In other words, Hurricane Ivan brought about the *de facto* destruction of Grenada.

The government required substantial financial resources to finance emergency and early relief measures and assist the country to return to normality. Emergency and early relief measures include the clearance of debris and the restoring of what is commonly described as 'critical infrastructure' such as hospitals, power plants, sanitation systems, communication and transport systems like roads, bridges and (air)ports. The provision of this public service serves society and the economy to return quickly to normality and enables humanitarian assistance to affected households in the first place. However, Grenada ran short of liquidity. The *de facto* destruction of Grenada and the income-generating economies brought about substantial decline in public revenues. The tax revenue shortfall was an estimated 5% of GDP between September 2004 and December 2004 (Ghesquiere et al., 2006, p. 2).

Due to limited financial reserves, the impossibility to access affordable credit—Grenada had been highly indebted—the Grenada faced serious problems not only to finance this post-disaster bill but also public salaries. Taken together it had difficulties to secure the continuation of public key services (Ghesquiere et al., 2006, p. 2; Iyahen & Young, 2011, pp. 11–12). Grenada was close to insolvency.

To finance the 'post-disaster bill', address public revenue shortfalls and prevent a state insolvency Grenada embraced a number of measures: First, it launched an international appeal and successfully secured US$ 150 million in donor assistance; yet, only US$ 12 million of these pledges were immediately available to finance urgent needs. The other US$ 138 million was earmarked by donors for reconstruction projects and thus subject to political preferences of donors. Second, it approached its creditors and offered them (or better: forced them by threatening them with a default) to restructure 85% of its commercial debts. Third, it increased consumption taxes and income taxes. The measures included a 45% rise on the retail price of fuel, an increase in taxes on alcohol and tobacco and a special levy on incomes over US$ 375 for a five-year period. Fourth, it strengthened its tax enforcement measures through an administrative reform (Ghesquiere et al., 2006, p. 2). Yet, despite these encompassing efforts to secure state revenues, "Grenada's fiscal situation remained challenging and the country still faced a financing gap of 4.5 per cent of GDP for 2005 with total debt projected to increase to 150 per cent of GDP" (ibid., p. 2).

The impact of Hurricane Ivan and associated fundraising problems raised concerns at the CARICOM and the World Bank, the prime international organization for the financing of post-disaster reconstruction efforts, about the country's long-term macroeconomic stability. In an emergency meeting held in November 2004, Grenada, the CARICOM and the World Bank discussed the political consequences the region could draw from these experiences. It noted the difficulties of the region in accessing rapidly financial resources and grasped the problem as a 'liquidity problem'. In search for a solution, the CARICOM mandated the

World Bank to develop an 'insurance solution'. With the particular situation in Grenada in mind, the World Bank designed a regional catastrophe insurance scheme for the Caribbean. The model of reference was private *business interruption insurance*—an insurance that covers short-term revenue shortfalls after an incident. I will deepen the history of CCRIF in Part II.

2 *To mitigate long-term negative macroeconomic impact of disasters*

The second goal of CCRF is to mitigate the negative long-term macroeconomic impact of disasters by empowering countries to accelerate rebuilding efforts. This goal is closely connected to the first goal and related to the consequences of a belated beginning of recovery efforts in the aftermath of Hurricane Ivan in Grenada 2004.

The financing problem in the aftermath of Hurricane Ivan bounded Grenada's administrative forces. Running short on revenues, the government set the priority on closing the emerging 'liquidity gap' to ensure the provision of basic public services. Yet, the search for funding distracted the administration from focusing on initiating other disaster-related tasks (Ghesquiere et al., 2006, p. 2).

The reconstruction of destroyed areas usually requires careful planning (Ghesquiere & Mahul, 2010, p. 9) as in some areas a certain infrastructure might not make sense anymore. Once an administration has decided what it wants to rebuild, "engineers need to design new infrastructure, projects have to be tendered and contractors have to mobilize" (Ghesquiere & Mahul, 2010, p. 19). In other words, administrations need some time to assess these post-disaster situations, create plans, establish and source the costs.

The fact that Grenada was occupied with the search for funding to cover the most urgent needs led to delays in kicking off the larger (and also more costly) reconstruction efforts (Ghesquiere & Mahul, 2010, p. 7; Ghesquiere et al., 2006, p. 2). The problem resulting from the sluggish launch of reconstruction measures in Grenada was that it contributed to prolonging the negative fiscal and macroeconomic impact of the hurricane.[2] When the "commercial infrastructure is damaged or destroyed, business slows or ceases" (Brooks, 2011, p. 2). As long as the economy is not back on its feet household incomes and thus tax revenues suffer from financial shortfalls. The welfare of the state and its people is tightly connected to the welfare of the economy.

Accordingly, what the instrument designers had in mind when they developed CCRIF was that a country's administration would need to have funds *quickly* available, so that the administration does not have to bother with covering short-term financing needs. The idea was that countries would utilize insurance payouts to finance immediate early relief and emergency measures to allow for a quick(er) fiscal and macroeconomic recovery. The vision was to mitigate the long-term financial impact of disasters.

However, and as I will elaborate later, there does not exist a built-in mechanism or designated actor that monitors how countries utilize insurance payouts. CCRIF is based upon a *laissez-faire* approach: Once a country gets an insurance payout, it can do whatever it wants to do with it. It must not all spend the payout on early relief and emergency efforts. As one of the instrument designer explained to me, the guiding idea was that countries would know best what is good for them (Personal Interview, World Bank official, 2015, April 28). That countries would use a payout to kick off early emergency and relief measures is thus rather an expectation (not say a hope) held by the instrument designers.

From a rule-driven Eurocentric perspective, such a *laissez-faire* approach looks dubious at first. It opens doors to fraud, misuse and corruption in a situation where financial resources are already scarce. And, yet, such a *laissez-faire* stance is thoroughly in line with recommendations from disaster risk managers. Disaster management professionals are not getting tired of emphasizing the emergency situation on the ground and thus the flexibility that is needed; that emergency needs cannot be known in advance and, thus, be planned for, is commonly argued; an effective response requires local knowledge (Handmer & Dovers, 2013, p. 166).

The insurance arrangement seeks to mitigate the problem of fraud by transferring the insurance payout to national treasuries. This agreement might, however, paradoxically also generate delays in the actual use of funds and thus somehow contradict its purpose (Ghesquiere & Mahul, 2010, p. 12). It touches complex administrative and legal questions typically associated with contingency plans, questions such as 'Who appropriates the budget?', 'Who is in charge?', 'How are the funds transferred?'. The clarification of these deceptively simple questions may not be underestimated.

3 To supply tools and knowledge for disaster risk management

Following contestations after a non-payout to Jamaica in the insurance policy year 2008–2009 (see Part II), CCRIF revised its mission and introduced the "Technical Assistance" Program. Since September 2009, CCRIF offers not only financial services but also support in the fields of meteorology, insurance and disaster management. CCRIF finances a scholarship program, (the organization and attendance of) workshops, conferences and studies to build up and disseminate disaster-relevant knowledge in the Caribbean region and provides small grants for risk reduction measures. It commissioned, for instance, an "Economics of Climate Adaption" study "to develop fact-based adaptation strategies that can be incorporated into national development plans to increase resilience against climate hazards" (CCRIF, 2010a). Over the years, CCRIF partnered with a number of specialized international and regional organizations in the field of meteorology and disaster management. This enabled CCRIF to offer satellite-based weather forecasting maps to members and non-members.

Summing up, the primary goal behind CCRIF is the *securitization of public liquidity*. CCRIF provides countries rapidly with limited funds so that they can bridge short-term disaster-related tax revenue shortfalls and cover explicit and implicit public service liabilities.[3] The secondary goals are to mitigate the long-term macroeconomic impact of a disaster by empowering countries to accelerate the launch of rebuilding measures and to provide countries with tools and knowledge in the field of disaster risk management.

Operational structure

Who provides the insurance? And how does the insurance work on a technical and practical level?

A regional state-owned insurance company

Insurance coverage is provided by CCRIF. CCRIF is a regional, state-owned insurance company that is public-private in nature. Legally speaking, CCRIF is a *captive* ('segregated portfolio company', SPC) that is owned by its insurance policyholders[4]—the Caribbean countries (including Nicaragua and Panama). A captive is in principle an insurance company established by an "industrial, financial or commercial group and has the exclusive aim of covering those risks to which the group is exposed" (LFF, 2017). In other words, a captive is essentially a self-insurance vehicle. It usually allows an entity to directly access the reinsurance market "to cover specific risks which are difficult to insure in the direct insurance market" (ibid.)

CCRIF is registered in the Cayman Islands. Captives cannot be licensed everywhere for various reasons. Typically they can only be registered in 'offshore jurisdiction'. Offshore jurisdictions are known for their low tax and regulatory standards. CCRIF is benefitting from the Cayman Islands' relaxed regulatory regime insofar as it has been granted an exemption from taxation until May 2027 (CCRIF SPC, 2015a, p. Annex). CCRIF is structured as a *not-for-profit* company. This implies that all profits must be given back to its owners in some form (e.g., in the form of dividends, reduced premiums).

From an institutionalist perspective, CCRIF is a 'virtual' company, that is, a letterbox company. CCRIF owns no physical buildings or staff *per se*. All operations are executed by private service providers under the guidance of a Board of Directors, composed of public representatives. It is in this sense that CCRIF is a public-private insurance company.

THE CLUB CHARACTER OF CCRIF

CCRIF offers its members individually *tailored insurance contracts* against the financial consequences of the region's main perils: *Hurricanes*, *earthquakes* and *tropical rainfall*. Countries can choose within certain parameters how much insurance coverage they want to obtain against each peril.

In order to be able to buy insurance coverage a country must, first of all, become a member of CCRIF. To become a member, countries must apply and pay a one-time membership fee; CCRIF labels this levy a "participation fee" (cf. Pearson & Young, 2009, p. 28). So far, membership was open only to CARICOM countries; however, since April 2015 membership is also open to COSEFIN countries. The level of the membership fee is equivalent to the (annual) amount of premium paid by a country. It thus varies from country to country (and must be retained at the level of a country's annual premium payment).[5] The membership fee is non-refundable. It serves to capitalize the facility and gives the countries access to its services.

As of September 2019, CCRIF counts 21 member countries: Anguilla, Antigua and Barbuda, Bahamas, Barbados, Belize, Bermuda, British Virgin Islands, Cayman Islands, Dominica, Grenada, Haiti, Jamaica, Montserrat, Panama, St. Kitts and Nevis, St Lucia, St. Maarten, St. Vincent and the Grenadines, Trinidad and Tobago and Turks and Caicos Islands; Nicaragua was the first Central American government to join CCRIF for the 2015/16 season. The British Virgin Islands, Montserrat, Panama and St Maarten are the most recent members. They joined CCRIF during the 2018/19 insurance season.

CCRIF OFFERS COVERAGE AGAINST HURRICANES, EARTHQUAKES AND TROPICAL RAINFALL

With the payment of the membership fee, member countries are entitled to buy a *limited* insurance coverage against the financial impact of *major* hurricanes, earthquakes and excessive rainfalls. 'Limited' refers to the fact that the payout is triggered and capped at a certain loss level (see what follows); it represents a fairly small amount and is *not* intended to cover all costs of a disaster. 'Major' refers here to loss events, which were statistically modeled to occur with a frequency of 1:10 years (hurricane), 1:20 years (earthquake) or 1:5 years (excessive rainfall).[6] That means, CCRIF offers largely (limited) catastrophe insurances.

INSURANCE CONTRACT(S)

CCRIF offers insurance contracts for one year. The insurance policy year runs thereby from 1 June—30 May. For every peril (hurricane; earthquake; excessive rainfall) there exists a separate insurance contract. Countries can choose to buy one, two or all three insurance policies. Every insurance contract has its own terms. The *Hurricane Policy*, for instance, covers only wind and storm-surge related costs. Costs related to heavy tropical rainfall—in fact a close companion of hurricanes and more often than not responsible for a great degree of hurricane-related damage—must be covered through a separate insurance contract, namely the *Excessive Rainfall Policy*. All insurance contracts operate independent of each other (CCRIF SPC, 2015b, p. 15). This implies also that a hurricane may trigger two insurance payouts if a country has bought hurricane plus excessive rainfall coverage.

Every contract is structured as an *excess-of-loss arrangement*. This implies that CCRIF members receive an insurance payout only if the losses exceed a certain threshold. The level at which they are entitled to a payout is determined by the *attachment point*. The attachment point is associated with a certain return period of a weather event and reflects a certain value in dollar terms. Countries are free to choose where they want to set this attachment point within certain parameters. Under the parameter of the 2016/17 season this could not be more frequent than a 1:10 years event for hurricanes, 1:20 years event for earthquakes or 1:5 years event for excessive rainfalls. No matter which attachment point a country chooses, losses below that level must be paid by the country.

The level at which the payout is capped is determined by the *exhaustion point*. Analog to the attachment point, the exhaustion point is associated with a certain return period of a weather event and reflects a certain value in dollar terms. The exhaustion point gives rise to the maximum payout. Losses beyond that level must be covered by the countries again. While there are in principle no parameters where the exhaustion point may be set, CCRIF has set the *coverage limit* at US$ 100 million. That means, a country may not receive more than US$ 100 million per peril *and* year as (aggregated) payout. However, there is "no limit in terms of how many events per year that policy can cover" (CCRIF SPC, 2015b, p. 39).

CCRIF offers very flexible terms and allows countries to tailor their coverage according to their needs and (financial) possibilities. Countries are free to determine the coverage by selecting the attachment and exhaustion point in each contract.

PREMIUM LEVEL

Policies are priced individually, according to the *risk profile* of the country and the characteristic of the insurance contract, that is, the coverage selected (CCRIF SPC, 2015b, p. 39). The 'risk profiles' are maps that are based on actuarial analyses and describe the country's exposure to losses. They are based on their (historical and simulated) hazard, vulnerability and financial exposure data (for details see next section). Simply put, some countries are more exposed than others to hurricanes, earthquakes and tropical rainfalls; some have more, other have less valuable assets. All these characteristics come with a different price tag.

That means, the premiums reflect the financial exposure of the countries. The rule of thumb of this 'fair premium' approach is 'the higher the coverage, the higher the premium'. That means, if a country selects the maximum coverage and is highly exposed to hurricanes and/or has a lot of highly valuable assets (e.g., new airports, bridges) it pays in principle a higher premium than a country with lower coverage, respectively a country with less exposure to perils and/or less valuable assets. The approach implies there is essentially *no cross-subsidization* between Caribbean countries. Richer countries do not support poorer countries.

Yet, as pointed out in Chapter 2, the precise premium levels depend not only on the contract terms and a country's risk profile but also on a number of other dynamic issues such as CCRIF's operating cost, professional fees, the cost of reinsurance as well as the premium contributions paid by other insurance policyholders (World Bank, 2010). Annual aggregated premium payments revolved between US$ 19,5—US$ 22 million.

A MODELED-LOSS TRIGGER

To calculate losses and determine if it needs to make a payout CCRIF uses a *modeled-loss* approach. That means, in contrast to 'traditional' insurers CCRIF does not dispatch loss-adjusters to 'ground zero' after a disaster to assess damages and establish claims. Instead, it uses a probabilistic model to estimate losses and establish claims *as the event happens*. CCRIF promises (and so far it has kept that promise!) to make a payment to national treasuries within 14 days once it has obtained the final results from running the model—and conditional upon the breach of the attachment point.

The model underpinning the calculation uses real-time hazard data (e.g., wind-speed, ground-shaking, volume of rainfall) derived from a network of weather stations and satellites that measures the intensity of a weather event and overlays it with geophysical, population and economic data. The losses are thereby determined by the exposure of people and "national assets" (CCRIF SPC, 2015b, p. 29) relative to the hazard intensity in a certain location. The model incorporates information on geophysical characteristics (e.g., characteristics of the land surface) and certain assumptions about the value and vulnerability of certain objects (e.g., cheap and highly vulnerable dwellings vs expensive and stable brick-and-stone buildings). 'National assets' refers hereby to public infrastructure (such as roads, bridges, ports) as well as public and private assets (such as public and private buildings).

The statistical equation informing the model weights in particular the exposure of 'valuable' national assets and population density to the location of the hazard. That means, if an area contains particular expensive assets and is densely populated it is assigned a high value in the model's equation (Brooks, 2011, p. 4). The equation of the model and the accuracy of the data informing the model are therefore of decisive importance for the level of a payout. The model and related data must therefore be seen as politically highly contentious issues and be closely looked at (see the following).

To prevent domestic political fraud, hazard data is supplied by US-based providers, namely the National Hurricane Center NOAA (for hurricanes), US Geological Survey (for earthquakes) and US-Japanese Tropical Rainfall Measurement Mission (TRMM), a joint space mission between the US National Aeronautics and Space Agency (NASA) and the Japan Aerospace Exploration Agency (JAXA) that provides satellite rainfall data. This implies also that CCRIF is highly depended on a functioning IT-infrastructure. A satellite or software failure may cause severe interruptions and challenge insurance arrangement.

The technical basis of CCRIF

CCRIF has done extensive modeling to establish premium levels and compute payout levels. Subordinated to the core goal of securing liquidity, the model values in particular the economic key centers of a country. The model consists of five components: (1) a hazard module (2) an exposure module (3) a vulnerability/damage module (4) a loss module and (5) an insurance module. These components follow, in principle, an industry-wide standardized catastrophe risk methodology:

1 Hazard module

The hazard module contains information on the intensity and extent of a peril. The hazards considered are *wind-speed* and *weave-height* for hurricanes, *ground acceleration* for earthquakes and *rainfall* for excessive rainfall. The hazard module draws thereby on *real-time* data obtained from a network of weather stations from the National Hurricane Center NOAA (for hurricanes), US Geological Survey (for earthquakes) and US-Japanese Tropical Rainfall Measurement Mission (TRMM). To estimate the hazard value areas without measurement point, the module considers land characteristics (e.g., land surface, soil properties, topographic slopes) to approximate the value.

2 *Exposure module*

Based on economic, demographic and satellite data, the exposure module provides financial estimates on the 'values at risk' (e.g., public infrastructure such as roads, bridges, ports as well as private assets such as houses, dwellings). As asset-specific data was largely not suitable for disaggregating assets in classes (see exemplarily CCRIF, 2013b, p. 40), CCIRF used a proxy approach and divided the countries into various 'asset zones'. They broke down a country into three zones: 'residential, 'agricultural/commercial' and 'institutional' area. Every asset zone was assigned a specific financial value. CCIRF assigned areas with a high economic activity ('agricultural/commercial' assets) a higher value. The compilation was performed at resolution of 1 km^2 grid cells (CCRIF, 2013b, p. 40).

3 *Vulnerability/damage module*

These two bodies of data are then merged in the vulnerability/damage module. The vulnerability module estimates the 'hazard footprint', that is, "the degree of to which an asset might be damaged when subjected to the hazard forces exerted by the peril" (CCRIF, 2013b, p. 13). The damage is a function of the hazard level: The greater the hazard, the greater the damage. It assumes different vulnerabilities for the various assets.

Damage functions translate the peak of the hazard at a certain site into a likely damage of assets. The damages functions utilized by CCRIF vary not only according to asset classes (or zones) but also to perils. They are based on established models from the academic literature on disaster economics (CCRIF, 2013b, p. 15).

4 *Loss module*

The three components (hazard, exposure, vulnerability/damage) are then combined to estimate the ground-up losses across the country (see Figure 3.1). To estimate the ground-up losses, the estimated damage is multiplied by the values at risk. The calculation provides estimates of repair or replacement costs and is done for each grid cell. Losses are then aggregated across the country. In more recent module updates, CCRIF incorporated also secondary costs such as evacuation costs (CCRIF, 2013b, p. 57).

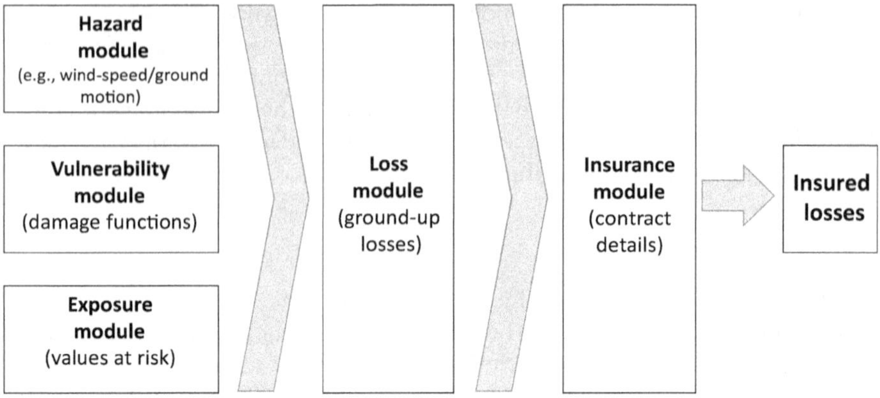

Figure 3.1 CCRIF's loss assessment framework

Source: simplified from CCRIF SPC, 2017, p. 24

5 *Insurance module*

The insurance module contains specific information on the insurance contract, namely the cost level at which a payout is due (attachment point), the level at which the payout is capped (exhaustion point) and the ceding percentage (the percentage of total losses that are covered). These details are selected by the countries themselves. If a certain loss level is breached a certain payout it due.

The (re-)insurance strategy of **CCRIF**

How does the rearrangement work? What is the relationship between CCRIF, re-insurers and the capital markets? CCRIF utilizes the practice of *risk layering* to ensure it has enough reserves to settle claims. CCRIF finances payout up to a certain level out of its own reserves *(self-retention)*. Claims exceeding that level are paid by commercial re-insurers and the capital markets (*excess-of-loss arrangement)*. That means that CCRIF is able to provide catastrophe insurance by removing the costs of 'mega-catastrophes' from its balance sheet. It does so by transferring the upper-loss levels to re-insurers respectively to capital market investors (that is, to institutionalized investors such as hedge funds).

CCRIF's self-retention level has varied throughout the insurance seasons and is outlined in its annual reports. It might be noteworthy, however, that in each year the total value of all insurance policies written exceeded CCRIF's own financial and reinsurance capacity. Hence, CCRIF has faced constantly the risk of insolvency. This is, however, no greater cause to worry—at least if one believes in the appropriateness of the gold standard in capital requirements for insurers. CCRIF's insolvency risk—modeled to be a millennial event—still falls comfortably within the EU's Solvency II Directive for insurers, which is commonly assumed to reflect the new 'best practice'.[7] And, yet, an extremely busy season—characterized, for instance, by one or a number of extreme hurricanes that hit the Caribbean's main economies Jamaica, Barbados, Trinidad, Cayman and Bahamas—or a larger series of smaller events could, in fact, have led to CCRIF's insolvency (CCRIF SPC, 2015b, p. 42). However, CCRIF modeled the occurrence of such an event with a probability of less than 1:1000 years; a probability "comfortably falling within CCRIF's guidelines for financial security and was substantially letter than any of its peers in either the public or private sectors" (CCRIF SPC, 2015a, p. 51).

How does the reinsurance structure look like in practice? To illustrate CCRIF's reinsurance structure, I will focus here exemplarily on the reinsurance structure 2016/17 for hurricane/rainfall insurance policies: For the insurance season 2016/17, CCRIF wrote 38 insurance policies for its 17 members at that time.[8] The aggregated value of these contracts totaled to about US$ 623 million (CCRIF SPC, 2017, p. 37). In other words: CCRIF promised to pay for claims aggregating to US$ 623 million. The arrangement foresaw that CCRIF would pay for all losses up to US$ 25 million out of its own capital reserves.[9] Losses from US$ 25 million up to US$ 163 million would then be covered by re-insurers and capital market investors via a cat bond. Losses greater than US$ 162,7 million would then be covered out of CCRIF's capital reserves until depletion (see Table 3.1).

How does this arrangement look like in terms of event probabilities? CCRIF would have paid for aggregated payouts that would occur 1:6 years, according to CCRIF's own financial modeling (see Figure 3.2). Aggregated claims exceeding US$ 25 million would then have been covered via traditional commercial re-insurers and the capital markets. To accommodate the diverging 'risk appetite' of re-insurers, CCRIF structured reinsurance into three risk layers. The first reinsurance layer covered aggregated payouts with a 1:10 years chance of occurring, the second reinsurance layer covered aggregated payouts with a 1:20 years chance of occurring and the third reinsurance layer covered aggregated payouts with a 1:625 years chance of occurring. The latter referred to an aggregated annual payout of US$ 163 million. Reinsurance was bought from a number of international commercial re-insurer companies.

In parallel to buying 'traditional' reinsurance coverage, CCRIF bought also alternative reinsurance coverage by issuing cat bonds. The cat bonds were placed in parallel to risk layer 2 and 3 and secured CCRIF another US$ 30 million in reinsurance coverage. In this context it is notable that these cat bonds were not sold by CCRIF but by the World Bank. The World Bank's treasury functioned

Table 3.1 CCRIF's reinsurance structure in 2016/17

Retained/Reinsured	*Losses*
Retained (CCRIF)	0—US$ 25 million
Reinsured (via re-insurers, capital markets)	25—US$ 163 million
Retained (CCRIF)	163—US$ 623 million

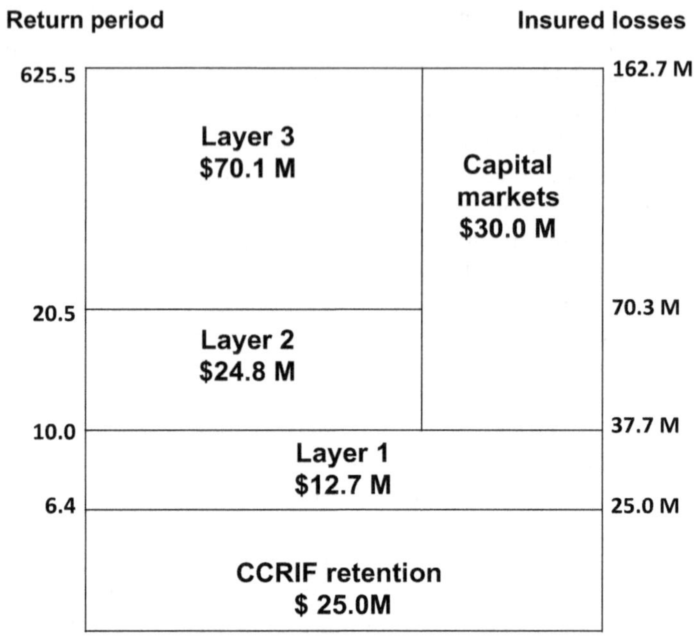

Figure 3.2 CCRIF's risk transfer structure (TP/EQ) in 2016/17
Source: based on CCRIF SPC, 2017, p. 37

here as intermediary: It sold the cat bond on CCRIF's behalf. The 2014/15 cat bond was the first cat bond ever (!) issued by the World Bank. The issuance of cat bonds via the World Bank was therefore quite a remarkable event in the World Bank's own history. The 2014/15 cat bond has a maturity of three years and was structured and placed at the capital markets with the aid of highly specialized capital market actors, namely GC Securities, Munich Re and Swiss Re Capital Markets (CCRIF SPC, 2015a, pp. 51–52).

Capitalization

Since 2015 CCRIF is structured as a 'segregated portfolio company'. Segregated portfolio companies, while sharing some operative costs, segregate assets and liabilities. The establishment of a segregated portfolio company was necessary to enable CCRIF's expansion into Central America and prevent a cross-regional subsidization. For each region and insurance product CCRIF runs separate accounts.

If these portfolios are added up, CCRIF held about US$ 120 million in shareholder equity,[10] as of 31 May 2016. US$ 120 million on the positive side of the balance sheet may sound a lot. Yet, by putting the sum into relation with the shareholder equity of global re-insurers such as Munich Re—the company's shareholder equity typically revolves around US$ 10–11 billion (Munich Re,

2014)—it becomes evident that CCRIF, as investor, is a fairly small actor. And yet, if one puts CCRIF's financial power into relation to the GDP of some small Caribbean islands—the GDP of Montserrat, for instance, which revolves largely around US$ 0,1 billion (IMF, 2013)—CCRIF holds quite a sum of capital.

CCRIF's capital has been sourced from the one-time membership fees which summarized to about US$ 20 million (CCRIF, 2008, p. 22), retained earnings from premium payments (annual premium incomes which revolved between US$ 19,5—US$ 22 million), investment income as well as donor contributions via a disbursement agreement between CCRIF and a Multi-Donor Trust Fund. The grant agreement enabled CCRIF to claim reimbursement for certain costs (e.g., operational costs, including reinsurance costs). The Multi-Donor Trust Fund was established in 2007 and financed by Canada, the European Union, the World Bank, the UK, France, the Caribbean Development Bank, Ireland and Bermuda. By 2012, the contributions summarized to US$ 67 million (World Bank, 2012, p. 7). Its original function has been to enable lower, affordable premium levels.

The Central America Segregated Portfolio was established in 2015 with funding from a second Multi-Donor Trust Fund which was supported by the US (US$ 10 million) and Canada (US$ 14 million). In May 2017, this second Multi-Donor Trust Fund received additional funding from the EU (US$ 14,8 million) and Germany (US$ 12,5 million). The funds from the EU were dedicated to facilitate the provision of geophysical and climate-related products for COSEFIN countries, while the funds from Germany were dedicated to provide new climate-related products to Caribbean members.

The governance structure of CCRIF

CCRIF is a 'virtual' organization and has a lean operational structure. In other words, CCRIF does not own physical buildings or employ own full-time staff; it is a letterbox company that is run through various service provision agreements with private subcontractors. The rationale to renounce brick-and-mortar facilities and outsource all financial, administrative, communicative and regulatory tasks was to keep the operation costs as low as possible. The private service providers are mandated by a 'Board of Directors' to conduct the day-to-day operations of the facility. The organigram of CCRIF designates five roles (Communications Manager, Risk Management Specialist, Insurance Manager, Insurance Manager, Placement Broker), which are performed by six different private service companies (see Figure 3.3). The responsibilities, duties and tasks are outlined in an Operations Manual. The manual also sets out policies and procedures (World Bank, 2010, p. 17).

The Board of Directors

The Board of Directors runs CCRIF on behalf of the countries and the donors, respectively (see the following). The Board of Directors oversees and approves all policies related to the strategy and operations of the facility (CCRIF SPC, 2015b, p. 8). The Board develops, sets and approves CCRIF'S insurance policies, sets the

Figure 3.3 Organigram of CCRIF

Source: simplified from CCRIF SPC, 2018b

vision of the future of the organization, develops and implements strategic plans while being at the same time financially responsible for company. Hence, its main decisions are strategic in nature.

The Board of Directors is composed of no more than five representatives. One represents the donors of the multi-donor trust fund and is nominated by the Caribbean Development Bank (CDB); one represents the member countries and is nominated by CARICOM; two board members are jointly appointed by CDB and CARICOM on the basis of their expertise in finance and insurance respectively; the chairperson is selected by the other four members. The four Directors are appointed for a three-year term and may not serve longer than two consecutive terms (World Bank, 2010, p. 44).

What is remarkable about the Board's composition is not so much that *multilateral actors* act somehow paternalistically *on behalf of the Caribbean countries* but that it is composed solely of multilateral actors that *represent financial interests*. The CDB and CARICOM have a substantial stake in the facility. The sole representation of financial interests is remarkable insofar as CCRIF has the explicit aim to act "*beyond* its core mandate of the provision of liquidity" (CCRIF, 2012, p. 9 my emphasis) and "assist member governments and their communities in understanding and reducing the socioeconomic and environmental impacts of natural disasters" (CCRIF SPC, 2015a, p. 3), as outlined in its mission statement. Also Pearson, Board Director of CCRIF, argued in a speech held at the UN in 2010, "*we are trying to do more* than simply provide parametric insurance policies to our members. We continuously *work towards engaging our members* and supporting them in the development and implementation of strategies for disaster risk management" (Pearson, 2011, pp. 55–56 my emphasis). Yet, despite these ambitions 'to do more' CCRIF has not systematically integrated representatives from disaster management agencies (such as, for instance, the Caribbean Disaster Management Agency) in its governance structure.

While it cooperates with disaster management agencies on an institutional level, CCRIF provides solely financial interests with structural power to shape the overall direction of the facility. This one-sided governance direction is further perpetuated through the individuals that have taken up the positions. All current Board Members (Timothy Antoine, Ken Blakely, Desriée Cheriben, Faye Hardy and Saundra Bailey) are economists by training and have worked in finance, either on the private or public side (CCRIF SPC, 2018a). This is not to discredit CCRIF or the Board *per se* but to underscore the apparent tension between a self-acclaimed vision and reality. And yet, this structural bias towards finance interest might simply be explicable due to compliance issues and thus with reference to regulations in the Cayman Islands. As a World Bank report pointed out: "Directors must be acceptable to the Cayman Islands Monetary Authority (CIMA), which is CCRIF's regulator" (World Bank, 2010, p. 44).

The Chief Executive Officer and the Chief Operations Officer

CCRIF is managed under the guidance of a Chief Executive Officer, a Chief Risk Management Officer and a Chief Operations Officer. Together they guide and monitor the operational functions of CCRIF, as laid out in CCRIF's operational manual. The Chief Risk Management Officer and the Chief Operations Officer oversee thereby five roles performed by six private service companies (there are two asset managers involved): The Placing Broker; Risk Management Specialist; Insurance Manager; Asset Manager; and the Communication Manager.

The Communication Manager is responsible for disseminating internal and external information related to CCRIF. The tasks involve strategic knowledge-building and capacity building—that is, enhancing the understanding of the working of CCRIF—among its stakeholders (participating states) and a broader, wider public. It also aims to advertise CCRIF and engages in marketing activities to consolidate its standing. It does so through direct contacts, use of print, radio, and television media, participation in conferences, and development of its own website. News bulletins, press releases and quarterly and annual reports are issued and made available on CCRIF's website, along with a variety of technical publications on specialized topics (World Bank, 2012, p. 17).

The Risk Management Specialist, Asset Manager, Insurance Manager and Placing Broker take over various commercial and financial tasks. Simply put, the Risk Management Specialist and the Placing Broker, for instance, develop, implement and manage the risk transfer strategy; this means that they report and advise the Board on new insurance and capital market developments and appropriate (re-)insurance strategies. The Placing Broker manages the reinsurance claims processes on behalf of CCRIF PSC, including verification of outputs from the models when they are triggered. The Insurance Manager and the Asset Manager take over financial tasks. The Insurance Manager does primarily accounting. It reports and updates the Board of Directors on the financial status of the facility by issuing quarterly financial statements. The Asset Manager, in turn, is responsible for investing the CCRIF's capital in accordance with CCRIF's investment guidelines.

An assemblage of eight private companies makes CCRIF happen: *Sustainability Managers*, a consultancy company based in Jamaica, play the Communication Manager/Technical Assistance Manger role. A consortium composed of *Evalucacíon de Riesgos Naturales* and *Risk Engineering and Design*, as well as *Kinetic Analysis Corporation* perform the perform risk management role. *Will Re* performs the role of the reinsurance broker. *Sagicor Insurance Manager*, a financial service organization based in Jamaica, plays the role of the insurance manager. *London & Capital* and *Butterfield Asset Management Ltd* perform the asset manager role (CCRIF SPC, 2017, p. 17).

Part II: the genesis of CCRIF

Ideational origin

The idea of creating a regional insurance instrument in the Caribbean was not totally out of the box. By contrast, the idea of developing a regional insurance instrument to support local insurers has been discussed within the Caribbean for years (Lalor, 1996, p. 23). The main rationale of these regional discussions has been to overcome the local insurance industry's dependency on international re-insurers and to develop a regional Caribbean reinsurance industry.

Throughout its recent history, the Caribbean has had a poor insurance penetration and the provision of insurance coverage largely has been depended on the provision of foreign capital and its pricing. Partly this has been the result of the region's colonial legacy. Historically seen, property was largely owned by persons who lived abroad in Europe—"mainly in England, where the majority of property insurance was placed" (Lalor, 1996, p. 2). Accordingly, local insurance has been mainly provided by insurance companies external to the region, companies which were mostly based in the UK (ibid.). The local insurance industry gained its own identity only in the 1970s with the end of colonization and the adoption of various Insurance Acts across the region (ibid.). Yet, the local markets remained highly fragmented and struggled due to their thin capitalization, low level of capital and solvency requirements, the lack of regulatory effectiveness in solvency vigilance and the comparatively weak technical infrastructure and database (Lalor, 1996; Pollner, 2001). The regional insurance business has been highly dependent on international market dynamics as local insurers have ceded most of their risks—that is, their payment obligations—to re-insurers outside of the Caribbean (Lalor, 1996; Pollner, 2001). As elaborated in the previous chapter, re-insurers demand a premium to take over insurers' payment obligation. Generally speaking, the lower the level of retention of the primary insurer the higher the premium re-insurers charge.

The dependency of Caribbean insurer on the pricing of foreign capital provided by re-insurers was particular evident during the late 1980s/early 1990s. A number of large-scale catastrophes across the world[11] drained substantial capital from global (re-)insurers. (Re-)insurers sought to recover by tightening coverage and increasing premium payments worldwide. (Re-)insurance became hard to obtain and more expensive. Hurricane Andrew (1992) stands exemplarily for the development of this 'hard market'. Hurricane Andrew brought major devastation to Florida and the Gulf of Mexico and created about US$ 15–17 billion in insured losses—a sum which was unprecedented by that time. It brought several insurers down to their knees. As a consequence, several re-insurers withdrew from the Caribbean, among them one of the largest players, Munich Re.

While insurance rates rose globally, they soared in the Caribbean: Within two years Caribbean countries experienced an increase in insurance premiums between 200–300%, despite any change in the countries' underlying risk (Pollner, 2001, p. 2). In fact, many Caribbean countries were not at all affected by the consequences of Hurricane Andrew. The sharp price increase reflected not only shortage in capital following Hurricane Andrew's devastating impact on the US but also "a changed perception by re-insurers on the potential costs of hurricanes" (ibid., p. 13). Re-insurers concluded that their models were underestimating the actual costs so that they were pricing too cheaply (see Chapter 2, *The impact of Hurricane Andrew*).

In the aftermath of Hurricane Andrew, as a consequence of the high insurance prices, on 1 July 1994 the CARICOM Heads of Government established a working group, led by Dennis Lolar, chairman of the Jamaican Insurance Company of the West Indies (ICWI), to assess the financial situation of CARICOM's insurance industry and develop recommendations to improve the financing possibilities of the local industry. The working group, composed of representatives from both the public and private sector, proposed, among others, the development of an own regional catastrophe pool, a Caribbean re-insurer for the local insurance industry. The guiding vision was an emancipatory one, one that reflects 'empowerment *from* global financial markets'.

The proposal by Lalor (1996)

Lalor proposed the creation of a regional re-insurer, a so-called Caribbean Trust Fund, to reduce the premium payments by Caribbean *primary* insurers to globally-acting re-insurers. He envisioned that this "Caribbean Trust Fund" should work as a buffer against the volatility of international reinsurance premiums (Lalor, 1996, p. 4). The working group imagined the creation of a regional reinsurance entity as a collaborative public-private undertaking. It proposed that the facility should be capitalized with contributions from local insurance companies and governments. The working group also recommended that a "multinational institution" should provide further financial support over the first years of operation, namely through a contingent line of credit "until additional premium income and/or contributions would pay this off" (ibid., p. 23).

It is noteworthy that, in contrast to the later CCRIF, the proposal had the health and development of a Caribbean insurance industry at the core. The objective of the "Caribbean Trust Fund" would be "to increase the availability of affordable catastrophe insurance protection and/or risk funding mechanism" (ibid., p. 23). The working group hoped that the availability of cheap reinsurance coverage would encourage the local industry to offer more and better insurance contracts (ibid., pp. 24–26).

The report emphasized the need for this market development to go hand-in-hand with regulatory reforms "in order to ensure the fiscal ability of insurance companies to meet their commitments" (ibid., p. 6). In fact, the report demanded nothing short of a tremendous regional institutional reform. A reason was that the working group ascribed the insurance industry a key role in the region's economic and physical development (ibid., p. 2).

As a next step the working group recommended the conduct of a feasibility study, commissioned by a multinational institution equipped with sufficient financial and technical capacities. It proposed to experiment with a sub-regional reinsurance entity for a smaller grouping of CARICOM countries, namely the countries of the Organization of Eastern Caribbean Countries (OECS). It is here "where there are fewer large insurers and where risk and risk sharing could be more beneficial" (ibid., p. 26), the report concluded.

The proposal by Pollner (2001)

Although it is unclear what precisely happened with Lalor's proposal, in 2001 the World Bank took up the idea to create a regional insurance arrangement for the Caribbean. In the context of a review of the Caribbean's insurance markets, the World Bank demonstrated the technical and financial feasibility of a sub-regional insurance instrument for the Eastern Caribbean countries and suggested its implementation. Although the World Bank report was neither commissioned by the CARICOM nor did it refer to the Lalor report explicitly, it clearly did the work the CARICOM Working Group demanded back in 1996.

In his comprehensive report *Managing Catastrophic Disaster Risks Using Alternative Risk Financing and Pooled Insurance Structures*, Pollner, a World Bank disaster economist, examined "the constraints and opportunities in implementing a catastrophe insurance system" (Pollner, 2001, p. 8) in the Caribbean context. He provided not only a comprehensive review of the Caribbean insurance markets and its relation to the global market but also a review of recent innovations in "catastrophe risk management, ranging from private securitization of insurance risk to public/private partnerships" (ibid., p. 9). He ultimately made the case for a sub-regional Caribbean insurance pool "to diminish economic volatility in countries subject to shocks which are unrelated to international market movements, or where countries affected by single natural events are disproportionately affected" (ibid., p. 83).

The historical context of the Pollner report was the 1999 hurricane season, a very busy season, similar to the 1992 season. The Atlantic hurricane season counted five Category 4 hurricanes—a number that was only tied in 2005. Similar to the impact of Hurricane Andrew in 1992, (re-)insurance prices rose globally and in the Caribbean particularly (World Bank, 2002b, p. 1). Yet, in contrast to the Lalor report, the starting point of the Pollner report was *not* the problem of high-level of Caribbean (re-)insurance premiums *per se* but the negative macroeconomic impact of disasters on the Caribbean's development.

In his report, Pollner emphasized the rising post-disaster bill worldwide due to "increasing urbanization, population growth, rising growth of both number and value of properties in catastrophe prone areas, as well as climatological and environmental changes" (Pollner, 2001, p. 24). He underscored that most of the costs were not insured. Pollner pointed out that natural disaster-related costs in the Caribbean were mainly borne by private households and only partly by insurers. This would reduce private wealth (ibid., p. 17)—and thus implicitly prevent investments: Small businesses would have to use their accumulated financial reserves to rebuild their properties.

To overcome this disaster-related 'poverty trap', Pollner envisioned insurance as *the* key solution. A (functioning) insurance industry could spread the costs over a large group of people and/or over a time and territory. He concluded that the World Bank's objective must be "to develop domestic financial industries and insurance-linked mechanism which can help to increase the private sector capacity to better absorb and spread these risks" (ibid., p. 83). To that end, it would be necessary to hedge the volatility of catastrophe insurance premiums and to increase the local industry's financing capacity. Pollner suggested that the Caribbean industry should make use of "the latest risk financing and risk transfer technologies" and tailor them to their needs to "ensure more sustainable development" (ibid., p. 83) in the region.

The Pollner Report created a link between disasters, insurance and sustainable development (see also Grove, 2012). It conceived the provision of liquidity through the insurance industry as key to reduce the macroeconomic impact of disasters and thus the overall regional macroeconomic volatility. It assumed that the provision of liquidity would allow for a quicker reconstruction (Pollner, 2001, p. 83), support the accumulation of wealth and thus help people out of poverty. The guiding vision to create a regional insurance pool was again an emancipatory one; yet, this time an emancipatory vision that meant 'empowerment *through* global financial markets'.

Pollner quantified and reviewed the cost and benefits of a sub-regional catastrophe insurance pool for the East Caribbean states (Antigua and Barbuda, Barbados, Dominica, Grenada, St Kitts and Nevis, St Lucia, St Vincent and the Grenadines, and Trinidad and Tobago) vs domestic catastrophe reserve funds. He concluded that a (sub-)regional catastrophe insurance pool would be financially the better solution as it would require less capital for the same amount of risk, permit higher level of loss levels as well as a higher retention level and thus cheaper premiums (ibid., p. 95).

Pollner suggested that local insurers could underwrite hurricane policies but then 'cede'—that is, sell—the 'catastrophe risks' to the pool (ibid., p. 80). He proposed that the regional catastrophe pool should be capitalized with financial contributions from both the local insurance industry and governments to make sure that it is appropriately capitalized. Regarding its reinsurance structure, Pollner suggested that the pool should buy classic reinsurance coverage and make use of new financial market innovations by utilizing cat bonds based on parametric payout triggers to source additional reinsurance capital (ibid., p. 111). Pollner proposed that the entity should be governed by an independent entity.

For the World Bank, Pollner envisioned the (new) role of a *facilitator* and *market-maker*: "The multilateral development bank role in this context would be to support the credit quality of such schemes as well as organizing and/or financing the appropriate technical assistance required" (ibid., p. 83). He argued that the World Bank should finance the start-up costs and provide a low-cost contingent line of credit on top (ibid., p. 105). Pollner emphasized that the "multinational institutional support could only be considered as a catalytic function since permanent support for such a scheme would be expected to be borne by the private markets" (ibid., p. 110).

What comes out of this short review is that Pollner's proposal envisioned in large parts what would at a later stage become CCRIF. However, before it came to CCRIF, the World Bank would have the chance to test the viability of the idea to create a *sub*-regional insurance pool.

The Eastern Caribbean Insurance Project (2001–2003)

From 2001–2003 the World Bank attempted to implement a modified version of the Pollner proposal together with the Caribbean Development Bank as implementing agency. The project was pretty ambitious.

The *Eastern Caribbean Insurance Project* sought to support the development of an Eastern Caribbean *financing strategy* to cope with macroeconomic costs of disaster ("to put in place comprehensive country and sub-regional risk management ex ante funding strategies to reduce fiscal, economic and financial instability in the wake of natural disasters", World Bank, 2002a, p. 1) and to facilitate the provision of insurance ("to strengthen the domestic insurance industries and their arrangements with the international reinsurance and credit markets, to augment the capacity of domestic insurance markets to manage and transfer large risk exposures, while building up the requisite reserves to pre-fund and insure against future natural disasters", World Bank, 2002a, p. 1). In other words, the World Bank sought not only to sensitize governments for financial planning but also to promote a regulatory 'strengthening' of the insurance law of the East Caribbean by providing suggestions to harmonize (which meant to liberalize) the local insurance laws and enhance capital and solvency requirements (World Bank, 2002a, p. 2).

The goal of these measures was the implementation of a sub-regional insurance pool. The sub-regional insurance pool was thought of as an instrument to protect *public assets and infrastructure:*

> a public assets insurance pool to be leveraged using reinsurance and credit support for ex ante coverage of critical buildings (government offices, schools, hospitals) and critical infrastructure (e.g.: water, ports, roads, bridges) as determined necessary for protection by the client Governments.
>
> (World Bank, 2002a, p. 2)

Hence, in contrast to the original Pollner proposal, which had the viability of the local insurance industry at the core, this project had the viability of public assets at its center.

The design of the sub-regional pool was assigned to a unit within the Caribbean Development Bank in consultation with a steering committee consisting of participating governments, CDB, ECCB and the World Bank. However, in 2003, after two years, the project failed. In a 2007 report, the World Bank concluded that the project failed because of the

> fairly mild hurricane season 2001–2003; lack of preparatory funds to support the preliminary studies required to implement the Facility, including

> the development of regional catastrophe risk models; worries about possible cross-subsidization among participating countries; insufficient political support for the insurance solution proposed at that time.
>
> (World Bank, 2007b, p. 5)

Time had not yet come for a Caribbean insurance scheme. There was no political momentum. The World Bank and the implementing agency, the Caribbean Development Bank, overestimated need or demand. It lacked a disaster and the goal of an insurance arrangement was not 'correctly' identified. The insurance solution was in search of a problem. It should take another year until problem and goal were correctly identified.

The World Bank as incubator

As the Pollner Report and the failed Eastern Caribbean Insurance Project demonstrate, the World Bank had apparently a strong interest in mobilizing the idea of a regional insurance scheme. But why was the World Bank so keen on setting up a regional insurance scheme? What were the potential rationales for the World Bank's support of a regional catastrophe insurance arrangement?

Organizational change and the accumulation of new knowledge

One reason was arguably the enhanced understanding of the economic consequences of natural disaster within the World Bank. Until the 2000s there was simply little understanding within the World Bank how disasters impact on economies and thus contribute to (re-)produce poverty, as a senior World Bank officer who was a leading figure within the development of CCRIF revealed (Personal Interview, senior World Bank officer, 2015, March 13). It was only by the end of 1990s and in the context of a small internal organizational change, a rising post-disaster bill and more orchestrated and global efforts to deal with disasters (see the following) that knowledge of the 'economics of disasters' was consequently built-up within the World Bank. This new knowledge prepared the grounds for new ideas and policy proposals and a more active engagement of the World Bank.

Key here was an internal organizational change, namely the creation of the Disaster Management Facility in July 1998. The World Bank established the Disaster Management Facility "to provide proactive leadership in coordinating efforts to introduce disaster prevention and mitigation practices in the range of development-related activities" (Kreimer et al., 1999, p. 6).

The Disaster Management Facility—which would later become the Global Facility for Disaster Reduction Recovery (GFDDR)[12]—was a small unit within the Bank and mobilized various members from country departments, the financial sector development department and the international finance corporation; but it partnered also with academic institutions (Virginia Polytechnic Institute and the International Institute for Applied Systems Analysis) and private insurers (the Aon Group, Gerlin Global Reinsurance, Munich Reinsurance, and Renaissance

Reinsurance Ltd) to "create public-private partnerships to advance disaster risk awareness and to provide effective incentives for institutional and individual investment in disaster mitigation measures" (Kreimer et al., 1999, p. 6). The unit created a number of reports and technical papers—basically country case studies—on disaster exposure, insurance coverage and the use of mitigation practices. It was important insofar as it fostered a new understanding of the macroeconomic costs of disasters and generated new policy ideas for implementation.

In an interview, a World Bank officer emphasized the key role of the Latin America country unit and the Disaster Management Facility in setting the stage for CCRIF:

> Especially the Latin America team started dealing these issues [financial impacts of natural disasters; N.S.] more. I think one element that was particular instrumental was the Global Facility for Disaster Recovery. That really had given impetus within the World Bank and outside then.
>
> (Personal Interview, senior World Bank officer, 2015, March 13)

This is evidenced by the increasing number of World Bank reports and technical papers by the end of the 1990s and early 2000s that quantify the macroeconomic costs of natural disasters in the Caribbean (see exemplarily Auffret, 2003; Benson, Clay, Michael, & Robertson, 2001; Pollner, 2001).

More frequent and expensive disasters

The creation of the Disaster Management Facility, the subsequent new understanding of disaster impacts on economic development and the increasing involvement of the World Bank in disaster-related economic policymaking came not out of the blue. It was driven by the rising numbers of large-scale disasters worldwide (see Figure 3.4)—in particular, the increasing and more intensive hurricane activity in the Atlantic basin since the late 1980s (Collymore, 2011, p. 8)—the rising post-disaster bill worldwide (see Figure 3.5) and the increasing international efforts to deal with disasters in a more coordinated and organized way that put mitigation first.

As the countries had little financial resources at their disposal to finance the rebuilding of their countries, the World Bank often financed (and still finances) the post-disaster bill. The World Bank typically diverted funds from previously planned development projects. It wanted to get the private sector involved to pay part of the costs and "save development resources and protect future investment" (Kreimer et al., 1999, p. 6).

The fact that the World Bank sought a solution with the private sector (in particular with insurers and thus the finance industry) must also be understood against the background of the political trends of the late 1990s. The late 1990s saw the rise of a political philosophy termed 'Third Way'. The 'Third Way' was a loose way of thinking, prominently put forward by social democrats across the transatlantic; it promoted the use of market practices and instruments for social

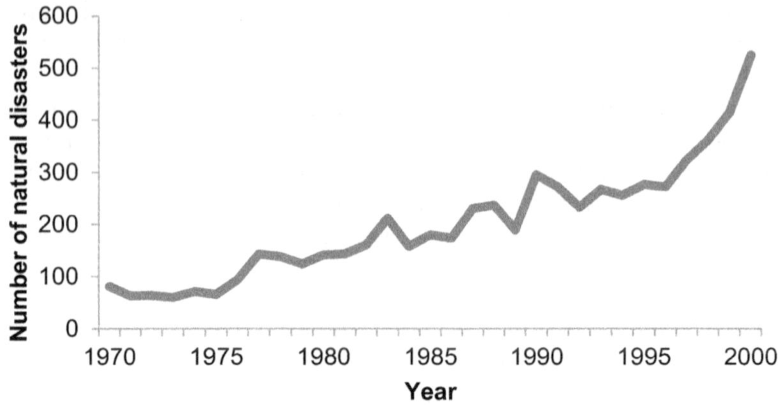

Figure 3.4 Number of natural disasters 1970–2000

Source: data taken from EM-DAT: The Emergency Events Database—Université catholique de Louvain (UCL)—CRED, D. Guha-Sapir—www.emdat.be, Brussels, Belgium

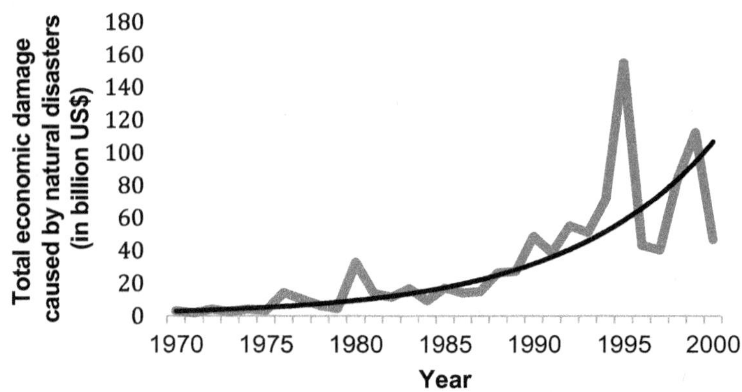

Figure 3.5 Total economic damage caused by natural disasters 1970–2000

Source: data taken from EM-DAT: The Emergency Events Database—Université catholique de Louvain (UCL)—CRED, D. Guha-Sapir—www.emdat.be, Brussels, Belgium

goals and goods. A result of this powerful ideology was the increasing formation of public-private partnerships across the OECD world. In other words, by proposing to involve financial market actors into disaster policies, the World Bank just went with the political fashion during that time.

International institutionalization of disaster management

The World Bank's leading involvement in developing a Caribbean insurance scheme must, however, also be understood against the background of more orchestrated global and regional effort to deal with disasters and thus with global institutionalization processes. Until the beginning of the 1990s, there existed no global and systemized approach on how to deal with natural disasters—not even in the Caribbean, despite its large exposure to natural disasters throughout history (Schwartz, 2015). This changed only in the course of the 1990s as various regional and global bodies started to politicize and thus question the management of disasters. In this context, the World Bank intensified its efforts to reduce the impact of disaster.

Until the 1980s disasters were dealt with largely on an *ad-hoc* basis in the Caribbean; efforts concentered mainly on compensation and relief distribution (Collymore, 2011, p. 9). Once a disaster had occurred, emergency and relief measures were provided in an "improvised and uncoordinated way" (PAHO, 1994, p. 31). No guidelines and plans existed on how to deal with the problems a disaster typically causes, such as rescuing survivors, treating the injured, providing shelter, distributing water and food, clearance of roads, reestablishing communication structures, etc. Disaster response was also typically organized by agencies responsible for internal and external security, such as the Ministry of Defense or Interior or the Armed Forces (PAHO, 1994, pp. 31–32) or even handed over directly to the heads of government (Poncelet, 1997). The institutional location of disaster management within agencies of defense has its deeper, intellectual and emotional origin in the context of the Cold War. Civil defense planning, pioneered by the United States in the early Cold War years, was informed by the fear of a nuclear strike and organized around the goals of reestablishing key infrastructure and industries and as well as a securing a functioning political command chain to enable a (military) response (Collier, 2008).

It was only in the wake of a number of high-profile disasters in the late 1980s—such as floods in the Sudan and Bangladesh, typhoons in the Philippines, Hurricanes Hugo and Gilbert in the Caribbean basin, all in 1988/89—that disaster management became a subject of civil 'high politics' and was subsequently institutionalized through the UN system and regional political entities such as CARICOM.

Following the rising number of disasters, in 1987 the UN General Assembly declared the 1990s the International Decade for Natural Disaster Reduction, a decade in which the international community would pay special attention to fostering international cooperation in the field of natural disaster reduction. In 1994, the first World Conference on Disaster Reduction was held in Yokohama, Japan. It adopted the Yokohama Strategy and thus the first global guidelines that put prevention, early warning systems and the development of contingency plans at the forefront of international efforts how to deal with disasters. Throughout the 1990s and in various resolutions, the UN General Assembly (re-)called to strengthen concerted collective efforts to improvements in early warning (see, for example, UN Resolution 51/185) and to facilitate the full integration of disaster reduction into the substantive efforts for sustainable development and environmental protection by the year 2000 (see, for example, UN Resolution 50/117).

In 2000, and thus after a decade of discussions, the UN General Assembly adopted the International Strategy for Disaster Reduction. The International Strategy for Disaster Reduction reflected a major shift from the traditional emphasis on disaster response to disaster reduction, and, in effect, sought to promote a culture of prevention. In 2005, the second World Conference on Disaster Risk Reduction was held at Kobe, Hyogo, Japan. It endorsed the Hyogo Declaration and the Hyogo Framework for Action 2005–2015. The Hyogo Framework identified the need to "promote the development of financial risk-sharing mechanisms, particularly insurance and reinsurance against disasters", as a priority action for "building the resilience of nations and communities to disasters" (UNISDR 2005).

In his review of Caribbean disaster management, Collymore argues that the 1990–2000 epoch stands for a significant departure from the previous country-based and *ad-hoc* way of dealing with disasters to a more collective, that is regional, and holistic *ex ante* approach (Collymore, 2011). This development was facilitated by repeated disasters in the mid/late 1990s (e.g., Hurricane Andrew 1992, the extremely busy hurricane season 1999 with five Category 5 hurricanes in the Atlantic basin) with high levels of loss and damage to sectors, which were of vital importance to the Caribbean economies, namely the agricultural and tourist sector. The 1990–2000 epoch has seen the regionalization of disaster management with the creation of the Caribbean Disaster Emergency Response Agency (CDERA) in 1991 and the increasing integration of NGOs, multilateral, regional and international financial institutions such as the Caribbean Development Bank or the World Bank into regional and national disaster-related policymaking processes and projects. With the integration of various stakeholders disaster management became more holistic, strategic and long-term in outlook. It received also a more economic touch, as organizations such as CDERA linked disasters increasingly to the concept of economic development (Collymore, 2011, p. 14; Lavell & López-Marrero, 2014, p. 240). It is also against this background of these increasing international and regional efforts that the World Bank's 2007 CCRIF Project Appraisal report claims:

> The World Bank has been working with the Organization of Eastern Caribbean States (OECS) over the past nine years on the development and implementation of the Emergency Recovery and Disaster Management Projects, aimed at reducing the countries' physical vulnerability and strengthening their emergency management capacities.
>
> (World Bank, 2007a)

To sum up, the World Bank's activism to develop an insurance proposal could be explained by its new unit, the Disaster Management Facility and the production of new knowledge on the relation between disasters and poverty. This organizational development should be seen in the wake of a rising number of disasters and thus a rising post-disaster bill and the increasingly orchestrated institutional developments at the global and regional level to reduce the impact of disaster. Disaster management gained increasingly importance.

The beginnings of CCRIF

So how did CCRIF then come about? As illustrated previously, since the mid-1990s ideas to develop a Caribbean insurance scheme took on more of a concrete shape. A number of large-scale disasters led to soaring insurance premiums and a rising post-disaster bill that affected states and the World Bank alike. In response, both the CARICOM and the World Bank worked out concrete proposals on how to strengthen the region's resilience against natural disasters. In 1996, CARICOM's Lalor Working Group developed the idea to set up a public-private regional re-insurer entity, backed by a multilateral financial institution, to ease and smooth the volatility of international reinsurance premiums. In 2001, the World Bank came up with the proposal to set up a sub-regional, public-private insurance entity in those Caribbean countries which are most affected by disaster to buffer macroeconomic and fiscal volatility, utilizing newest financial market innovations such as catastrophe bonds to source sufficient reinsurance capital. The World Bank pushed the idea further and, together with the Caribbean Development Bank, initiated a pilot project. The Eastern Caribbean Insurance Project (2001–2003) had the development of a sub-regional insurance entity to insure public infrastructure and assets as its goal. The project, however, failed due to fairly mild hurricane seasons and insufficient political support. There was no political momentum. This changed, however, the subsequent year.

Agenda-setting: how CCRIF made it to the table of decision-makers

Trigger for what would then become CCRIF was the destructive impact of Hurricane Ivan. Hurricane Ivan caused billions of losses and killed about 60 people across the Caribbean. It was the most powerful hurricane to hit the Caribbean in ten years and only the destructive and deadly highpoint of a highly devastating hurricane season 2004.

Hurricane Ivan had a substantial impact on several Caribbean countries (Bahamas, Cayman Islands, Cuba, Dominican Republic, Dutch Antilles, Jamaica, Grenada, Haiti, St Lucia, St Vincent and the Grenadines) of which Grenada and the Cayman Islands were most affected. Hurricane Ivan made a direct landfall on Grenada as a Category 4 hurricane with a sustained wind-speed of 220 km/h. Hurricane Ivan caused financial losses in Grenada and the Cayman Islands close to 200% of their annual national GDP. Grenada's economy was severely affected, if not totally "wiped out" as put by Grenada's prime minister in a press conference (CARICOM, 2004a). Nearly every single citizen of Grenada was affected. The Red Cross reported about Grenada that

> At least 39 people died and approximately 90 per cent of the country's homes sustained damage. Water, electricity and telephone services were all cut off. Approximately 60,000 people [of ca. 100,00 inhabitants; N.S.] were in need of assistance and approximately 5 thousand people were accommodated in 145 official and ad hoc shelters. The hurricane caused structural damage to nearly every major building in the Grenadian capital of Saint George's,

> including the island's emergency operations centre, prime minister's residence, several schools, the main hospital and a nearby prison.
>
> (IFRC, 2005, p. 3)

The impact and financial losses for other Caribbean countries such as Jamaica (8% of GDP) and the Bahamas (7% of GDP) were significant lower and, yet, they signified a substantial economic burden (MCII, 2013, p. 14).

Instrument selection: the CARICOM's heads of government mandate the World Bank

Following the destructive impact of Hurricane Ivan the CARICOM Heads of Government held an emergency meeting on 15 September 2004 in Port-of-Spain, Trinidad and Tobago to discuss the region's response to the damaging impact of the hurricane season in general and the regional financial assistance to Grenada in particular. The meeting was the official kick off for CCRIF.

While the heads of governments agreed to assist Grenada "in a form of grants and aid and not loans" (CARICOM, 2004b) they also established the need for a regional catastrophe insurance system. The CARICOM formally asked the World Bank for technical assistance to develop a regional insurance instrument to reduce and transfer natural hazard-related financial risks.[13] According to an interviewee, the proposal to develop a regional insurance instrument and to task the World Bank with it came thereby from one of the Prime Ministers who remembered the failed mid-1990s initiatives and the World Bank's efforts in the Eastern Caribbean to set up a sub-regional catastrophe insurance scheme (Personal Interview, former World Bank Consultant, 2015, May 18). Hence, it was a government, respectively a long-serving prime minister, who brought the idea to develop a regional insurance instrument on the region's agenda.

The creation of CCRIF was, thus, ultimately demand-driven and a subject of 'high politics'. Interestingly, there was no real instrument selection process in the sense that the prime ministers weighted different instruments and/or mandated an agency to develop a number of instruments. The idea to develop a catastrophe insurance instrument was apparently immediately accepted and considered to be an issue that should be realized with supreme priority. The CARICOM simply mandated the World Bank to work out a detailed proposal for a regional catastrophe insurance instrument. Moreover, it is also noteworthy that the CARICOM asked the World Bank and not, for instance, the Caribbean Development Bank. I claim this must be understood against the background of the special situation, the previous discussions as well as the analytical work the World Bank had already put into the development of a catastrophe insurance instrument.

Overall the CARICOM meeting was an *emergency meeting* and emergencies take politics typically beyond the 'normal' process of exploring and carefully weighing pros and cons of different measures in favor of immediate action (Buzan, Waever, & Wilde, 1998). The dramatic situation of Grenada during that period might have facilitated this somehow depoliticizing move of foreclosing public debates and more intensive discussions with national executive branches. During that period Grenada found itself in quite a critical situation. According to CCRIF documents, "Grenada's treasury had great difficulty in meeting its debt payments and paying its civil servants, income having dried to a trickle due to Ivan's impact on business activity" (Pearson & Young, 2009, p. 16). Also the Cayman Islands had substantial difficulties in paying the post-disaster bill (see the following). Moreover, the Caribbean had just experienced a busy hurricane season. That put every government under pressure to meet public expectations 'to do something'.

In these critical situations of uncertainty or disarray policymakers typically turn to a range of actors, in particular to 'epistemic communities' (that is, specialized experts/professionals) in search for assistance and policy instruments (Hall, 1993). As Peter Hall claims, a consensus in the epistemic community of relevance can then inform a policy (instrument) change (Hall, 1993). Central to CARICOM's move to ask the World Bank was the ascription of intellectual authority. The World Bank was granted this status because of its previous intellectual and practical groundwork in disaster financing. Moreover, as elaborated previously, in the Pollner Report (2001) the World Bank positioned insurance confidently as *the* key solution for the region's economic development. There was already a consensus. This may have foreclosed any further discussion.

Also interviewees involved in the development of CCRIF referred to the World Bank's expertise and reputation through their engagement in previous disaster management projects to explain why they were mandated to develop this instrument:

> It's not that the Caribbean Development Bank lacks knowledge, it is just that the World Bank has a much better reputation for technical expertise. So I think this is why the CARICOM governments came to the World Bank: Because they figured that was the best chance of getting some solution.
>
> (Personal interview, former World Bank officer, 2015, April 28)

It seems that positive *experiences* and, thus, *trust* in its expertise were key to why the CARICOM government conferred the instrument development process to the World Bank.

Instrument development

The World Bank's subsequent instrument development process can be subdivided into five key activities[14]: (1) assessing the needs (2) formulating the goals (3) mobilizing support (4) coordinating design activities (5) overcoming challenges.

Assessing the needs: the fiscal impact of Hurricane Ivan on Grenada and the Cayman Island

For the goal and subsequent instrument design the World Bank drew heavily on the experiences in Grenada and the Cayman Islands. The World Bank conducted a comprehensive analysis of the situation on the ground; it engaged with government officials and relevant key decision-makers to identify a core problem and to establish the goal and the design of the insurance arrangement (Pearson & Young, 2009, p. 16). So what happened in Grenada and the Cayman Island? Which needs were identified?

EXPERIENCES IN GRENADA: DESTRUCTION AND A POLICY INSTRUMENT FAILURE

As already illustrated previously, Hurricane Ivan brought about the *de facto* destruction of Grenada. In terms of its consequences, Hurricane Ivan was unprecedented in Grenada's history. It destroyed an estimated 89% of Grenada's housing stock and damaged about 80% of Grenada's public and commercial building (Francis; Ghesquiere & Mahul, 2012). Large numbers of people were in need of assistance and large parts of the economy destroyed.

Given the immense impact on the island, the government required substantial financial resources to fund early relief and recovery measures to assist the country to return to normality. Yet, it had serious difficulties to source enough funds. Grenada did not budget *ex ante* for a disaster; it had, for instance, no catastrophe reserve fund in place, an instrument frequently used in the Latin American context, but could only reallocate budget. Yet, the scope of budget reallocations is typically in the order of only 1% of GDP for small economies such as Grenada (World Bank, 2012, p. 1).

As a consequence of this shrinking financial space, Grenada struggled to fund the public service bill—including salaries of public employees—and thus secure the provision of early relief and recovery measures (Ghesquiere et al., 2006, p. 2). These key services include the provision of emergency assistance, emergency shelter, food and medical attention, as well as the clearance of debris, the restoration of critical services (such as water supply and electricity) and critical infrastructures such as roads, bridges, ports and airports, and the provision of material support for people to restore their destroyed private assets (e.g., shelter materials, tools, etc.) (Pearson & Young, 2009, p. 16). All of these services cost money. Given the regional impact of Hurricane Ivan, there was also a high regional demand for building materials, which dynamically drove up the costs thereof, making the restoration process particularly costly (IFRC, 2005, p. 8). As a consequence, the government of Grenada quickly ran out of financial resources; it simply did not have enough reserves to finance these immense disaster-related costs.

At the same time the government faced a significant reduction in state revenues as the income-generating sectors of the economy were destroyed (Iyahen & Young, 2011, pp. 11–12). Public revenues shortfalls were estimated to be around 5% of GDP between September and December 2004 (MCII, 2013, p. 15). In other words, at a time of great financial need Grenada was running out cash. It lacked liquidity to deal with an event of that magnitude.

In an immediate effort to secure public funding, Grenada launched an international appeal and was able to fundraise about US$ 150 million in international financial assistance. Grenada was supported by international agencies such as the United Nations (UN) and the Pan American Health Organization (PAHO) (PAHO, 2004). Yet, despite these relatively high international pledges, only US$ 12 million thereof were immediately available to finance early relief and recovery measures. What is more, the bulk of financial assistance was earmarked for later reconstruction projects, in other words, tied to specific expenditures at a later stage. This left Granada little room for maneuver.

Struggling with liquidity to finance immediate emergency and relief measures and meet its liabilities to uphold the public service, Grenada raised and imposed several taxes and stressed tax enforcement measures, yielding over 2% of GDP (MCII, 2013, p. 15). These measures included "(i) an increase of about 45% in the retail price of fuel, (ii) an increase in excise taxes on alcohol and tobacco, (iii) a special levy on incomes over US$ 375 per month for a five-year period, and (iv) greater investment in improved tax administration" (Iyahen & Young, 2011, p. 12). Moreover, Grenada approached its creditors for a 'voluntary' restructuring about 85% of its commercial debt extending Grenada's debt service payments by 20 years (World Bank, 2012, p. 39).

Despite this all-encompassing policy instrument mix, the public budget for 2005 entailed a gap of 4,5% of GDP with a projected total debt increase to 150% of GDP (Ghesquiere et al., 2006, p. 2). What is more, the effort to secure funding bound administrative forces and distracted Grenada from recovery and reconstruction efforts and led to a delay in the rebuilding of the country.

THE EXPERIENCES IN THE CAYMAN ISLANDS: DESTRUCTION AND POLICY INSTRUMENT FAILURE

The Cayman Islands found itself in a similar challenging situation although the country was much more prepared for a large-scale event such as Hurricane Ivan in terms of availability of funds and stability of infrastructure. The Cayman Islands are among the richest countries of the Caribbean in terms of GDP and income per capita. Over the last 30 years they have made substantial investments in tourism, commercial, government and residential buildings, road networks, water and sewerage infrastructure, ports and airports (Nixon, 2009). Moreover, the Cayman Islands have a relatively high private sector insurance penetration, one that approaches the insurance penetrations in the OECD-countries (cf. Swiss Re). In other words, most assets are insured.

Hurricane Ivan killed two people and created an estimated loss of US$ 3,4 billion or 183% of GDP. In contrast to Grenada, the Cayman Islands had two policy instruments in place to deal with the public costs resulting from catastrophes: An (traditional) indemnity-based insurance coverage and a catastrophe reserve fund, the National Disaster Fund (NDF). Since the mid-1980s the insurance instrument provides all-risk insurance coverage against physical loss of *major* government buildings and assets. In 2011, it insured more than US$ 500 million in assets across the three Islands (Nixon, 2009, p. 65). However, the program did not provide coverage for all public assets. Elements such as, for example, roads, seawalls, piers, etc. were excluded in the terms. Such a limited coverage is not unusual as full catastrophe coverage is financially unaffordable (MCII, 2013, p. 15). Moreover, the program did not provide coverage for secondary public service cost resulting from emergency and relief measures as elaborated previously. With the goal to secure a budget for these post-disaster activities, the Cayman Islands created the NDF in 1999. It has been funded through an annual allocation of the public budget.

When Hurricane Ivan made its ways through the Cayman Islands, the fund had accumulated reserves of US$ 3,01 million. Following the passage of Hurricane Ivan the government reported claims of about US$ 120 million and withdrew US$ 2,989 billion from its catastrophe "to fund emergency relief measures and emergency house repairs for displaced and uninsured residents" (Nixon, 2009, p. 66).

The insurance program did not work as intended: Due to the high claims from the public and private sector the insurer was unable to meet its obligations and paid out only a portion of the claims. What is more, the expenditures for social relief measures far exceeded the accumulated funds. The problem of the resulting 'liquidity gap' was, however, far less dramatic than in Grenada: The Cayman Islands simply engaged in deficit spending. It 'just' moved from a small budget surplus of US$ 0,5 million to the largest public deficit in the country's history totaling US$ 36,7 million (Nixon, 2009, p. 65).

The experience demonstrated that although the Cayman Islands had an *ex ante* financing instruments in place, not to say a multilayered safety net, it faced some serious financing problems after the event. Yet, in contrast to Grenada this resulted from a *policy failure*. The Cayman Islands sought to learn its lesson. It reformed both economic instruments, the indemnity insurance and the reserve pool. In addition, the Cayman Islands adopted a multi-year Capital Development Program (CDP) to upgrade its buildings and critical infrastructure to reduce its vulnerability to future disasters. Moreover, after considering several alternatives (among others, cat bonds and a national parametric insurance solution), the Cayman Islands decided on the basis of a (1) comparative cost-benefit analysis (2) reduced financial maneuver—also due to the higher insurance premiums of its own property insurance programs—and (3) in recognition of an increase in frequency and intensity of hurricanes in the Caribbean basin, to adopt CCRIF as an additional instrument (Nixon, 2009, pp. 70–71).

Summing up, while Grenada did not plan for a disaster *ex ante* the Cayman Islands did not plan effectively. Both countries were in *need of short-term liquidity* to cover early relief and recovery measures.

Formulating goals: the 'lesson learned'

Having reviewed the situation in Grenada and the Cayman Islands, the World Bank concluded that the *lack of short-term liquidity* was the key problem; it would prevent the smooth functioning of government services by slowing them down, rendering government activity and the rebuilding of the country 'inefficient'. As a consequence, the World Bank identified the *rapid injection of limited funds* following a major disaster as key goal of a regional insurance arrangement. The lesson drawn was thus 'rapid and limited funds are needed to keep government activities running'.

Importantly, the World Bank envisioned CCRIF from the very beginning as a (contingent) *budget support tool* and not as a comprehensive financial instrument that covers all disaster-related costs; instead, it conceived it as an instrument that would *supplement* others. In its 2012 CCRIF Completion Report, the World Bank explains:

> CCRIF was established *to add* to its members' revenues in the short-term by providing a cash infusion to help them fund the initial phase of their disaster response and avoid interruption of their basic business of government. CCRIF was never conceived of, nor designed as, an instrument for avoiding longer-term impacts of the disaster on the government's debt and balance sheet. As already indicated, attempting to insure the entire costs of the covered disasters would be prohibitively expensive. Nor was CCRIF conceived or designed to be a complete and comprehensive disaster response strategy for its members. Instead, CCRIF was explicitly seen as only one instrument in the tool chest for a comprehensive disaster risk management strategy.
>
> (World Bank, 2012, p. 11 my emphasis)

To understand the 'need' of *immediate liquidity* was key for the later successful adoption of a regional insurance scheme, as a senior World Bank officer argued:

> the main breakthrough was *to identify the needs* of the islands. Many of the ideas that had been floated were more about how do we pool risks; to provide, in fact, insurance . . . but the main breakthrough was to understand that need was *immediate liquidity* after a disaster. . . . I think the crux was to identify that that was at the core of the problem, the main urgent problem to fix. And that was actually a very simple problem to fix.
>
> (Personal interview, senior World Bank officer, 2015, May 13, my emphasis)

After years, the World Bank's search for a problem that would fit a solution was finished.

Mobilizing support: pooling donor resources and topping-up

The World Bank was quick in coming up with a preliminary idea to meet the CARICOM's demand and thus use the political momentum generated by Hurricane Ivan. Already in January 2005—that is, less than four months after the CARICOM's formal approach—the World Bank launched the *Caribbean Catastrophe Insurance Initiative.*

The World Bank economist Francis Ghesqiuere and insurance expert Olivier Mahul played a key role in envisioning the tool. They were the 'task managers' and could be described as *the instrument's inventors*, the intellectual minds behind CCRIF. The World Bank's previous analytical work, namely the Pollner Report

(2001) and an earlier draft thereof[15], provided intellectual underpinnings for the initiative (World Bank, 2012, p. 8).

Francis Ghesquiere, Olivier Mahul and in particular Caroline Astney, at that time Country Director for the Caribbean within the World Bank, were also central in securing *internal* support for the realization of the project and thus the development of the insurance instrument. This was not an easy process, as the instrument proposal required to bring together a various departments, expertise and resources. A World Bank official pointed out

> I remember we had to have clearances from close to 20 departments on different aspects and dimensions of the proposal. This was not a simple, easy, straightforward process within the World Bank. But surely we have very strong support from the hierarchy, especially from the country director.
>
> (Personal interview, senior World Bank official, 2015, May 13)

In contrast to the failed 2001 initiative, securing funding was not a problem. By January 2005, the World Bank secured about US$ 2,2 million to cover the start-up costs. Funding, however, did not come from the Caribbean countries but from various donors. Main sponsor was the Japanese government through its Policy and Human Resource Development Program (PHRD). Contributing US$ 1,8 million, it financed the largest share.

To provide funding, Japan diverted earmarked budgets "for closely related" (World Bank, 2012, p. 7) development projects in Jamaica (worth US$ 1 million) and the Eastern Caribbean, namely Antigua and Barbuda, Dominica, St Kitts and Nevis, St Lucia, St Vincent and the Grenadines (worth US$ 800,000). The World Bank administered this budget and contributed another US$ 400,000 to make sure the instrument is really developed this time (World Bank, 2012, p. 7). This means, in conclusion, that *key* for the realization of CCRIF was the *political willingness of donors to pool resources*, in other words *external financial sponsorship*.

Coordinating design activities: the World Bank as a hub

The subsequent technical design work took about 1,5 years and lasted until October 2006. The World Bank set out the initial vision and outsourced the technical process providing guidance and supervision. Consequently, it launched an international tender process for the project execution, which was won by an international consultancy consortium for the technical design (CGM Gallagher Group, one of the world's largest consulting firms specialized in risk management, insurance and reinsurance broking) and by EQECAT (one of the three leading catastrophe modeling agency) in January 2005.

While the consortium did the technical work, the World Bank collected and brought together the input and feedback from myriad actors. The design phase lasted for about 1,5 years. The World Bank drew

> on data and knowledge accumulated in the Caribbean, North America, Europe, and Asia by catastrophe risk insurance entities; insurance regulators;

> actuaries; probabilistic catastrophe risk modeling companies; private sector insurance and reinsurance companies, brokers, and associations; disaster emergency management organizations; and meteorological, hydrological, and seismic monitoring and research centers such as the World Meteorological Organization (WMO), U.S. National Oceanographic and Atmospheric Agency (NOAA), National Hurricane Center (NHC), California Earthquake Authority (CEA), the Florida and Hawaii hurricane funds, US Geological Survey (USGS), and SRC.
>
> (World Bank, 2012, p. 8)

The coordinative and communicative tasks involved in holding such a loose, heterogeneous network together may not be underestimated. The network members came from numerous backgrounds and were highly specialized; in other words, they produced specialized knowledge and used their own language. This required substantial 'translation' skills. In fact, the World Bank's efforts to explain and feed relevant information back and forth were massive: Activities included

> consultations, workshops, conferences, and briefings for CARICOM Heads of Governments, Finance Ministers and Permanent Secretaries, insurance supervisors and regulators, and other senior government officials; the CARICOM Secretariat and CARICOM organizations such as CDEMA and CIMH; U.N. bodies, technical specialists; donors; research institutes; and private sector entities involved in probabilistic hazard modeling and catastrophe insurance and reinsurance.
>
> (World Bank, 2012, p. 12)

These activities vividly illustrate the World Bank's organizational capacity but also its convening power, that is, its power to bring actors together.

While the World Bank outsourced the technical work, it concentrated itself on the political process. Throughout the project phase the World Bank worked close with the Caribbean countries to make sure their needs, demands, doubts are addressed in a potential instrument structure and that project progress is communicated (World Bank, 2012, p. 12). To keep the momentum and secure the political buy-in, the World Bank took care that potential stakeholders even at highest political level ('political champions') were constantly informed about recent developments, as illustrated, for instance, by the regional introductory workshop for the CARICOM Council for Finance and Planning in Kingston, Jamaica in April 2006 and the briefings for the CARICOM Heads of Government in July and October 2006—organized in collaboration with the Caribbean Development Bank (World Bank, 2012, p. 11).

Overcoming challenges

The process from the initial idea to a detailed concept note was, however, not a straightforward one, as several interviewees highlighted. Major technical and political challenges characterized the development process. So what were the key challenges?

The biggest technical challenge in the beginning was that the World Bank had only a *rough idea of the precise institutional and legal structure* of a regional catastrophe insurance scheme. A World Bank officer highlighted that the World Bank envisioned "a political animal owned by CARICOM and representing the countries but at the same time is not too influenced by the political process and can function rather as an independent entity" (Personal interview, senior World Bank officer, 2015, May 13). Yet, it was not clear to the World Bank team how that could look like in practice. "It was something off the shelves" as a consultant involved in the design put it; he continued, "We were basically breaking new grounds, nobody within the Bank really knew, you know . . . there weren't really any examples"; also, "there wasn't any inherent structure within the [insurance] industries that we could piggy-beg . . . so we really had to start from the bottom-up" (Personal interview, former World Bank consultant, 2016, May 16). As a result, for the consultants a lot of work revolved around

> trying to get people within the Bank, especially the legal team within the Bank, to come up with some ideas what the structure could look like, whether you it would be a mutual . . . whether you wanted to form some kind of charity . . . whether you something the IFC could get involved.
>
> (Personal interview, former World Bank consultant, 2016, May 18)

Also a World Bank official underscored that clarifying the "legal and governance dimensions" (Personal interview, senior World Bank officer, 2015, May 13) was the most complicated challenge.

Another technical challenge was ensuring the development of an *operational and institutional structure that is affordable* for Caribbean countries. As already indicated previously, the Caribbean countries typically lack financial resources. Overall, the recognition of their limited financial capacity was the main reason for embracing a regional pooling concept. The World Bank discussed intensively how operational costs could be kept at a low level. The World Bank finally opted not only for parametric insurance policies, which meant lower administration costs, but for a 'virtual' organizational structure. Problematically, a parametric payout model was "alien not only to the sovereign themselves in the ministries of finance but also the insurance industry in the Caribbean" (Personal interview, former World Bank consultant, 2016, May 18). It required substantial 'capacity building', that is, knowledge transfer and thus additional work. The decisions to keep CCRIF as cheap as possible and limit overhead costs drove the World Bank to propose the creation of a letterbox company. As highlighted by Young and Pearson, both involved in the instrument design process: "it was paramount that any *operational structure be minimalist and flexible*. A 'bricks and mortar' facility with permanent staff was discarded as being too costly and inflexible"(Pearson & Young, 2009, p. 23). As a result, CCRIF has neither physical buildings nor own staff. All services are run by contracted service providers operating in their own facilities—in London, Washington DC and Kingston (Jamaica), among others. They make up the organization's 'virtual staff' (World Bank, 2012, p. 11).

Another challenge resulted from the fact "that it is *very difficult to set up trans-border deals within the Caribbean*" (Personal interview, former World Bank consultant, 2016, May 18). While the EU, for instance, liberalized its insurance markets in 1994 in the course of the creation a single European market (see Chapter 2; *ART in historical context*, pp. 86), the Caribbean largely resisted liberalizing tendencies to boost cross-regional businesses. Moreover, once the World Bank decided to propose a captive—a self-insurance vehicle—it still had to find a jurisdiction. As pointed out in Chapter 3, captives cannot be licensed everywhere. This limited the geographic space where to set up an insurance vehicle. An internal legal analysis within the World Bank suggested setting up CCRIF in St Lucia or Bermuda. Yet, as the World Bank finally decided to play an indirect role in the arrangement in the form of a trustee by providing funding to set up CCRIF and reimburse operative costs (see the following), St Lucia and Bermuda had to be ruled out. Both jurisdictions were not members of the two main international associations of insurance supervision; the World Bank guidelines, however, required that the jurisdiction was a member of one of the two main international associations of insurance supervision for their regulatory standards (Personal interview, former World Bank consultant, 2016, May 18). So, CCRIF ended up in the Cayman Islands.

Another technical challenge was the *lack of data and resulting consequences for the payout model*. Due to the lack of available data new data had to be collected and many proxies had been used where this information was simply not available (Pearson & Young, 2009, p. 20). EQECAT, which developed the hazard and loss estimation models, subcontracted other agencies such as the University of the West Indies to collect data for the hazard and risk mapping for all 20 CARICOM countries. It created new datasets containing geographic, geological as well as economic information. It might be noteworthy to underline that "[m]ost of the Caribbean countries had never undertaken any major effort to collect this information" (Iyahen & Young, 2011, p. 16). Hence, this was already a challenge in itself. Yet, closely related to the lack of data and/or the partly poor data quality was the problematic performance of the (first generation) model, as an interviewee pointed out:

> The first model we used was barely fit for purpose . . . barley! You know, we were really. . . . We ran the model back and forth, back and forth and said 'OK, should we delay another year? Because this isn't quite what we wanted it to be'. As a group we decided now that we need to get this out . . . this will not gonna be perfect or probably will never be perfect, we can certainly make it better work . . . so we commit[ed] to make it work better as were launching with this and I think that's what we had to do. If we hadn't done that, CCIRF wouldn't have happened.
>
> (Personal interview, former World Bank consultant, 2015, May 18)

A major political challenge in devising the instrument and ultimately realizing CCRIF was, however, *to keep the CARICOM countries on board*, as several interviewees highlighted, politically but also in terms of understanding. To maintain the political momentum, assure local ownership and make itself understood, the World Bank held extensive workshops with high-level country representatives as well as with experts from the region from the very beginning onwards. The World Bank concluded that these consultations "served to assure the CARICOM Heads that the World Bank was being responsive to their request" (World Bank, 2012, p. 12).

To summarize the section, the World Bank defined a problem, provided a vision, recruited a consultancy and a catastrophe modeler to fledge out details, mobilized financial resources and local input to feed it back into the design process. It held extensive discussions with the Caribbean countries and other relevant policymakers. This section suggests that the World Bank's participatory design process was crucial to the implementation of CCRIF. The participatory design process helped to tailor the instrument according to local needs, created a sense of local ownership and thus helped to secure the political momentum. The World Bank did not persuade the Caribbean countries to adopt an instrument but it involved them *to make them want to implement it*. I will develop this argument further in the next section.

Instrument implementation

On 13 October 2006 the World Bank presented the concept note to the CARICOM Ministers of Finance in a conference in Barbados, fulfilling its mandate. It recommended the creation of CCRIF—a regional catastrophe insurance facility. CCRIF would allow the CARICOM countries to purchase insurance coverage for immediate liquidity needs in case of a *major* hurricane or earthquake. CCRIF was designed akin to a business interruption insurance and would provide the countries swiftly with *limited* funds—just enough to keep government activities going. To achieve this goal it would be based on a simple parametric trigger. A parametric trigger does not require a lengthy loss assessment process to check out claims; instead it pays out based on the *representation of losses*, calculated via a model.

CCRIF would be owned by the insured themselves ('member countries') and provide buyers with tailor-need coverage. Through pooling of financial resources on a regional scale, this form of catastrophe insurance would be cheaper, as if countries would build up these reserves on their own to sustain a catastrophic event and/or purchase such reinsurance coverage on individual scale from a reinsurer and/or the financial markets (Ghesquiere et al., 2006, p. 2). Countries, however, would retain a certain loss level. CCRIF, in turn, would use reserves to finance a payout up to certain level and buy reinsurance coverage via reinsurer and financial markets participants for larger claims via a *cat swap*. Simply put, CCRIF was designed as a "financial intermediary" (World Bank, 2007d) between the CARICOM and the financial market actors. Figure 3.6 illustrates the concept:

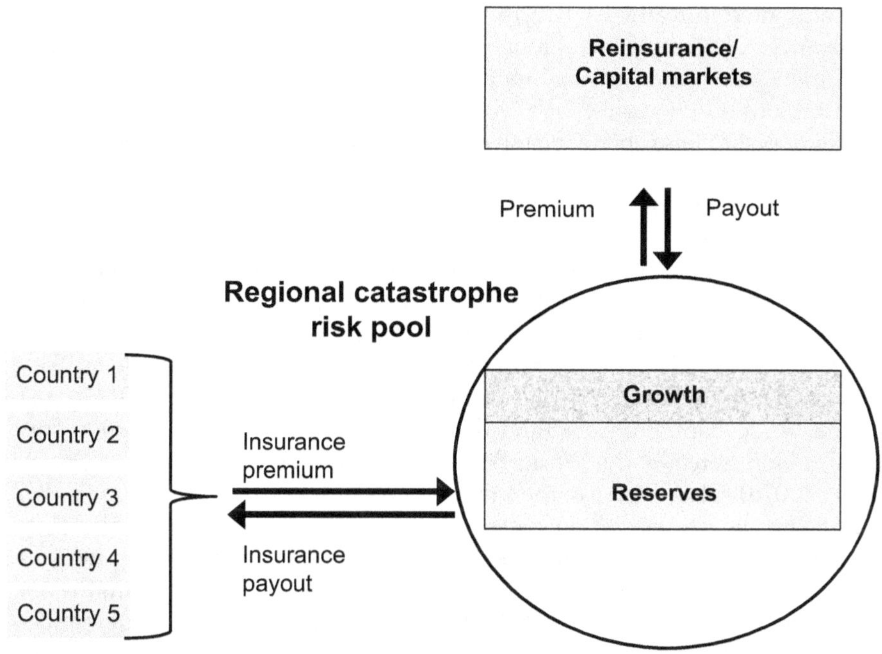

Figure 3.6 Concept of CCRIF

Source: simplified from World Bank, 2007c, p. 7

Since the World Bank worked from the very beginning on with the countries, *there was no formal act of adoption*. The development phase evolved rather smoothly into the implementation phase in early 2007. Crudely put, the World Bank basically offered the CARICOM countries to realize the initiative. The World Bank asked the CARICOM countries for a formal declaration of intent by 31 January 2007 as its goal was to set up CCRIF for the 2007 hurricane season starting on 1 June. The World Bank would then engage in formal country negotiations to tailor the insurance contracts according to individual coverage preferences.

While the development phase has brought about a template with all technical information needed to allow for the operation of a regional insurance facility, the implementation phase still had to overcome some challenges: (1) the capitalization of the facility (2) the precise design of the ownership and governance structure (3) putting together the service providers to run the operations of this 'virtual organization' (4) convincing re-insurers to provide reinsurance coverage for the facility (5) securing that a critical mass of governments stick to their intent to buy insurance coverage. In the following, I will briefly highlight the capitalization and the securitization of a 'critical mass' of governments as they were of crucial importance to the adoption of CCRIF.

To capitalize the facility, allow for affordable insurance premiums for the impoverished Caribbean governments and enable the financing of the first years of operation, *the World Bank used its convening power to collect and leverage donor resources* and thus realize CCRIF. In February 2007 it organized a donor pledging conference. The conference brought together ministerial representatives from 18 Caribbean governments and representatives from international and regional organizations (Ghesquiere et al., 2006, p. 3).

It managed to raise approximately US$ 30 million through contributions from the Canadian International Development Agency (CIDA) (US$ 17,7 million), the UK Department of International Development (DFID) (US$ 7,5 million) and the Agence Française de Développement (AFD) (US$ 6,3 million) (World Bank, 2007a). The World Bank and the Caribbean Development Bank contributed another US$ 17,5 million to realize CCRIF so that by June 2007 CCRIF was capitalized with US$ 47,5 million. Moreover, as mentioned previously, CCRIF member governments paid a participation fee equal to their premium in the first year which added to the donor contributions (McGee, Phelant, & Wenta, 2014, p. 18).

The World Bank proposed (and then also established) a Multi-Donor Trust Fund, administered by the World Bank to channel the donor resources (World Bank, 2007d). The purpose of the Multi-Donor Trust fund was to support CCRIF in the start-up and operation costs, including potential insurance payouts, to allow for cheaper premiums and accelerate the building up of financial reserves. The World Bank expected CCRIF would become self-sustainable within the first five years of operation (until 2012). Hence, administrative fees, banking fees, registration fees, audit cost, operational expenses and insurance payouts to the extent that they were not covered by the reinsurance coverage were reimbursable under a grant agreement until funds were ceased. The World Bank would reimburse CCRIF in tranches, namely every quarter (ibid.).

This grant agreement with the World Bank as trustee of donors may not be underestimated. It gave the World Bank a supervisory, not to say, a fiduciary role through the backdoor. It allowed the World Bank to secure a role for itself and shape CCRIF policies for the first years of operation by drawing on the conditionality principle, that is, providing or not providing funds for certain activities. This subtle power was further secured through its 'observer status' as representative for the donors in meetings of CCRFI's Board of Directors.

By the end of March 2007 the CCRIF was legally established in the Cayman Islands as a non-profit organization based on a four-year business plan (World Bank, 2012, p. iii). The CCRIF obtained its license as insurer on 23 May 2007 with the legal capacity to sell insurance coverage to its participating countries, to buy reinsurance coverage and to issues financial securities such as cat bonds to transfer its financial risks to the international capital markets.

Major political challenges: premium payments

The World Bank established that ten countries would be the critical number for CCRIF to be a viable tool. As the smaller OECS countries (Dominica, Grenada, St Lucia, St Vincent and the Grenadines) and Haiti were not willing or able to budget for the participation fee and insurance coverage the World Bank jumped in. It sponsored the participation fee and the premium of Haiti and granted the OECS countries a loan to secure their participation[16] (World Bank, 2007a). Also the Caribbean Development Bank made some funds available and financed the participation fee and the premium of some CARICOM countries (World Bank, 2007a).

The fact that the World Bank as well as the Caribbean Development Bank sponsored and/or helped to fund the participation of CARICOM countries in CCRIF indicates either that the political momentum was already gone because countries saw little value in such arrangement, and/or that the premium were still not affordable. In any case, it fairly safe to assume that CCRIF would not have seen the light if the World Bank, the Caribbean Development Bank and donors would not have joined forces. The pooling of donor resources, organized by the World Bank, was a little but decisive piece on the way to CCRIF.

As can be inferred from the documents, the World Bank interpreted its role as a catalyzer. It was sure that at least one large-scale catastrophe might realize in the designed four-year business plan. It 'hoped' that the occurrence of such a (insured) catastrophe would demonstrate its functioning, secure and stimulating demand and thus secure the longevity of the arrangement: "This will confirm the utility of the risk financing instrument. It is expected that this demonstration effect will reinforce the commitment of Dominica, Grenada, St. Lucia, and St. Vincent & the Grenadines to continue their participation" (World Bank, 2007a, p. 9).

The first year

CCRIF came into operation on 1 June 2007 as envisioned. In the first year CCRIF collected US$ 19 million in premiums and sold 19 insurance policies to 16 countries (CCRIF, 2008). As elaborated in the previous part, each country selected an individual attachment point, exhaustion point and coverage limit. While CCRIF guaranteed to cover all insured losses up to $10 million *per annum*, a number of re-insurers (Munich RE with Paris RE and Hiscox, a Lloyd's of London Syndicate) and capital market investors agreed to pay the next US$ 110 million in insured losses. The capital markets were incorporated through a cat swap-arrangement with the World Bank's Treasury. That means, the CCRIF placed cat *swaps* in the international capital markets via the World Bank Treasury as intermediary. In other words, the CCRIF ceded the top financial risks to the World Bank, which, in turn ceded them to the international capital markets in the form of a tradeable swap.[17]

That issuance of cat swaps implies that the World Bank used the trust it enjoys among financial market actors to create a business between poor countries and hedge funds and other affluent speculators. Put differently, it used its reputation to bring distinct late twentieth century capitalist products to poor countries. Doing so, it provided financial market actors with fresh capital for speculation. That means also the World Bank helps not only to facilitate the expansion of the financial market activities into poor areas south of the OECD but also to contribute to legitimizing the *financialization of capitalism*.

Instrument modification: the evolutionary phase

The first year of operation (2007/08)

In its first year of operation the Caribbean experienced 17 tropical cyclones, of which 6 attained hurricane status, and 10 earthquakes. In other words, it was quite a busy season. The World Bank saw its desired 'demonstration effect' realized, but probably not in the way it had wished for. The season demonstrated both the potential and flaws of CCRIF.

Hurricane Dean crossed the Caribbean in August 2007 and revealed some severe problems within the insurance instrument: Hurricane Dean had a substantial impact on Jamaica, where it made a direct landfall as a Category 4 hurricane (wind-speed > 220 km/h) and on other Caribbean islands, notably Dominica, St Lucia and Haiti. Hurricane Dean was thus the weather event for which CCRIF was actually designed: a major catastrophe. However, no payout was made to one of these countries.

One reason was that most of the damage occurred due to excessive rainfall and storm floods. Damages resulting from these perils were, however, explicitly excluded in the terms of the insurance contract. CCRIF provided insurance coverage only against wind-related damages.

The other reason was related to the payout model itself and rather a design failure: The statistical equation to estimate losses weighted in particular areas with 'valuable' infrastructure, namely areas which provided high fiscal revenues—basically economic key centers. Yet, most of the wind-induced damages occurred in the Southern part of Jamaica, where exists only little *valuable* state infrastructure and no economic key center but largely small-scale farmers. Since the Southern part of Jamaica is not a fiscal stronghold, no massive financial losses were modeled and consequently no payout was made (CCRIF, 2008)—despite, however, the substantial damage local farmers and the poorer segments of society living in that area experienced.

How did this come about? Simply put, to estimate losses the instrument designers sub-classified a country into more and less valuable zones. Depending on the value of public assets in a certain area and the contribution of this region to GDP, a region was assigned a low or high value in the statistical equation. If such a predefined region comprised few or invaluable public assets and contributed little to fiscal revenues, the level of government exposure was assigned a low value. Logically speaking, this design was only consequentialist. Again, CCRIF was designed akin to a business interruption insurance—an insurance that provides liquidity for *fiscal* (!) revenue shortfalls.

Yet, the non-payout triggered a media discussion in Jamaica about the usefulness of the insurance instrument (Nixon, 2009, p. 76). Jamaica had paid US$ 4 million in insurance premium to the facility, that is about eight times (!) the amount of the budget of its national disaster management agency, to secure a maximum coverage of worth US$ 94,4 million (Grove, 2012, p. 151) and, yet, did not receive a single dollar. The problem was not that CCRIF's model was wrong; quite the contrary, the model performed quite well as the comparative *ex-post* analysis conducted by CCRIF revealed (see the following). Overall, CCRIF pays out not against the losses on the ground but against modeled losses, that is, the *representation of losses*. Yet, Hurricane Dean just demonstrated that the way the insurance was designed caters primarily government loss and *not* private loss. In other words, Hurricane Dean simply showed the insurance was not so much of help for the poorer and, thus, the most vulnerable segments of society.

The earthquake in Dominica and St Lucia in November of the same year demonstrated another flaw. They were the most severe earthquakes in terms of ground acceleration in the Eastern Caribbean over the last 30 years. The parametric equation indicated a severe economic loss to both countries. Subsequent payouts were made to the treasuries within the set time frame of 28 days. This time, however, reality showed that the earthquakes brought about only minimal damage.

Yet, despite of the minimal damage, Dominica and St Lucia received a payout. The payouts may be interpreted as a "lucky strike" (Christian Aid, 2009, p. 34) in favor of these countries. Following the insurance contract, countries were free to allocate the funds where they needed them. Dominica, for example, used the insurance payout of US$ 500,000 to finance the rehabilitation of infrastructure and home construction for people not affected by the earthquake but by the still unattended damage due to Hurricane Dean earlier that year (ibid., p. 34). People in need due to Hurricane Dean were therefore paradoxically the main beneficiaries of this payment.

CONSEQUENCES OF HURRICANE DEAN: DEBATES AND MODIFICATIONS

The events around Hurricane Dean stimulated a debate within CCRIF. CCRIF compared its economic loss calculations with the calculations made by the Planning Institute of Jamaica. Although CCRIF was satisfied with the overall performance of its model, it reacted to the critiques and engaged in some major revisions following the first insurance year: First, CCRIF decided to develop a flood/excessive rainfall product and partnered with the Caribbean Institute of Meteorology and Hydrology (CIMH) in Barbados to do so. Second, it decided to modify its payout structure by lowering the attachment point (the level at which a payout is triggered) and by increasing the exhaustion point (the maximum payout level). For the policy year 2008/09 CCRIF offered coverage for events with a modeled 1:15 chance of occurring (down from 1:20 chance) and a maximum coverage of US$ 100 million (up from US$ 50 million). Third, CCRIF decided to develop a new and more precise model, for which it hired Kinetic Analysis Corporation (CCRIF, 2008). In addition, CCRIF lowered the premium price (yet, is to note that the reduction in insurance premium was planned from the early beginning on and not a reactive development to the events in Jamaica), topped-up its capital reserves through additional donor grants from Ireland and Bermuda so that donor capital summarized now to about $54 million (World Bank, 2012,

p. iv) and accelerated the payout process by guaranteeing the finalization of payout within 14 days (instead of 28 days).

Apart from these changes CCRIF moved on with discussions to revise its original goal, which was to provide insurance coverage. Already during the first insurance policy year, the Board discussed second order changes. Some people within CCRIF feared it might be called into question if does not provide additional services to its members if there is no insurance payout. Given that the insurance contracts last for only one year, contract renewals were (and still are!) of central importance for the organizational survival; less than ten renewals would possibly put an end to CCRIF.

The discussions continued in the second policy year and resulted in the adoption of the Technical Assistance Program (TA Program) in late 2008 and a number of other cross-regional collaborations and projects (see the following). An interviewee remembers the discussions as follows:

> I lobbied very hard to make sure that [something else outside of pure insurance; N.S.] was in place because, you know, I have been involved in that disaster management and insurance space in the Caribbean for a long time and it seemed to me that we won't have—even across a group of countries—we won't gonna have policies paying out every year. . . . We got a lot of momentum, we got 16 countries on board, which is fabulous, but we won't be able to keep that level of engagement *just through insurance*. So *we needed to show that CCRIF was supportive to a broader set of objectives*. . .
>
> (Personal interview, former World Bank consultant, 2015, May 18, my emphasis)

The same interviewee also pointed out that CCRIF held lots of capital

> . . . and defending hundred million dollars of capital in the Caribbean, which was capital committed to disaster management—I mean, that would have paid the disaster risk management budgets of all the disaster risk management agencies in all of the CARICOM countries for many years—I mean this was two or three times orders of magnitude more than normal budgets; so it was a very big number. It is pretty well ring-fenced in a legal sense but ultimately, you know, [if] the CARICOM countries all got together [and] said 'We want the money', then, CARICOM could have . . . could have caused the twine down . . . my thinking was *we needed to look like we are relevant so that people don't start asking why do you have so much money on the balance sheet when you never pay out.*
>
> (Personal interview, former World Bank consultant, 2015, May 18, my emphasis).

The comments indicate the internal discussions (and the subsequent adoption of the TA program) centered around taming unfriendly critics and providing additional legitimacy to CCRIF so that no-one even asks 'why was the organization formed?'. The idea to add a number of services was to encourage contract renewal, foster commitment and reassure CCRIF members that the organization is valuable to them in many regards. This strategy is frequently used by entrepreneurs when it comes to the foundation of new organizations (Aldrich, 1999, p. 99). The symbolic nature of adding services can further be demonstrated with the following quote:

> And I was kind of, say, '*scholarship is a really cheap way of getting PR*' . . . not that I want to devalue the value of a scholarship, I think it is a very positive thing to do, it's hard to argue against a few thousands of dollar, you know, to students to go to a university in the Caribbean to study the subject . . . 'and then, you know, we build some other projects around that'. *It was really so that the CCRIF could be seen as active in the countries and investing in some research but also investing in people and supporting initiatives of the countries and were supportive otherwise.*
>
> (Personal interview, former World Bank consultant, 2015, May 18, my emphasis)

A CASE OF POLICY (INSTRUMENT) LEARNING: DEBATES AND MODIFICATIONS AFTER HURRICANE DEAN

The modifications and discussions demonstrate a *high degree of responsiveness* to negative feedback and *effective collective action*. CCRIF sought to learn and address concerns of its members swiftly. This is best illustrated by CCRIF's willingness to develop a second-generation model—despite its good performance—and the commitment to develop a parametric excess rainfall product following requests of some Caribbean governments. Again the purpose was to demonstrate to its members that the facility is on the right track.

This willingness to learn from trial and error experiences and the swift move to mobilize resources and knowledge to act upon negative feedback is somehow remarkable given the numerous stakeholders involved. It indicates an organizational culture characterized by playfulness, experimentation and a 'willingness to learn'. CCRIF did not silently negotiate away critique but embraced it. This 'willingness to learn' is reflected in the various social interactions CCRIF sought with its members. Just in 2007/08 CCRIF held about ten conferences, seminars and workshops, which involved representatives from the regional disaster management and Caribbean Ministry of Finance (CCRIF, 2008, p. 16). Frequent social interactions create trust; trust, in turn, creates the necessary space for critical comments.

These numerous internal discussions indicate that from the very beginning the Board thought of CCRIF as an instrument whose goal and structures should be flexible—even only to secure policy renewal and thus organizational survival.

Key developments 2008–2018

Also the following years CCRIF worked as intended. The Board continued to be very responsive to demands and concerns raised by its members. CCRIF substantially expanded and professionalized its activities. CCRIF established itself as regional hub for disaster management and started to marketize itself as 'role model' on international climate policy platforms. Since 2012/13 CCRIF offers an Excessive Rainfall Policy to cover tropical rainfall and since 2015/16 COSEFIN countries may apply for membership and purchase insurance coverage. CCRIF continues to evolve. In the following section I will highlight some key events and developments since the 2007.

DEVELOPMENT OF PAYOUTS: A TENDENCY OF UNDERINSURANCE?

Between 2007 and December 2018 CCRIF has made 38 payouts to 13 countries (see Table 3.2). Total payouts summarize to US$ 138,815,479. Most of the payouts were derived from claims related to tropical cyclones (see Table 3.3 and Figure 3.7)

Eleven of these 38 payouts were made in the insurance policy year 2017/18, following the passage of the Hurricanes Irma and Maria in September 2017. The two hurricanes were responsible for CCRIF's highest aggregated annual payout

Table 3.2 Payouts made by CCRIF until December 2018

Policy year	*Country*	*Payout*	*Hazard*
2007/08	Dominica	US$ 528,021	Earthquake
	St Lucia	US$ 418,976	Earthquake
2008/09	Turks and Caicos Islands	US$ 6,303,913	Tropical Cyclone Ike
2009/10	Haiti	US$ 7,753,579	Earthquake
	Anguilla	US$ 4,282,733	Tropical Cyclone Earl
	Barbados	US$ 8,560,247	Tropical Cyclone Tomas
2010/11	St Lucia	US$ 3,241,613	Tropical Cyclone Tomas
	St Vincent and the Grenadines	US$ 1,090,388	Tropical Cyclone Tomas
2014/15	Anguilla	US$ 493,465	Excessive rainfall
	Anguilla	US$ 559,249	Excessive rainfall
	St Kitts and Nevis	US$ 1,055,408	Excessive rainfall
	Barbados	US$ 1,24,882	Excessive rainfall
2015/16	Dominica	US$ 2,402,153	Excessive rainfall
2016/17	Nicaragua	US$ 5,000,000	Earthquake
		US$ 1,110,193	Tropical Cyclone Otto
	Barbados	US$ 975,000	Tropical Cyclone Matthew
		US$ 753,277	Excessive rainfall
	St Vincent and the Grenadines	US$ 285,349	Excessive rainfall
	St Lucia	US$ 3,781,788	Excessive rainfall

Policy year	Country	Payout	Hazard
	Belize	US$ 261,073	Excessive rainfall
	Haiti	US$ 20,388,067	Tropical Cyclone Matthew
		US$ 3,020,767	Excessive rainfall
2017/18	St Kitts and Nevis	US$ 2,294,603	Tropical Cyclone Irma
	Anguilla	US$ 6,529,100	Tropical Cyclone Irma
		US$ 158,823	Excessive rainfall
	Turks and Caicos Islands	US$ 13,631,865	Tropical Cyclone Irma
		US$ 1,232,769	Excessive rainfall
		US$ 419,372	Tropical Cyclone Maria
	Dominica	US$ 19,294,800	Tropical Cyclone Maria
		US$ 1,054,022	Excessive rainfall
	Antigua and Barbuda	US$ 6,794,875	Tropical Cyclone Irma
	The Bahamas	US$ 163,598	Tropical Cyclone Irma
	Barbados	US$ 1,917,506	Tropical Cyclone Maria
	St Lucia	US$ 671,013	Tropical Cyclone Maria
	St Vincent and the Grenadines	US$ 247,257	Tropical Cyclone Maria
	Trinidad and Tobago	US$ 7,007,886	Excessive rainfall
2018/19	Barbados	US$ 5,813,299	Tropical Cyclone Kirk
	Trinidad and Tobago	US$ 2,534,550	Excessive rainfall
Total		**US$ 138,815,479**	

Source: data taken from CCRIF SPC, 2018a, pp. xiii–ix

Table 3.3 Payouts made by CCRIF until December 2018 according to peril

Peril	*Total payouts*
Tropical cyclone	US$ 100,730,068
Earthquake	US$ 9,200,576
Excessive rainfall	US$ 32,163,584

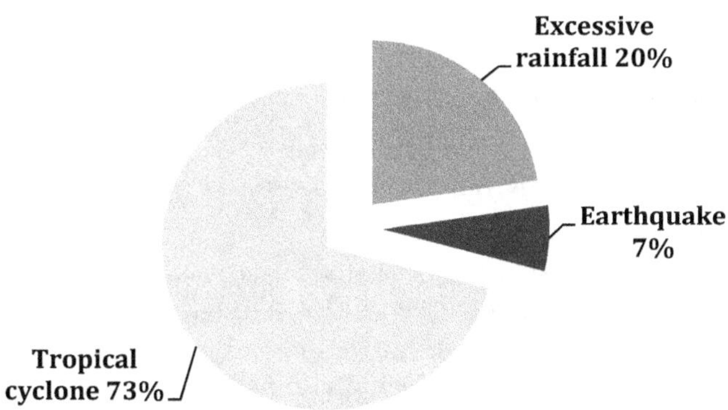

Figure 3.7 Payouts made by CCRIF until December 2018 according to peril

so far (US$ 48,838,880). Already in the policy year 2016/17, CCRIF made nine payouts to six countries totaling to about US$ 31 million, among them the largest single payout so far: Haiti received US$ 20 million on its tropical cyclone policy following the passage of Hurricane Matthew in September 2016.

All insurance payouts were transferred within less than three weeks, in some cases even within a week (CCRIF, 2013a, p. 1). Taken together, it appears that CCRIF works as intended: It provides Caribbean governments with immediate liquidity in the aftermath of a *major* disaster. Accordingly, the CCRIF has gained a first-class reputation and attracted the attention of other regions of the world as a model to draw on (see next case studies).

Yet, what is often overlooked is that there is a slight tendency of underinsurance due to flaws in the model or the insurance contract. Having already illustrated the experiences of 2007—a non-payout to Jamaica and an unexpected payout to Dominica and St Lucia—it might also be worth to pointing to Anguilla's experience in 2010. Although CCRIF's developed a new and more precise second generation model that estimated the country's losses at US$ 38,194,203, Anguilla received a payment of just US$ 4,282,733 due to specific characteristics of the insurance contract (CCRIF, 2011, p. 15). Jamaica had a similar experience the following year. In 2011, Tropical Cyclone Nicole led to estimated damages of about US$ 150 million, which reflect roughly 1,9% of the country's GDP. The public road network, drainage system, river protection and associated infrastructure were substantially damaged. Yet, no payout was made since most of the damage resulted from tropical rainfall (CCRIF, 2011, p. 19). In 2010, Belize also suffered from a substantial loss of its agriculture and public infrastructure as well as its tourist key infrastructure in Belize City due to Tropical Cyclone Richard. The total damage was estimated at US$ 25 million, but, again, no payout was made as the trigger level was too low. The government had to release immediate emergency response and relief efforts worth US$ 1,5 million as thousands of people lost their home and economic income basis (CCRIF, 2011, p. 21).

While an index-based insurance contract entails unavoidable basis risk since, by definition, it only *estimates* the *real* costs it is hard to deny that there is a slight tendency of a *negative* basis risk, forcing compromised government to cope with insufficient or absent payouts. The case of Dominica and St Lucia in 2007 was, so far, the expectation as they benefited from a payout despite minimal damage.

INCREASING EXPANSION OF ACTIVITIES: THE CCRIF AS HUB FOR REGIONAL DISASTER POLITICS

Despite or possibly precisely because of these flaws, over the years CCRIF has done much more than providing 'just' insurance. It has increasingly established itself as a hub for regional disaster mitigation and management and provides countries with early warning tools, data, disaster education and assistance to its member countries.

Since 2009, for example, CCRIF offers its members a storm impact forecast tool. The *Real Time Hazard and Impact Forecasting System*

> provides access to hazard maps in Google Earth showing wind speed over terrain, wave height in open water, storm surge height and inundation along the coast, cumulative rainfall over duration of storm and wind effects on vegetation, structures and electrical power
>
> (CCRIF, 2009, p. 10)

The instrument enables the national disaster management agencies to prepare for potential disasters and minimize the impact. The underlying data is also integrated into the Caribbean's DEWETRA platform, and decision-making early warning platform for hydro-meteorological and wildfire events. Moreover, in the aftermath of the devastating earthquake in Haiti 2010, CCRIF supported the country in its long-term reconstruction and rebuilding measures by making its hazard data and tools available to the Caribbean Institute for Hydrology and Meteorology (CIHM). CCRIF's goal was to help "planners and relief workers in Haiti to make better decisions about where to re-settle the citizens of Haiti and rebuild infrastructure to minimize people's exposure to flooding and landslides" (CCRIF, 2010b).

Another "flagship" (CCRIF, 2010a, p. 23) is CCRIF's Technical Assistance Program (TA). The TA Program was launched in 2009 and consists now of three components: a Scholarship/Professional Development Program, Regional Knowledge Building and a Small Grant Program to support local disaster risk reduction initiatives. Since the inception of the Scholarship/Professional Development Program, CCRIF has provided 36 scholarships to students interested in disaster risk management, environmental management, actuarial science, geography, climate studies and related studies. Under this label and since 2015 CCRIF also offers a Regional Internship Program designed to provide opportunities for students to complement their academic studies in the disaster risk management field with practical work experiences. Under the Regional Knowledge Building component, CCRIF has set up a number collaborative programs "to improve disaster risk management, increase climate resilience and reduce vulnerabilities in the Caribbean" (CCRIF SPC, 2017, p. 44). It has provided, among others, grants for technical projects, hazard and risk assessments, research studies, trainings, workshops and conferences to support the development and implementation of local disaster management policies. To that end it has concluded a number of Memoranda of Understanding with various actors in the region. It partnered, among others, with the Caribbean Community Climate Change Centre (CCCCC), the Caribbean Institute for Meteorology and Hydrology (CIMH), the Caribbean Disaster and Emergency Response Agency (CDERA), the OECD International Network on Financial Management of Large-Scale Catastrophes and the United Nations Economic Commission for Latin America and the Caribbean (UNECLAC), the Organisation of Eastern Caribbean States (OECS) the

University of the West Indies (UWI) and plans to engage in further collaborations. In March 2015, CCRIF launched the Small Grants Program. The Small Grants Program provides funds (in the range of US$ 5,000-US$ 25,000) for small disaster risk reduction projects in local communities by NGOs, grassroots organizations, charities and research institutions across CCRIF's Caribbean member countries.

This apparent trend to strategic knowledge building and institutionalization demonstrates not only the increasing professionalization of CCRIF but also signifies an increasing accumulation of power. In fact, these days CCRIF provides not just cash: instead, it has become a hub for regional disaster management. CCRIF contributes to mainstreaming disaster management in the Caribbean. Doing so, it has joined more systematic and global efforts to institutionalize (that is, to standardize) measures how to deal with recurrent disasters and stands next to regional actors such as CDERA or PAHO.

INCREASING PROFESSIONALIZATION OF COMMUNICATION AND MARKETING ACTIVITIES: CCRIF'S SELF-CONFIDENT CONSTITUTION AS 'ROLE MODEL'

The expansion of activities CCRIF has been accompanied by a more offensive and increasingly professional communication and marketing strategy to inform about its products and activities, manage its relationships with its stakeholder, build up and strengthen the facility's reputation.

Over the course of time CCRIF has produced an annually increasing number of reports, technical papers, leaflets, brochures, training manuals, press releases; it has set up a mailing list and various social media accounts (Twitter, Facebook and a YouTube channel) and professionalized its website. Just during 2016/17 CCRIF produced about 30 publications and two videos to facilitate better understanding of CCRIF and promote and its products and services (CCRIF SPC, 2017, p. 61). Since 2009 CCRIF has been carefully documenting what and how the media reports about CCRIF, analyzing the press coverage it gets by media source, geographic origin and thematic focus as well its number and origin of website visitors (cf. CCRIF, 2010a, pp. 31–41). During 2016/17 the CCRIF website was accessed 22,790 times, a 175% increase over the 8,300 visits in CCRIF's first year of operations (CCRIF SPC, 2017, p. 62).

CCRIF has also started to attend international conferences and workshops to increase understanding about the facility, its role in disaster risk financing, share lessons learned and stimulate discussions about disaster risk management, climate change adaptation, sustainable development and the Agenda 2030. The annual reports carefully document the about 150 international meetings and conferences CCRIF has attended or sponsored since 2009 (CCRIF, 2010a, 2011, 2012, 2013a). CCRIF participated thereby in high-profile conferences such as the UN General Assembly, UNFCCC Climate Change Conferences (COPs), the UN Conference on the Last Developed Countries and Meetings of the Board of Governors of the Caribbean Development Bank (cf. CCRIF, 2011).

In short, CCRIF has started to regularly develop and disseminate materials and knowledge to national, regional and international stakeholders to facilitate better understanding of CCRIF and its products and services. In doing so, it has taken to care to build up a positive reputation and presenting itself as a potential model for others to draw on.

IMPROVING AND EXPANDING THE PRODUCT LINE: A NEW PAYOUT MODEL AND THE LAUNCH OF NEW INSURANCES

Apart from the expansion of its activities and its increasing outreach activities CCRIF continued to be responsive to local concerns and critics—also from its re-insurers (CCRIF, 2011, p. 27).

In the insurance policy year 2010/2011, for instance, CCRIF introduced for some countries the second generation model, the 'CCRIF Multi-Peril Risk Estimation System'. The model incorporated some improvements in modeling techniques and utilizes fewer proxies as datasets were updated. The upgrade to the second generation model became fully effective with the policy year 2012/13. In its most recent annual report CCRIF announced to replace the current model with a new third generation model for the insurance policy year 2018/19. The envisioned SPHERA (System for Probabilistic Hazard Evaluation and Risk Assessment) models will be designed produce *ex ante* estimates of future earthquake and hurricane-induced losses and thus improve insurance policy pricing; moreover, they will provide loss estimates in near real time. Unlike the current 'CCRIF Multi-Peril Risk Estimation System', the new exposure database will thereby provide the economic value of the exposed assets disaggregated by sector (residential, commercial, industrial, hotels, education, healthcare, public, airports, ports, power facilities, road networks, etc.) *and* by building class (e.g., reinforced concrete, masonry, etc., of different age and height) (CCRIF SPC, 2017, p. 22). In short: CCRIF promises to make substantial improvement in its modeling.

CCRIF also expanded its product line. In 2012/13 CCRIF launched the Excessive Rainfall Policy. The insurance policy was designed in collaboration with Swiss Re. As elaborated in the first part of this chapter, the Excessive Rainfall Policy is not linked to hurricanes or earthquakes and must be bought separately. The Excessive Rainfall Policy complemented the shortcomings from the hurricane insurance and was developed in response to demands from Barbados, Jamaica, St Lucia, St Kitts and Nevis, Trinidad and Tobago and Guyana (CCRIF, 2013a, p. 27).

Over 2010–2014 and with funding from Germany, CCRIF partnered with the "Climate Risk Adaptation and Insurance in the Caribbean" project, led by the Munich Climate Insurance Initiative (MCII), an advocacy network for climate risk insurances, to pilot two new micro-insurance products: the Livelihood Protection Policy and the Loan Portfolio Cover. Based on CCRIF's catastrophe model, the Livelihood Protection Policy has been designed to provide low-income individuals (such as farmers and day laborers) with swift cash payouts following a hurricane or excessive rainfall. It is offered through local insurance companies and financial institutions in St Lucia, Jamaica and Grenada. The second product—the Loan Portfolio Cover—has been targeted at financial institutions. It offers hedge for lending institutions with credit portfolios exposed to natural disaster risk. The products are currently under review for improvement and a larger rollout (CCRIF SPC, 2017, p. 28).

CCRIF expanded, however, not only its product line but also its membership base. In 2015, with financial support from the US and Canada, CCRIF opened membership to the Central American COSEFIN countries. Nicaragua was the first country to join. Only very recently, during the policy year 2018/19 the British Virgin Islands, Montserrat and St Maarten joined CCRIF. In the 2019/20 insurance season, CCRIF consists, for the first time, of 21 members. In other words: CCRIF continues to expand and evolve.

Part III: summary

CCRIF is a **regional, not-for-profit insurance company** ('captive'/'segregated portfolio company') that offers 21 states from the Caribbean and Central America (policy year 2019/20) with **insurance coverage against the financial consequences of** catastrophic events in exchange for annual premium. States receive rapidly limited funds following the occurrence of a ***major*** **hurricane** or **earthquake** or **excessive rainfall.** The level of the payout is modeled on real-time hazard and archived exposure data and capped at a certain, pre-defined level. Insurance payouts flow into national treasuries and are not tied to specific expenditures. Governments are free to decide where they want to allocate the payout.

The **key goal** behind CCRIF is to **provide countries rapidly with short-term liquidity.** CCRIF provides governments with *limited* funds to bridge disaster-related tax revenue shortfalls and cover explicit and implicit public service liabilities. The secondary goals are to mitigate negative macroeconomic impact of disasters and to supply countries with tools and knowledge for disaster risk management.

CCRIF is a 'virtual organization', **a letterbox company**, and is registered in the Cayman Islands. CCRIF is formally owned by the Caribbean countries (including Nicaragua) and **operated as public-private partnership.** While the public stakeholders—the countries—legally own the facility, all services are outsourced and thus delivered by private service providers. The 'virtual existence' is enabled through a number of service provision agreements. This implies also that CCRIF exists 'in reality' just on paper, respectively online.

CCRIF was created in 2007 and the **first multi-country insurance pool in the world**, providing governments with parametric-based insurance policies against natural disasters backed by re-insurers and capital market investors. **Between its creation in 2007 and December 2018,** CCRIF has made **38 payouts totaling to about US$ 138 million** to 13 governments following major disasters. CCRIF work as intended although there is a slight tendency to 'underinsurance'.

Historically seen, **CCRIF emerged out of a number of *ideas*, an *exogenous event* and a subsequent *policy failure*.** Even though the idea to develop a regional insurance arrangement had been discussed in the Caribbean since the mid-1990s, the idea did not become viable until 2004. Trigger for the creation of CCRIF was the destructive impact of Hurricane Ivan in 2004. Hurricane Ivan brought about the *de facto* destruction of Grenada. Large numbers of people were in need of assistance and large parts of the economy were destroyed. Grenada required substantial financial resources to fund early relief and recovery measures and rebuild the country. Yet, it had substantial difficulties in sourcing these funds. Grenada had only limited financial instruments in place to cover disaster-related expenses; moreover, it faced a significant reduction in state revenues as the income-generating sectors of the economy were destroyed. It sought international assistance, raised taxes and restructured its commercial debts to make ends meet. Faced with Grenada's struggling situation, the CARICOM held an emergency meeting to discuss political consequences for the region. They called for a policy instrument that would prevent a future liquidity crunch and mandated the World Bank to develop a catastrophe insurance instrument.

The **CARICOM mandated the World Bank** because of its newly developed technical expertise in disaster financing and its conceptual 'readiness'. Faced with an exceptional, critical situation, policymakers typically turn to epistemic communities and/or specialized experts and thus to agents to whom they grant intellectual authority in a certain issue area (Haas, 1992). The examination indicated that the World Bank enjoyed this authority due its previous engagement in the region. Since the mid-1990s the World Bank had played around with the idea of setting up a Caribbean insurance scheme as it conceived insurance as *the* solution for the region to cope with disasters financially. The fact that the CARICOM simply mandated the World Bank also shows that there was **no instrument selection process** in the sense that policymakers considered and weighted different policy options how to deal with the consequences of Hurricane Ivan. The CARICOM, it appears, did not consider seriously any alternative.

The **World Bank** may be understood as ***instrument constituency.*** From the very beginning on, it was keen on realizing CCRIF. The World Bank did not just deliver a proposal. It made *tremendous efforts* by involving the broader regional policy community and relevant key actors in the design process. It held frequent exchanges with the CARICOM countries and other relevant actors in the insurance and disaster management field to make sure the instrument matches local needs. This provided the instrument with substantial input legitimacy. Moreover, the World Bank did not just devise the instrument but it also organized the necessary funds for both the instrument design and implementation process. It helped to pool donor resources and provided its own funds, where necessary, to make sure the instrument would also become reality. It organized a donor conference to raise the funds to sufficiently capitalize the facility and (together with the Caribbean Development Bank) sponsored the premiums for some countries. The World Bank's ***participatory approach*** and its ***sponsoring activities*** were **crucial** for the **successful implementation** of CCRIF. It secured a buy-in.

Since its creation in 2007 **CCRIF has changed form and purpose**. The Governing Board has turned out to be highly responsive to accommodating various demands. In the context of a non-payment to Jamaica and tendency to 'underpayment' CCRIF did not only update its model but to started to provide much more than insurance. To encourage annual contract renewal and reassure its members that the company is valuable, CCRIF substantially expanded its products, scope of activities and geographic reach. Today, by buying into CCRIF, member countries gain access to a number of disaster-related services such as satellite-based weather forecast services and technical assistance/capacity building measures. While these services have primarily a secondary legitimizing function (and thus secure organizational survival), they also demonstrate that CCIRF is not anymore a purely financial instrument. **In fact, it may also be understood as a low-cost tool for public investment in disaster-related infrastructure and education.** CCRIF has increasingly established itself as a hub for regional disaster mitigation and management. CCRIF, it appears, is a highly responsive and flexible instrument.

Notes

1 All data refers to the insurance policy year 2016/17 unless noted otherwise.
2 The fiscal impact must be understood as a subset of the overall macroeconomic impact. It might be noteworthy that the precise impact of a disaster on a government budget is hard to discern as most countries do not record emergency or post-disaster expenditures in separate budget lines (Ghesquiere & Mahul, 2010).
3 With implicit liabilities I mean liabilities out of moral obligations (based on public expectations or political pressure). By explicit liabilities I mean liabilities by law or contract.
4 This has changed over the years. CCRIF was initially formally owned by a World Bank-administered Trust Fund.
5 That means if, at some point, a country decides to increase its coverage (and/or has to pay a higher premium for whatever reason) it must top up its participation fee to the same level (Pearson & Young, 2009, p. 28).
6 The data refers to the insurance policy year 2015/16. 'Excessive rainfall' event occurs when a 5-day aggregate rainfall exceeds 50mm and ends on the day before the rainfall falls again under 50mm.
7 Simply put, the guideline foresees that insurers limit the insolvency risk to a 1:200 years event.
8 The British Virgin Islands, Montserrat and St Maarten joined only in the course of 2018/19.
9 The excess-of-loss arrangements for hurricane/earthquakes policies have different retention levels than the arrangements for excessive rainfall. For the sake of simplicity I just focus here on hurricane/earthquake policies. The principles are, however, the same. The difference is that reinsurance kicks in at a different loss levels.
10 A shareholders' equity is describes a company's total assets minus its total liabilities. It is commonly used a metric to describe a company's financial health.
11 I am referring here to the floods in Sudan and Bangladesh (1988), typhoons in the Philippines (1988) and the Hurricanes Gilbert (September 1988), Hugo (September 1989) and Andrew (August 1992) in the Caribbean.
12 "The Global Facility for Disaster Reduction and Recovery (GFDRR) is a global partnership that helps developing countries better understand and reduce their vulnerabilities to natural hazards and adapt to climate change. Working with over 400 local, national, regional, and international partners, GFDRR provides grant financing, technical assistance, training and knowledge sharing activities to mainstream disaster and climate risk management in policies and strategies. Managed by the World Bank, GFDRR is supported by 34 countries and 9 international organizations" (GFDRR, 2017).
13 More precisely speaking, the CARICOM asked the World Bank to conduct a feasibility study.
14 These five activities happened not necessarily in sequence; they partly overlapped.
15 The paper I am referring to is entitled "Insurance, Reinsurance and Catastrophe Protection in the Caribbean" and was published jointly by the World Bank, United States Agency for International Development (USAID), and the Organization of American States (OAS) in 1996.
16 Participation respectively premium fee for Dominica, Grenada and St Lucia was about US$ 1,286 million, for St Vincent and the Grenadines it was about US$ 0,2 million (World Bank, 2007b).
17 CCRIF made use of a cat swap-arrangement until 2014/15. It then changed to issue cat bonds.

References

Aldrich, H. (1999). *Organizations evolving*. London: Sage.

Auffret, P. (2003). *High consumption volatility: The impact of natural disasters* (World Bank Policy Research Working Paper 2962). Retrieved from http://papers.ssrn.com/sol3/papers.cfm?abstract_id=636324

Benson, C., Clay, E., Michael, F., & Robertson, A. (2001). *Dominica-natural disasters and development in a small island state* (Disaster Risk Management Paper Series No. 2). Washington, DC. Retrieved from http://documents.worldbank.org/curated/en/875391468770118094/Dominica-natural-disasters-and-economic-development-in-a-small-island-state

Brooks, L. (2011). The Caribbean catastrophe risk insurance facility: Parametric insurance payouts without proper parameters. *Arizona Journal of Environmental Law & Policy*, *2*(135), 135–152. http://doi.org/10.1525/sp.2007.54.1.23

Buzan, B., Waever, O., & Wilde, J. H. (1998). *Security: A new framework for analysis*. Boulder, CO: Rienner.

CARICOM. (2004a, September 16). *CARICOM heads give top priority to Grenada relief*. Port-of-Spain, Trinidad and Tobago. Retrieved from www.paho.org/disasters/index.php?option=com_content&view=article&id=921:caricom-press-release-16-september&Itemid=0&lang=en

CARICOM. (2004b, September 17). *Press release 144/2004: CARICOM leaders recommend grant aid for Grenada*. Georgetown, Guyana. Retrieved from www.caricom.org/jsp/pressreleases/pres144_04.jsp

CCRIF. (2008). *Annual report 2007–2008*. Grand Cayman. Retrieved from www.ccrif.org/publications/ccrif-annual-report-2008-2009

CCRIF. (2009). *Annual report 2008–2009*. Grand Cayman. Retrieved from www.ccrif.org/publications/ccrif-annual-report-2009-2010

CCRIF. (2010a). *Annual report 2009–2010*. Grand Cayman. Retrieved from www.ccrif.org/publications/ccrif-annual-report-2009-2010

CCRIF. (2010b). *CCRIF and CIMH provide support for Haiti in its reconstruction efforts*. Retrieved October 7, 2014, from www.ccrif.org/news/ccrif-and-cimh-provide-support-haiti-its-reconstruction-efforts

CCRIF. (2011). *Annual report 2010–2011*. Grand Cayman. Retrieved from www.ccrif.org/publications/ccrif-annual-report-2010-2011

CCRIF. (2012). *Annual report 2011–2012*. Grand Cayman. Retrieved from www.ccrif.org/publications/ccrif-annual-report-2011-2012

CCRIF. (2013a). *Annual report 2012–2013*. Grand Cayman. Retrieved from www.ccrif.org/publications/ccrif-annual-report-2012-2013

CCRIF. (2013b). *Country risk profile Dominica*. Grand Cayman, Cayman Islands. Retrieved from www.ccrif.org/sites/default/files/publications/Dominica_CountryRiskProfile_2013.pdf

CCRIF SPC. (2015a). *Annual report 2014–2015*. Grand Cayman, Cayman Islands. Retrieved from www.ccrif.org/sites/default/files/publications/CCRIFSPC_Annual_Report_2014_2015.pdf

CCRIF SPC. (2015b). *Understanding CCRIF—A collection of questions and answers* (Vol. 53). Grand Cayman, Cayman Islands.

CCRIF SPC. (2017). *Annual report 2016–17*. Grand Cayman, Cayman Islands. Retrieved from www.ccrif.org/publications/ccrif-spc-annual-report-2016-2017

CCRIF SPC. (2018a). *Annual report 2017–18*. Grand Cayman, Cayman Islands: Caribbean Catastrophe Risk Insurance Facility. Retrieved from http://ccrif.org/publications/ccrif-spc-annual-report-2017-2018

CCRIF SPC. (2018b). *Organisational structure*. Retrieved August 1, 2019, from www.ccrif.org/content/aboutus/ccrif-organisational-structure

Christian Aid. (2009). *The potential role of the Caribbean Catastrophe Risk Insurance Facility (CCRIF) as a tool for social protection, disaster risk reduction and climate change adaption—A civil society perspective*. Retrieved from www.cpdcngo.org/cpdc/index.php?option=com_docman&task=doc_download&gid=21&Itemid=27

Collier, S. (2008). Enacting catastrophe: Preparedness, insurance, budgetary rationalization. *Economy and Society*, *37*(2), 224–250. http://doi.org/10.1080/03085140801933280

Collymore, J. (2011). Disaster management in the Caribbean: Perspectives on institutional capacity reform and development. *Environmental Hazards*, *10*(1), 6–22. http://doi.org/10.3763/ehaz.2011.0002

EM-DAT: The Emergency Events Database. Brussels, Belgium: Université catholique de Louvain, CRED, D. Guha-Sapir. Retrieved from www.emdat.be

GFDRR (2017). Who we are. Retrieved from https://www.gfdrr.org/en/who-we-are

Ghesquiere, F., & Mahul, O. (2010). *Financial protection of the state against natural disasters—A primer*. Washington, DC: World Bank. Retrieved from http://elibrary.worldbank.org/content/workingpaper/10.1596/1813-9450-5429

Ghesquiere, F., & Mahul, O. (2012). *Caribbean Catastrophe Risk Insurance Facility (CCRIF)—Pooling risk to protect against natural disasters*. Washington, DC. Retrieved from http://documents.worldbank.org/curated/en/319251467999348409/Caribbean-catastrophe-risk-insurance-facility-CCRIF-pooling-risk-to-protect-against-natural-disasters

Ghesquiere, F., Mahul, O., Forni, M., & Gartley, R. (2006). *Caribbean catastrophe risk insurance facility—A solution to the short-term liquidity needs of small island states in the aftermath of natural disasters*. *IAT03–13/3*. Washington, DC. Retrieved from http://siteresources.worldbank.org/PROJECTS/Resources/Catastrophicriskinsurancefacility.pdf

Grove, K. (2012). Preempting the next disaster: Catastrophe insurance and the financialization of disaster management. *Security Dialogue*, *43*(2), 139–155. http://doi.org/10.1177/0967010612438434

Haas, P. M. (1992). Epistemic communities and international policy coordination. *International Organization*, *46*(1), 1–35.

Hall, P. (1993). Policy paradigms, social learning, and the state: The case of economic policymaking in Britain. *Comparative Politics*, *25*(3), 275–296. http://doi.org/10.2307/422246

Handmer, J., & Dovers, S. (2013). *Handbook of disaster policies and institutions*. Abingdon: Routledge.

IFRC. (2005). *Operations update: Caribbean—Hurricane Ivan*. International Federation of the Red Cross and Red Crescent Societies. Retrieved August 19, 2019, from www.ifrc.org/docs/appeals/04/210402.pdf

IMF. (2013). *Caribbean small states—Challenges of high debt and low growth: New directions for youth development*. Washington, DC: International Monetary Fund. Retrieved from www.imf.org/external/np/pp/eng/2013/022013b.pdf

Iyahen, E., & Young, S. (2011). Understanding the CCRIF mechanism and policies by. In *A natural catastrophe risk insurance mechanism for the Caribbean: A collection of papers, articles and expert notes* (Vol. 2, pp. 11–20). Grand Cayman: Caribbean Catastrophe Risk Insurance Facility. Retrieved from www.ccrif.org/sites/default/files/publications/CCRIFBookletNovember252011.pdf

Kreimer, A., Arnold, M., Barham, C., Freeman, P., Gilert, R., Krimgold, F., . . . Vogt, T. (1999). *Managing disaster risk in Mexico—Market incentives for mitigation investment*. Washington, DC: World Bank.

Lalor, D. (1996). *Report of the working party on insurance*. Leyshon.

Lavell, A., & López-Marrero, T. (2014). Disaster risk management in Latin America and the Caribbean—Four decades of evolution and change, 1970–2010. In *Disaster management: International lessons in risk reduction, response and recovery* (pp. 229–247). London and New York: Routledge.

LFF. (2017). *Reinsurance*. Retrieved February 24, 2017, from www.luxembourgforfinance.com/en/products-services/insurance-pensions/reinsurance

McGee, J., Phelant, L., & Wenta, J. (2014). Writing the fine print: Developing regional insurance for climate change adaption in the Pacific. *Melbourne Journal of International Law*, *15*, 1–29.

MCII. (2013). *Climate risk adaption and insurance—Reducing vulnerability and sustaining the livelihoods of low-income communities* (Report No. 13). Bonn.

Munich Re. (2014). *Munich reinsurance company annual report 2014*. Munich. Retrieved from www.munichre.com/publications/302-07807_en.pdf

Nixon, M. (2009). The Cayman Islands experience with the Caribbean catastrophe risk insurance facility. In *A natural catastrophe risk insurance mechanism for the Caribbean: A collection of papers and articles* (Vol. 1, pp. 60–81). Grand Cayman: Caribbean Catastrophe Risk Insurance Facility.

PAHO. (1994). *A world safe from natural disasters—The journey of Latin America and the Caribbean*. Retrieved from www.mona.uwi.edu/cardin/virtual_library/docs/1214/1214.pdf

PAHO. (2004). *Hurricane Ivan in the Caribbean a comprehensive appeal for health emergency needs*. Pan American Health Organization. Retrieved August 19, 2019, from www.paho.org/disasters/index.php?option=com_content&view=article&id=904&Itemid=904

Pearson, M. (2011). Caribbean partnerships for disaster risk reduction—The CCRIF experience. In CCRIF (Ed.), *A natural catastrophe risk insurance mechanism for the Caribbean: A collection of papers, articles and expert notes* (Vol. 2, pp. 52–57). Grand Cayman: Caribbean Catastrophe Risk Insurance Facility. Retrieved from www.ccrif.org/sites/default/files/publications/CCRIFBookletNovember252011.pdf

Pearson, M., & Young, S. (2009). Natural catastrophe risk insurance mechanisms for Asia and the Pacific: The CCRIF as a technical model. In *A natural catastrophe risk insurance mechanism for the Caribbean: A collection of papers and articles* (Vol. 1, pp. 14–37). Grand Cayman: Caribbean Catastrophe Risk Insurance Facility.

Pollner, J. D. (2001). *Managing catastrophic disaster risks using alternative risk financing and pooled insurance structures* (World Bank Technical Paper No. 495). Washington, DC: World Bank. Retrieved from http://documents.worldbank.org/curated/en/940891468766227622/Managing-catastrophic-disaster-risks-using-alternative-risk-financing-and-pooled-insurance-structures

Poncelet, J. L. (1997). Reports and comments disaster management in the Caribbean. *Disasters*, *21*(3), 267–279. http://doi.org/10.1111/1467-7717.00061

Schwartz, S. B. (2015). *Sea of storms: A history of hurricanes in the greater Caribbean from Columbus to Katrina*. Princeton: Princeton University Press.

UNISDR. (2005). *Hyogo framework for action 2005–2015. Building the resilience of nations and communities to disasters*. United Nations Office for Disaster Risk Reduction. Kobe, Hyogo. Retrieved from https://www.unisdr.org/2005/wcdr/intergover/official-doc/L-docs/Hyogo-framework-for-action-english.pdf.

World Bank. (2002a). *Organization of Eastern Caribbean States (OECS)—Barbados catastrophic risk management and insurance reform project*. Washington, DC. Retrieved from http://documents.worldbank.org/curated/en/637051468780955765/Organization-of-Eastern-Caribbean-States-OECS-Barbados-Catastrophic-Risk-Management-and-Insurance-Reform-Project

World Bank. (2002b). *Organization of Eastern Caribbean States (OECS)—Catastrophe risk management and insurance reform project*. Washington, DC. Retrieved from http://documents.worldbank.org/curated/en/2002/08/2000402/organization-eastern-caribbean-states-oecs-catastrophe-risk-management-insurance-reform-project

World Bank. (2007a). *Project appraisal document on proposed credits to Dominica, Grenada, St Lucia and St. Vincent & the Grenadines* (Report No. 38539-LAC). Washington, DC. Retrieved from http://www-wds.worldbank.org/external/default/WDSContentServer/WDSP/IB/2007/02/16/000020953_20070216094855/Rendered/INDEX/38539.txt

World Bank. (2007b). *Project Information Document (PID) appraisal stage—OECS-catastrophe insurance* (Report No. AB2661). Washington, DC. Retrieved from http://documents.worldbank.org/curated/en/832161468298482917/OECS-Countries-Catastrophe-Insurance-Project%0A

World Bank. (2007c). *Results of preparation work on the design of a Caribbean catastrophe risk insurance facility*. Washington, DC. Retrieved from http://documents.worldbank.org/curated/en/2007/02/8556209/results-preparation-work-design-caribbean-catastrophe-risk-insurance-facility

World Bank. (2007d). *Untitled report on IDA R/2007–0044*. Washington, DC: World Bank.

World Bank. (2010). *A review of CCRIF's operation after its second season*. Washington, DC. Retrieved from http://siteresources.worldbank.org/INTLAC/Resources/CCRIFReview2008-2009.pdf

World Bank. (2012). *Implementation completion and results report on CCRIF*. Washington, DC. Retrieved from www.ccrif.org/sites/default/files/publications/2012WBImplementationCompletionReportCCRIF.pdf

4 The African Risk Capacity (ARC)

I will now turn to the African Risk Capacity (ARC). Empirically speaking, ARC is an international organization that offers, through a financial affiliate, African countries insurance to finance drought-related contingency plans to prevent famine.

In the first part of this chapter (Part I) I will center stage the goal and operation of ARC. I will explore the problems ARC seeks to solve and illustrate how the insurance arrangement works on a practical and technical level to fulfill these objectives. To guide the reader through these complexities, I will first provide a short overview of ARC, which forestalls key takeaway points of this chapter. I will then elaborate on these points by zooming onto the goals, the operational structure, the technical basis, the governance structure and the eligibility requirements.

In the second part of this chapter (Part II) I will explore how ARC has come about. I will explore the ideational origin of ARC and illuminate the processes and activities leading up to the creation of ARC highlighting in particular political and technical challenges (1–5). I will also briefly review the first year of operation (6).

I will finish this Chapter 4 with a concluding summary (Part III).

Part I: understanding ARC

The African Risk Capacity (ARC) is a public-private self-insurance vehicle that provides **five** (insurance policy year 2017/18) **African countries** with **insurance coverage to finance response measures to mitigate drought-related famine** in exchange for an annual premium and the submission of pre-approved contingency plans.

ARC enables payouts *before* the event—a *severe* drought—**has realized**. It does so via a drought surveillance software labeled Africa Risk View (ARV). Combining rainfall, crop and population data, the software estimates how many people are 'at risk' (or have been affected) by an agricultural drought in a given season; the software then translates this information into response costs to ensure food security. If the estimated response costs cross a pre-defined threshold, a country receives a payout. ARC aims to make payouts within two to four weeks of the end of the rainfall season, that is, before the impact of the drought fully hits the exposed population.

The idea behind realizing funds *ex ante* is to enable African countries to kick off interventionist measures *timely* in order to protect **the most vulnerable populations from engaging in typical coping or survival strategies** when it comes to food shortage such as the selling of assets, the killing of their livestock, a reduction in food consumption and malnutrition. The **interventionist measures financed by ARC** are largely up to the countries; they **include cash transfers, the provision of food vouchers, targeted food distribution or feeding programs**. Yet, any measure requires the approval of ARC.

To be eligible for insurance payouts countries must first of all become a member of ARC and submit a 'contingency plan' that details how they would use an insurance payout. ARC is *a specialized entity* of the African Union (AU) established through an inter-governmental treaty. As an AU agency, membership is in principle open to all AU members. ARC members must then develop and submit contingency plans to the organization. The contingency plans must be targeted at the most vulnerable segments of society and are peer reviewed by the ARC Governing Board for approval. Once approved, countries are entitled to buy insurance coverage and can tailor the insurance contract according to their financial needs and possibilities.

The **primary goal** of ARC is **to prevent a drought-related famine** by ensuring the food security[1] of the most vulnerable segments of a population. ARC aims at protecting the lives and livelihoods of the most vulnerable populations, primarily small-scale subsistence farmers. It does so by providing state agencies with funds so that they can buy and distribute food to the poorest segments of society according to a previously established and approved contingency plan. The **secondary goal** is to **limit the long-term socioeconomic consequences** of a drought by accelerating the speed at which humanitarian assistance is mobilized and distributed. The quicker provision of support aims to prevent the most vulnerable segments of society from engaging in typical behavioral coping and survival strategies. These have long-term socioeconomic consequences such as health problems and income reduction.

ARC offers its insurance policyholders a **maximum insurance coverage of US$ 30 million** per season for drought events that were modeled to occur with a statistical frequency of 1:5 years or less.[2] The financial cap implies that the insurance covers only a part of a disaster—or, more specifically, a part of the overall drought-related costs—namely the sociopolitical emergency measures targeted to those in need, those poor segments of society living in the drought-affected area.

From an organizational-sociological perspective **ARC is a hybrid comprising two entities: A political (ARC Agency) and a commercial entity (ARC Ltd)**. While the inter-governmental ARC Agency provides general oversight and develops the insurance policies, ARC Ltd is the actual insurance vehicle that carries out the parent's insurance functions on a day-to-day basis (ARC, 2014f).

The political entity, the **ARC Agency**, is a **'Specialized Entity' of the African Union (AU)** and, as such, an international organization. It has been established through an inter-governmental treaty between 33 AU Member states[3] (Status: December 2018). It is located in Johannesburg, South Africa. ARC signatories are: Burkina Faso, Burundi, Central African Republic, Chad, Comoros, Republic of Congo, Côte d'Ivoire, Djibouti, Gabon, The Gambia, Ghana, Guinea, Guinea Bissau, Kenya, Liberia, Libya, Madagascar, Malawi, Mauritania, Mozambique, Niger, Rwanda, Sahrawi Arab Democratic Republic, Sao Tome and Principe, Senegal, Sierra Leone, Sudan, Togo, Zambia, Zimbabwe (see Table 4.1).

The **financial affiliate, the ARC Insurance Company Limited (ARC Ltd)**, is a mutual—an insurance company owned by its policyholders—domiciled in Bermuda and thus subject to its national jurisdiction. The ARC Ltd has been capitalized through donor contributions from the UK (DFID, UK's development agency) and Germany (KfW, Germany's development bank) and through insurance premiums.

ARC was **launched in 2014 and has provided a varying number of states with insurance coverage**. In its first year of operation (2014/15) ARC provided four of its members with **insurance coverage against drought**: Kenya, Mauritania,

Table 4.1 List of ARC member states

ARC member states		
Benin	Guinea	Nigeria
Burkina Faso	Guinea Bissau	Republic of Congo
Burundi	Kenya	Rwanda
Central African Republic	Liberia	Sahwari Arab Democratic Republic
Chad	Libya	Sao Tome and Principe
Comoros	Madagascar	Senegal
Cote d'Ivoire	Malawi	Sierra Leone
Djibouti	Mali	Sudan
Gabon	Mauritania	Togo
The Gambia	Mozambique	Zambia
Ghana	Niger	Zimbabwe

Source: data taken from ARC, 2018a

Mozambique and Niger. In its second season (2015/16), it provided coverage to seven countries (Kenya, Mauritania, Niger, Senegal, The Gambia, Malawi and Mali). In the third season (2016/17) it provided coverage to six countries (Burkina Faso, Niger, Senegal, The Gambia, Mali and Mauritania). In the most recent season (2017/18) this book considers ARC coverage to five countries (Burkina Faso, Mali, Mauritania, Senegal and The Gambia). While the number of countries varied between four and seven per year it is notable that most countries that have bought insurance coverage are located in the Sahel.

Between 2014 and December 2018, ARC has made **five payouts** to four different governments (Senegal, Niger, Mauritania and Malawi) totaling to about **US$ 36,8 million** to finance early intervention measures to prevent famine. ARC claims that at least 2.1 million people and 900,000 livestock could be assisted with the payouts. The largest payouts occurred thereby in ARC's first year of operation. In January 2015 three countries in the Sahel—Senegal, Mauritania and Niger—received an aggregated payout of US$ 26,3 million to address a major drought.

Generic goals of ARC

What are the goals and the rationales behind ARC? Three core objectives can be distilled: (1) to prevent a drought-related famine (2) to limit the (long-term) socioeconomic consequences of a drought (3) to enhance the cost-benefit ratio of international aid provision.

1 To prevent drought-related famine

The primary goal of ARC is to prevent a drought-related famine by ensuring food security—the access to food—*of those in need and living in a drought-affected area*. It intends to do so by providing national disaster management agencies with *limited* funds (maximal US$ 30 million) at the onset of a drought period to empower them to undertake early life-saving activities aimed at the most vulnerable segments of society. The most vulnerable must be classified by the government *ex ante* and are thus subject to its definition. The idea of the arrangement is that countries get the funds *at the onset* of a severe drought period to provide *timely* assistance and thus to prevent a deterioration of the situation of the most vulnerable.

Importantly, ARC has been designed in response to food insecurity triggered by a 'creeping' exogenous event, namely *a drought related to the shortage of rainfall*. This is important as the linkage between drought and food insecurity is, in fact, not so straightforward as it may seem on first sight (Clarke & Hill, 2012). The literature on food security indicates that the relationship is far more complicated: Food insecurity may have many reasons. For instance, it might be related to "failures of entitlement" (Sen, 1982, p. 457), as Amartya Sen famously put it, and thus to sociopolitical questions, to 'state failure' and thus to the ineffectiveness of public distribution systems, to 'trade disputes, market dynamics and speculations' (e.g., flow, production and price of goods) and thus to (polit-) economic questions, to 'wars and conflicts' and thus to political questions, to 'the use, access to and management' of available food resources (e.g., crop, water) and thus political, infrastructural and behavioral questions (Clarke & Hill, 2012). In other words, the reasons for food insecurity are, *in practice*, often not clear-cut. They may be endogenous or exogenous and are, in fact, often interlinked.

ARC is well aware of this inherent complexity; it acknowledges and defends the shortcoming of its design. In response to a cost-benefit analysis in preparation of ARC, the Secretariat argued

> As an insurance scheme, ARC cannot model for all risks to food security. Instead ARC has chosen one of the most predictable and greatest exogenous threats to food security in an Africa dominated by rain-fed subsistence agriculture—drought—as the initial risk factor against which to provide coverage. Indeed insurance is not an appropriate tool for many food security problems in a country—nor chronic or frequently occurring drought risks.
>
> (ARC, 2012a, p. 6)

Hence, the goal of ARC is clearly limited: It seeks to provide a response to 'drought-related famine'.

2 To limit the long-term socioeconomic consequence of a drought

The second generic goal of ARC is to prevent the long-term adverse socioeconomic consequences of a drought by speeding up the mobilization and distribution of aid. The arrangement seeks to mitigate the delays of the (inter)national aid contribution system by making funds to governments for living-saving, humanitarian emergency measures *readily* available, that is, *in the initial days and weeks* of a drought—in other words, at the beginning of the annual harvest season. ARC claims that the speeding-up of aid provision saves not only lives and money but also the long-term socioeconomic consequences of a drought:

> As currently structured, the international system for responding to natural disasters is not as timely or equitable as it could be. *Funding is secured on a largely ad hoc basis* after disaster strikes and only then can relief be mobilized toward the people who need it most. In the *meantime, lives are lost, assets are depleted, and development gains suffer major setbacks*—forcing more people into chronic destitution and food insecurity in the world's least developed countries.
>
> (ARC, 2016a, p. 6 my emphasis)

The claims are based on the insight that African countries are usually deprived of greater financial resources. Often they lack a public welfare net as well as appropriate administrative capacities and contingency plans to support disaster-affected households (ARC, 2014f; Kimetrica, 2014). African countries typically reallocate budgets by diverting funds from ongoing policy programs and rely on international donations to engage in humanitarian activities. This comparatively limited financial room for maneuver makes African states heavily dependent on the generosity of international disaster funds and donors and leads to delayed responses, as the mobilization of these international funds and the distribution of aid require time.

This time lag has a particular negative impact on the poorer segments of society (DFID, 2012, p. 12), especially on those whose livelihoods depend on (small-scale) farming. Left without public support, drought-affected households typically adopt their own coping and survival strategies (ibid., p. 12). These include malnutrition, reduced food consumption and the killing or selling of livestock (ARC, 2012a, p. 12). These strategies, in turn, have long-term consequences such as health problems and income reductions, as outlined by ARC. ARC hopes that by providing funds and organizing their distribution *timely* these coping and survival strategies can be prevented (see Figure 4.1).

A timely provision requires, however, functioning contingency plans and aid distribution systems (Clarke & Hill, 2012, p. 6). This is why ARC has made the submission (and approval) of contingency plans an obligatory eligibility criterion for insurance payouts. ARC suggests that countries use the funds to scale-up existing social safety nets. The overarching goal of a timely provision of funds and

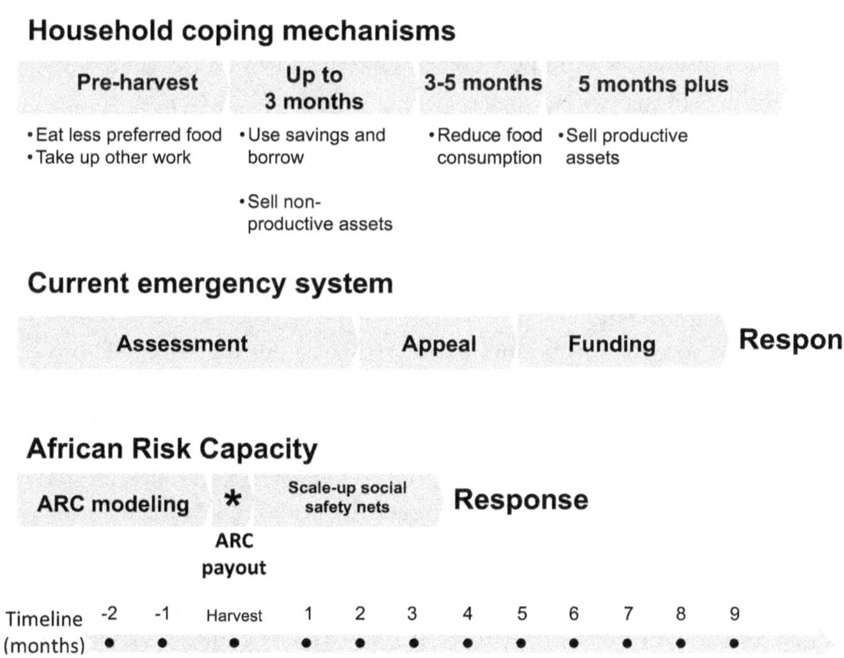

Figure 4.1 How ARC seeks to limit the socioeconomic impact of a drought

Source: adapted from ARC, 2019a

the development of contingency plans is to limit the long-term socioeconomic consequences of a drought.

It is important to underline that ARC understands insurance only as a supplementary financial tool. ARC emphasizes that it provides governments only with *limited funds* and does not intend to replace existing instruments:

> In particular, with an initial maximum payout currently modeled at USD 30 million per country per year, ARC is not intended to replace traditional aid channels, which will continue to be needed to finance the bulk of drought and other food security response cost needs on the continent for the foreseeable future.
>
> (ARC, 2012a, p. 6)

ARC claims that a payout covers less between 10–30% of the total disaster funding requirements "with the remainder largely remaining with the UN appeals process" (ARC, 2017c, p. 1). Hence, ARC conceives a drought insurance only as "one of several tools available to countries to respond to food security problems" (ARC, 2012a, p. 7). In an interview an ARC team member also expressed the hope that a payout would help countries to mobilize international aid (Personal interview, ARC team members, 2015, January 9).

The operational structure

Who provides insurance? And how does the arrangement work on a technical and practical level?

The hybrid character of ARC

In exchange for an annual premium, ARC offers African countries individually tailored insurance policies to finance social interventionist programs against drought. Countries can choose (within certain parameters) how much insurance coverage they want to buy.

ARC is based on an institutional two-tiered structure that is public-private in nature. That means, it is composed of a political (ARC Agency) and a commercial entity (ARC Ltd). The political entity (ARC Agency) is an international organization established as a *Specialized Entity* of the AU through an inter-governmental treaty between interested AU member states. It is located in Johannesburg, South Africa. The ARC Agency provides political guidance and supervision; it offers, negotiates and designs the insurance contracts and approves the contingency plans. The commercial entity (ARC Ltd) is the insurance company that carries out the insurance functions (e.g., investment of premiums, insurance payouts) on a day-to-day basis. It is a letterbox company domiciled in Bermuda. Akin to CCRIF, it exists just on paper. ARC Ltd is fully owned by the ARC Agency; the donors of the ARC Agency (the KfW on behalf of Germany and DFID on behalf of the UK) sit, however, on ARC Ltd's Governing Body.

ARC Ltd offers insurance only to the ARC Agency and thus to ARC members. Legally speaking, ARC Ltd is a *mutual*.[4]

The club character of ARC

In order to be able to buy an insurance coverage a country must, first of all, become a member of the ARC Agency. To become a member, states must sign the ARC Establishment Treaty. As *Specialized Agency* of the AU, ARC membership is in principle open to all AU members. No membership fee is required. The initial establishment treaty was negotiated between the 18 AU member states. By December 2018, 33 AU members have become part of the ARC Agency by signing the treaty.[5] Upon request the ARC Agency would then develop and 'customize' an insurance contract according to the demands (and financial possibilities) of the country. The ARC Agency is accountable to its signatories and has to report about its activities.

ARC offers coverage against drought

As member of ARC countries are entitled to buy *limited* insurance coverage against the emergency response cost of a *severe* drought. 'Limited' refers to the fact that the payout is triggered and capped at a certain level. 'Severe' refers here to drought events, which were statistically modeled to happen 1:5 years or less. That means, ARC offers essentially insurance for relatively frequent events and not necessarily for catastrophic events.

Conceptual basis: regional risk pooling

Similar to arrangement in the Caribbean, the operation of ARC is based on the concept of *regional* risk pooling. Since not every African country is subject to the same weather events at the same time—in other words it is highly unlikely that severe droughts will occur in all parts of the African continent at the *same* time in the *same* policy year—the countries can capitalize on the diversification of weather risks and pool their financial resources in a common fund. The pooling of financial resources allows for cost-savings since countries must not set aside the same amount of capital they would usually set aside in annual contingency budgets to attain the same level of financial protection. ARC estimated that the cost-savings revolve around 50% (ARC, 2014e).

The insurance contract

The ARC Ltd offers an insurance contract for one year. The insurance policy year runs from 1 May–30 April. The member states pay an annual premium reflecting the financial risk they want to retain, respectively transfer to the insurance facility. With 'financial risk' I refer here the modeled emergency 'response costs' of a public interventionist program in dollar terms. These costs have been estimated with the aid of a software called Africa Risk View (ARV).

ARV estimates the losses in crop yields due to rainfall deviations and the impact thereof on vulnerable households in drought-affected regions. It then translates this impact in equivalent financial terms, that is, it estimates the likely financial costs of the country's program aimed at preventing those vulnerable households from engaging in coping strategies and protecting them from famine. The countries are free to select the proxies for their drought-related response costs by defining the poverty line and the interventionist programs they intend to implement (e.g., cash voucher, food aid, social safety-net scale-up) (AU & WFP, 2014, pp. 23–24). The software assists countries to understand how much funds they require to realize their contingency plans and, thus, informs their decision select an appropriate insurance coverage. In the insurance policy year 2017/18, ARC provided drought insurance to Burkina Faso, Mali, Mauritania, Niger, Senegal and The Gambia.

The premium level

In their decision on the level of coverage, countries select the 'attachment point' (the response cost they want to retain using other financial resources), the 'ceding percentage' (the percentage of the response costs they want to transfer to the insurance facility in exchange for a premium) and the 'exhaustion point' (maximum payout they can receive from the insurance facility). This implies that the insurance contracts are highly customizable; they can be tailored around country preferences to reflect political circumstances and financial possibilities. The premium varies accordingly.

The maximum coverage a country may buy is, however, limited to US$ 30 million. For the insurance policy year 2016/17, only Niger decided to buy the maximum insurance coverage of US$ 30 million. The other countries' coverage limit was largely well below the upper coverage limit (Burkina Faso: US$ 10 million, Mali: US$ 15 million, Mauritania: US$ 10 million, Senegal: US$ 25 million, The Gambia: US$ 5 million). For the insurance policy year 2017/18, Senegal bought the highest insurance coverage (US$ 21,1 million); the other countries' coverage was well below the maximum coverage they could buy (Burkina Faso: US$ 7 million, Mali: US$ 13 million, Mauritania: US$ 7,5 million, The Gambia: US$ 7 million) (ARC, 2017b, p. 9).

In its first year of operation ARC Ltd collected about US$ 17,5 million in premiums (Wilcox, 2014, p. 3) and insurance premiums varied between US$ 500,000–4,5 million.[6] For the policy year 2016/17 and 2017/18 premiums summarized to US$ 11,3 million (2016/17) and US$ 8,7 million (2017/18) (ARC, 2017b, p. 20).

Modeled-loss approach

The ARC's insurance scheme is based on the performance of the Water Requirement Satisfaction Index (WRSI). The WRSI predicts crop yields based on the amount of rainfall. ARC incorporates the index in a model that estimates the number of people that are affected by a drought and associated costs of a social interventionist program to assist them. It does so via a software called Africa Risk View.

The technical basis of ARC

Africa Risk View quantifies the costs of the social interventionist program. A WFP-affiliated 'start-up team' developed ARV (see Part II). The software is based on three modules and integrates scientific insights from the natural and social sciences. It connects real-time rainfall data with assumptions about crop developments and combines it with socioeconomic assumptions and public finance data to estimate the likelihood and impact of a drought on a public budget. In this section I will illustrate the three modules on which Africa Risk View is based.

1 Hazard module

The first component provides an estimate on "the seasonal performance [of crop yields; N.S.] at the district, regional and national level under different rainfall conditions" (AU & WFP, 2014, p. 23). To do so, the software uses satellite-based rainfall estimates, produced by the Climate Prediction Center (CPC) at the US National Oceanic and Atmospheric Administration (NOAA), and feeds these rainfall data into a crop growth model, based on the Water Requirement Satisfaction Index (WRSI) and developed by the UN Food and Agriculture Organization (FAO) and the US Geological Survey NOAA (ARC, 2014b). WRSI is based on how much water is available for the crop and a widely used early warning indicator for famine. It proceeds by information on the timing, amount and distribution of rainfall and connects it to assumptions about the growth of annual rain-fed crops. This means that the hazard module provides satellite-based estimates on the impact of rainfall on the development of rain-fed crops and the availability of pasture on various spatial levels. Simply put, it anticipates crop failures and allows identification of drought-affected regions.

2 Vulnerability module

Africa Risk View then converts the spatial drought information into the number of people projected to be in need of assistance. It does so by overlaying the spatial drought information with population data. To identify those in need, that is, to distinguish the vulnerable from the non-vulnerable population, ARV uses two population characteristics: A household's percentage of a household income that comes from agricultural activities (e.g., production, casual labor and livestock) ('exposure') and a household's distance from the national poverty line ('resilience'). The latter can be redefined by the countries. To compute this vulnerability profile WFP made use of household surveys where possible and used proxy data from UNICEF's Multiple Indicator Cluster Surveys (MICS) or the Demographic and Health (DHS) as substitutes (ARC, 2014b, p. 23). The ARC classified a household as 'highly exposed' to drought when it would lose at least 12% of its livelihood because of drought.

3 Cost module

As a third step the number of affected people is translated into financial response costs. Modeled costs are standardized at US$ 50 per affected person for regions with bimodal rainfall seasons, that is, where there are two rainfall seasons per annum, and at US$ 100 per affected person for regions with uni-modal rainfall seasons. However, the costs per beneficiary should be adjusted by the countries to reflect their response programs (e.g., cash voucher, food aid etc.) to food insecure populations (AU & WFP, 2014, p. 24). In other words, countries must determine the actual costs per person on the basis of envisioned emergency measures and the historical experiences with response cost for past droughts.

Overall Africa Risk View may be understood as *a forecast* and *budget tool*. It provides estimates on geographic areas of low rainfall and on the number of people affected. Estimates on the impact of low rainfall allows to targeting and organizing public (or international) assistance more specifically and timely.

The governance structure of ARC

ARC comprises two entities: ARC Agency and ARC Ltd. The ARC Agency is located in Johannesburg and is an intergovernmental organization. It is the political branch of ARC. Its key mandate is to make African countries aware of the benefits of ARC, prepare and help member states to take out insurance. To its key activities belong strategic oversight and capacity building measures on risk modeling, disaster risk financing and contingency planning. ARC Ltd is a separate legal entity, namely a *not-for-profit* insurance company ('mutual'). It is legally owned by the ARC Agency and domiciled in Bermuda. The ARC Ltd is the financial arm of ARC. It carries out the day-to-day insurance functions. Activities include collecting the premiums, pooling and investing them, buying reinsurance coverage and organizing the payout (e.g., computing and verifying the claims).

The ARC Agency is led by a Governing Board that consists of eight members: Two AU Commissioners, one representative (plus one alternate each) for each of the five sub-regions of the African continent (Central Africa/East Africa/West Africa/South Africa/North Africa) and one Director General *ex officio*. According to Article 14 of the ARC Establishment Agreement, one AU Commissioner is appointed by the chairperson of the AU for "having experience in insurance matters" (AU, 2012, p. 8), the other AU Commissioner is appointed by the AU Chairperson and in collaboration with the Executive Director of WFP for "having expertise in the area of food security, extreme weather events and disaster risk management" (ibid., p. 8). The sub-regional members are elected by the countries that have bought or expressed intent to buy insurance coverage, taking into account geographical representation. The Director General is elected by the seven board members and has no voting right. The Director General acts *ex officio* and oversees a chief of staff and various program and policy officers to implement ARC. In 2017 and in the context of its further professionalization ARC introduced the position of M&E Advisor and Communication Advisor. All ARC Board members serve part-time (and in their 'personal capacity'); they are elected for three years in staggered terms to ensure continuity.

ARC Ltd is a *not-for-profit* insurance company ('mutual') legally owned by the ARC Agency and domiciled in Bermuda. It carries out the insurance functions. Not-for-profit implies that all profits must be given back to its owners in some form (e.g., in the form of dividends, reduced premiums). Its insurance functions include collecting the premiums, pooling and investing them, buying reinsurance coverage and computing and verifying claims. The Governing Board of ARC Ltd consists of representatives from the countries taking out insurance and representatives from donors. As of December 2018, donors on the board are the KfW (a German development bank) on its own behalf, and for the account of the Federal Republic of Germany (acting via the Federal Ministry for Economic Cooperation and Development, BMZ) and the United Kingdom (via the Department for International Development, DFID). The Governing Board oversees a Chief Executive Officer, a Chief Operating Officer, one Underwriter, an Executive Assistant and a few service providers. Like CCRIF, ARC Ltd is organizationally 'lean'; that means, it exists only in the form of a 'letterbox company'. There exists no physical building and staff *per se* in Bermuda. All back-office functions are outsourced to private service providers located somewhere else. The providers during the first season involved *Marsh Ltd* as insurance manager, *Willis* as reinsurance broker, *Price Water Coopers* as auditor—all leading actors in their respective fields (Syroka, 2014, p. 7).

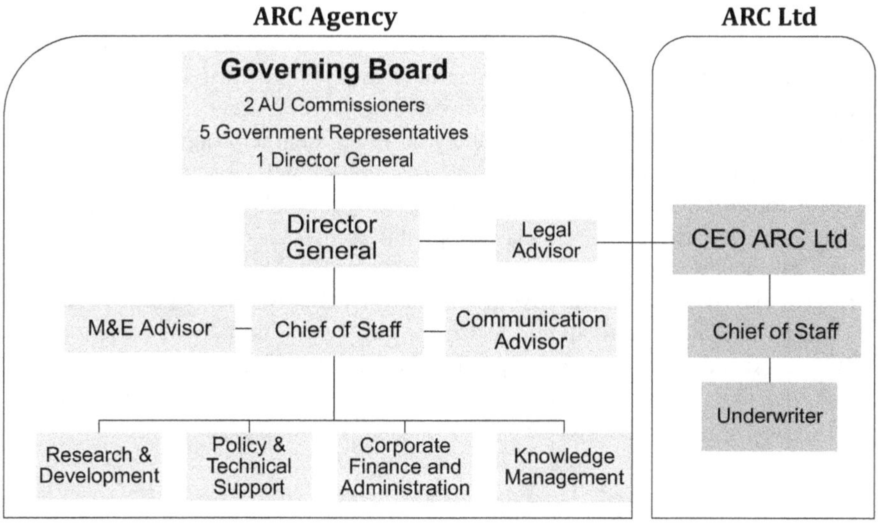

Figure 4.2 Organigram of ARC

Source: simplified from OPM, 2017, p. 6

Eligibility requirement: the development of contingency plans

A key feature of the ARC arrangement is that ARC member states must submit contingency plans to be eligible for insurance payouts. ARC's Agency Governing Board must approve these contingency plans.

The eligibility process entails two steps: First, countries must apply for a Certificate of Good Standing (CoGS). To obtain that certificate countries must submit a provisional "operation plan" exposing the measures they intend to implement in the case of an insurance payout. The plans are reviewed by the Governing Board of the ARC Agency according to a set of preestablished criteria (see the following). Second, shortly before an insurance payout, the countries must update this plan and submit a "Final Implementation Plan" (FIP) entailing "comprehensive information on how the payout will be deployed" (ARC, 2012b, p. 2).

While the operations plan entails a set of measures that *could* be implemented, the FIP lists the measures that *will* be implemented by taking into account the country's (fiscal, political, food security) situation at the time of the emergency situation (ibid., p. 9). Hence, in order to receive an insurance payout, ARC member countries must make two submissions to the ARC Agency's Governing Board. The measures are evaluated according to three criteria (ibid., p. 7):

1 *Time sensitivity and/or catalyst*: the measures must be implementable within 120 days and/or enable other activities
2 *Livelihood saving*: the measures must aim at saving livelihoods of beneficiaries. They must be not aimed at general investment
3 *Duration*: the measures must be completable within a six-month time frame

In its illustration of eligibility requirements, the ARC Agency list exemplarily some measures that fulfill these criteria. They include conditional or unconditional cash-transfer (to the most vulnerable), targeted food distribution, the issuance of food stamps, vouchers and coupons and feeding activities (ibid., p. 23). In more general terms, the ARC Agency urges its member countries to develop contingency plans that complement or *scale-up*—that is, expand in scope and target—potentially *existing social safety net programs*. The ARC Agency also underscores that the countries may adopt different solutions according to contextual circumstances: "It is possible that a country could find itself in a situation where the best use of an ARC payout is an activity that was not included . . . then the country should seek to amend its Operations Plan" (ibid., p. 31). Hence, the ARC Agency does not insist on a specific plan or set of instruments but on *a* plan in the first place. Once a payout has been made, the countries are required to monitor and evaluate the measures they decided to implement.

Voluntary 'capacity building' as add-on

As most of ARC members have never had contingency plans in place and therefore the lack the knowledge and administrative capacities to develop and implement these (Kimetrica, 2014), the ARC Agency offers ARC members a capacity building program in the run-up of the development of contingency plans and the customization of insurance policies. The program is intended

> to prepare in-country government staff for the prospect of risk transfer to the international market. Countries formalize their engagement with the ARC Agency by signing a Memorandum of Understanding (MoU), which commit resources both from the ARC Agency and the government to undertake the capacity development programme.
>
> (ARC, 2014f)

The ARC Agency developed and offers workshops, trainings and conferences with national and international actors involved in disaster risk management to transfer knowledge and build up and strengthen skills. The activities offered include support in the use of *Africa Risk View software;* support in the quantification of the country's drought risk and selection of appropriate tools for the different layers of the risk transfer model, including risk transfer parameters for ARC; support in the drafting of operations plans for the use of a potential payout; and support in the securing of funds for the premium payment.

The goal of the capacity building program is to build up knowledge—expertise and skills—on disaster risk management at various technical and political levels. The adoption of this capacity building program is not an eligibility requirement for taking out insurance but an offer. During the second season, in November 2014, nine ARC member countries (Malawi, Kenya, Niger, Lesotho, Senegal, Burkina Faso, Mozambique, Mauritania, Zimbabwe)[7] signed up for the capacity building program.

By making the submission of contingency plans an eligibility requirement, the ARC Agency has some power to influence disaster response policies and politics at the national level. By providing financial incentives to engage in disaster response planning the ARC Agency is, in fact, in a unique position to disseminate and harmonize public disaster management policies across the African continent. The power to disseminate and harmonize disaster management policies across the African continent manifests also in a more subtle way: In the context of the capacity building program, ARC teaches and familiarizes African countries with what counts as 'best practice'. It shows countries what other countries have implemented. This knowledge transmission reflects a 'soft' form of power. Yet, it is important to not overstate the power of ARC. The ARC Agency has no enforcing 'hard power' to institutionalize and harmonize disaster management across the African continent. It cannot (or does not want to) prescribe specific policies. Countries are still free in terms of how to govern disasters. Moreover, from a post-colonial perspective, it is also important to underscore that the ARC Governing Board, the Board which approves the contingency plans and issue the certificates of good standing, is composed primarily of political representatives from the African states. Hence the subtle structuring form of power is African-owned.

The (re-)insurance strategy of ARC

Akin to CCRIF, ARC utilizes the practice of *risk layering* to ensure it has enough reserves to settle claims. It financed payouts up to a certain level of its own reserves *(self-retention)*, claims exceeding that level are paid by commercial re-insurers *(excess-of-loss-arrangement)*.

ARC's self-retention level has varied throughout the insurance seasons and is outlined in the financial audits of ARC Ltd. For the 2016/17 insurance policy year, ARC retained US$ 7 million of the total underwritten exposure. Above this level of self-retention ARC had purchased reinsurance coverage for the next US$ 53 million. ARC pays again for (aggregated) annual payouts that exceed US$ 60 million until depletion. In respect of the 2017/18 insurance policy period, ARC retains US$ 6 million of the underwritten exposure. Above this level of self-retention, ARC had purchased reinsurance cover for the next US$ 41 million and pays again for (aggregated) annual payouts that exceed US$ 47 million (ARC, 2017c, p. 9).

In contrast to CCRIF, ARC's reinsurance arrangement secures sufficient funds for a 'black swan' event, an event in which all countries would receive the maximum payout. In other words: It did not write more insurance policies than it could finance.

As of 30 September 2017, ARC held about US$ 108 million in assets or shareholder equity (ARC, 2017b, p. 5). ARC's capital basis has been sourced from retained premium incomes (annually between US$ 8–17,5 million), investment incomes and donor contributions. In 2014 the UK (through DFID) and Germany (through KfW) had capitalized ARC by providing them with about US$ 200 million in risk capital to get it off the ground.[8] These interest-free loans must, however, be paid back within 20 years.

In 2016/17, ARC had invested around US$ 95 million of its assets. It has invested its income mainly in corporate and sovereign bonds (US$ 67 million and US$ 15 million respectively) of which US treasury bonds (US$ 5,8 million) make up the greatest share. ARC invested around US$ 7,7 million in more risky asset-backed bonds, namely debt securities (ARC, 2017b, pp. 14–16). Credit ratings for the bonds held by ARC ranged thereby from AAA to BBB- according to the credit rating agency Standard & Poor's.

Table 4.2 ARC's reinsurance structure in 2016/17

Retained/Reinsured	*Losses*
Retained (ARC)	0—US$ 7 million
Reinsured (via re-insurers)	7—US$ 60 million
Retained (ARC)	>US$ 60 million

Source: data taken from ARC, 2017b, p. 9

Part II: the genesis of ARC

Ideational origin

The idea to create insurance for disaster relief and humanitarian assistance in Africa was not out of the box. Quite in contrast, it has been discussed within the UN system at least since the late 1980s. The main rationale to consider the development of insurance was to overcome the inadequate financial resources of the UN and thus the shortcomings of the international aid system regarding the provision of relief and assistance.

Despite these early discussions, the idea to create an international insurance to finance the provision of disaster relief and humanitarian assistance never really gained prominence until the early 2000s, when WFP and the World Bank tested the idea of using an insurance mechanism to finance the provision of humanitarian relief and assistance in a number of pilot projects. WFP and the World Bank realized several projects of which I will highlight the ones in Ethiopia and Malawi. I do so because they not only envisioned some elements of what would later become ARC but also because it were those people who took part in these projects who would then develop the idea of ARC.

The Nordic Report (1991)

The idea to insure the provision disaster relief and humanitarian assistance gained prominence within the Nordic UN Project. The Nordic UN Project was set up in 1988 as a collaborative intellectual project led by the Scandinavian countries that collected and proposed ideas from various sources to reform the activities of the UN system in the social and economic policy field. In its final report from 1991 ("Nordic Report"), the authors, basically Scandinavian policymakers, criticized the scarcity of financial resources available to the UN agencies and the speed at which these resources are mobilized to deal with emergency situations such as drought-related famine and proposed the introduction of an 'emergency insurance':

> The solution may thus be an insurance system under which the funds can be used for the UN's initial cost in the event of emergency. The responsibility for this insurance could be entrusted to a UN organization or a private insurance company. UNDRO, WFP and UNICEF in particular would benefit from such a system.
>
> (The Nordic UN Project, 1991, p. 79)

The idea was not 'owned' by the authors. As any idea it was a social product. It was inspired from a scholarly contribution made by Ernst Michanek in a rather alternative international development journal from 1981 (Michanek, 1981). The authors of the UN report made a clear reference to the article. The reference to Michanek's scholarly intervention entitled *An International Disaster Relief*

Insurance indicates that the idea of using insurance as a mechanism to finance contingencies was already floating around in some epistemic development circles before it finally gained greater visibility through the UN report.

Although the Nordic Report was well received among the UN community it took a decade until that insurance idea was finally taken up by policymakers and realized in modified version. It was only in 2004/05, in the wake of more global efforts to deal with disasters (see Chapter 3, *International institutionalization of disaster management*) and the availability of new financial market instruments (see Chapter 2, *ART in historical context*), when WFP, the UN's agency for food aid, took up the idea. It partnered with the French re-insurer Axa Re and initiated a pilot project in Ethiopia to test the feasibility of an index-based catastrophe insurance for drought.

WFP's drought insurance for Ethiopia

Ethiopia was severely hit by a drought in 2003. In 2003, about 13,1 million people were in need of assistance. WFP provided food aid at a cost of about US$ 1 billion (WFP, 2012b). The drought affected in particular the poorest segments of society. As a result of this catastrophe, Ethiopia and the broader development community took action. Ethiopia revised its strategy to deal with droughts and introduced, among others, a new social policy program, the Productive Safety Net Programme (PSNP) in collaboration with WFP and other development actors, to assist the most vulnerable segments of its society and prevent future drought-related food shortages. The program's goal provides regular transfers such as food, cash or both "to help chronically food insecure people survive food deficit periods and avoid depleting their productive assets while attempting to meet their basic food requirements" (WFP, 2012a, p. 1).

In 2005–07 WFP sought to scale-up the program from the "chronically food insecure" (Hellmuth, Osgood, Hess, Moorhead, & Bhojwani, 2009, p. 19) to the "transiently food-insecure" (Wilcox & Arthurs, 2007, p. 5), those "who are able to feed themselves in a good season but suffer when the rains fail" (Hellmuth et al., 2009, p. 19). It did so through the creation of an index-based insurance program for Ethiopian farmers. The pilot project was designed to provide Ethiopian farmers with cash in the event of a severe crop shortfall to prevent them from selling their assets to stabilize income streams and protect their livelihoods (ibid., p. 18).

WFP offered the insurance in collaboration with the Ethiopian government. WFP signed the contract on behalf of the Ethiopian government and the United States Agency for International Development (USAID) and paid the premium of US$ 930,000. The maximum payout was capped at US$ 7,100,000. The contract was based on a newly developed index, the Ethiopia Drought Index (EDI). The index used historical and real rainfall data from the national meteorological agency and a crop–water balance model to track crop growth. As pointed out by Hellmuth et al, "The index had a 80% correlation with the number of food aid beneficiaries from 1994 to 2004, indicating that it is a good proxy for human need when a disaster strikes" (2009, p. 18).

WFP's risk of a potential insurance payout—in other words, the risk of a drought in Ethiopia—were packaged and sold in the form of a weather derivative to international capital investors via Axa Re as issuer. Since the insurance policy year 2006 was characterized by average rainfall, no payout was made to the government via WFP. The following year, the insurance contract was not renewed, because WFP did not manage to secure sufficient donor support for the annual premium payment. Ethiopia apparently decided that it could not afford the premium by itself. While the project came quickly to an end, it nevertheless demonstrated the feasibility of an index-based insurance mechanism for a public entity. It showed that re-insurers and financial market investors were willing to support such an insurance arrangement. They were ready to underwrite and buy weather related financial risks. In other words: African weather risks were ready for marketization.

Since the drought insurance was not renewed in 2007/08, WFP sought to draw lessons. Some basis risk was noted: In contrast to the model, some areas had experienced crop stress because the rainfall ended earlier, but according to the model the crop had already been harvested (Hellmuth et al., 2009, p. 19). WFP worked on improving the index and made the plan to reintroduce a massively scaled-up version of the drought insurance in 2010. The goal was to set up an index-based drought insurance that offers up to US$ 250 million in payout. WFP, together with the World Bank, developed a drought early warning software called LEAP (Livelihoods, Early Assessment and Protection). LEAP used rainfall data from stations on the ground and satellite data to develop crop production estimates in combination with vulnerable population distribution data to project the number of people in need of assistance due to anticipated crop production reductions. The software then estimated the financial needs to support livelihood-saving interventions for "transiently food-insecure" (Hellmuth et al., 2009, p. 18) people. LEAP was created in 2008 and by August 2019 was still operational. It serves as an early warning tool for severe drought and floods and provides a proxy for the funding needs to scale-up Ethiopia's interventionist *Productive Safety Net Programme* (PSNP). According to the website, "LEAP's ultimate objective is to link early warning, contingent planning and capacity building to a contingent fund managed by the World Bank" (WFP, 2017). Plans to reintroduce the insurance were, however, given up. Unfortunately, the reasons are not clear.

The World Bank's drought insurance in Malawi

In 2005, the same year the Ethiopian insurance was developed, the World Bank, through its Commodity Risk Management Group, set up a very similar index-based insurance pilot project in Malawi.

The World Bank designed an index-based crop insurance for small-scale famers. The objective of this pilot project was to prevent small-scale farmers from selling their assets to survive in the case of a major drought. The arrangement was designed in a way that farmers would receive an insurance payout when the measured rainfall at a pre-defined weather station in certain area was below a certain threshold in a certain time period. The money farmers would receive would then sustain them through the drought period and enable them to continue farming in the long run.

This insurance arrangement was very different from the Ethiopian one. The main difference was first of all that the insurance contract was designed to pay out farmers. This means that the arrangement was targeted at individuals and not at a sovereign entity, as it was the case with Ethiopia. Moreover, the insurance arrangement involved a range of stakeholders from the public and private sector, many of them from Malawi, and was, as such, much more complex in terms of the underpinning web of relations to make this arrangement function and in terms of who bears responsibility for what.[9] The main rationale for the World Bank to engage in such an insurance arrangement were mainly macroeconomically driven. A World Bank report concluded that Malawi's economy

> is highly vulnerable to adverse weather shocks not only because of the *direct impact on agricultural production* (and GDP), but also due to the *indirect impact of government finances* resulting from the unanticipated need for emergency intervention (which may translate into increased domestic borrowing), and *increased pressure on the current account* because of the need for exceptional food imports.
>
> (Syroka & Nucifora, 2010, p. 3 my emphasis)

The insurance as such should therefore provide a hedge against the volatility of the Malawian economic growth, which would hamper the country's ability to attract investment, create income generating jobs and stabilize state revenues (Syroka & Nucifora, 2010, p. 4). It should prevent a greater fiscal crisis due to losses in state revenues due to a shortfall of taxes following a shortfall of crop production in the private sector and thus secure the country's ability to repay debts (ibid., p. 3). Malawi had only a limited budget line for "unforeseen expenditures" amounting to about 1% of its GDP and relied therefore mainly on international assistance for major disasters.

This macroeconomic not to say fiscal rationale of the Malawian project can be contrasted with the WFP's rationale to create drought insurance for Ethiopia. Here, the motivation was mainly driven by considerations around food security. This difference must, however, be seen in the context of diverging organizational mandates of the World Bank and WFP.

A scientific article

The projects in Ethiopia and Malawi represented the playground for rethinking the provision of disaster aid. Key to the creation of the ARC was a scientific article published by Johanna Syroka and Richard Wilcox in the *Journal of International Affairs* in 2006. The vision set out there inspired the development of ARC. The authors of article, Johanna Syroka and Richard Wilcox, were development professionals and involved in the WFP project in Ethiopia, and Syroka later in another index insurance project realized by the World Bank. Syroka and Wilcox may be understood as the 'intellectual minds' behind ARC. Both authors would later play a key role in setting up ARC.

The article *Rethinking International Disaster Aid Finance* called for a conceptual shift on how humanitarian aid should be financed. It argued for a move from a reactive, ad-hoc "emergency aid business model to a proactive risk-management investment model" (Syroka & Wilcox, 2006, p. 197). The lack of timely and sufficient funds hampers the effective operation of emergency relief measures. Syroka and Wilcox proposed to reinterpret WFP, the world's largest food aid provider, as "insurer of last resort for vulnerable populations facing food shortages in the developing world" (ibid., p. 199). The authors suggested reading them as financial market actors who hold a portfolio of risks:

> They hold portfolios of contingent liabilities—inherent to their mission of aiding vulnerable populations in times of crisis—that are triggered not by movements in the financial markets, but, like any other agriculture or property/casualty (re)insurer, by drought, earthquakes, floods and conflict. These financial liabilities manifest themselves not in cash claim settlements, but in the mobilization of aid workers, shelter, water, food, clothes, health care and all the other aspects associated with emergency relief.
>
> (ibid., p. 199)

Syroka and Wilcox proposed that aid organizations like WFP should make use of the newest risk management and tools available in the finance sector to make cash more readily available for emergency relief operations. To be able to do so, they should make use of the newest technology, quantify the risks, that is, estimate the likelihood of a natural disaster and the likely financial costs thereof in terms of necessary emergency relief operations to prevent a greater humanitarian disaster. This information could be based on existing data and collected from the experiences of different aid actors. Using probabilistic tools and having cash readily available, Syroka and Wilcox claimed, would lead to a more sustainable solution to address the lack of donor funding to finance relief measures associated with food shortage.

From a historical, politico-economic perspective, the vision to employ strategies and practices used in the finance world reflected very much the *zeitgeist* of the millennium. The late 1990s and early 2000s saw the rise of the financial sector. In a range of OECD countries it even became the most important sector (Assa, 2012). The success of this 'new economy' inspired many political actors to propose or utilize corporate or financial market practices for the greater social good (see in the context also Chapter 3, *Part II: The genesis of CCIRF*).

The beginnings of ARC[10]

A financier: the Rockefeller foundation

While the visionary article by Syroka and Wilcox was published in 2005, it did not receive much attention until 2008 when the Rockefeller Foundation and

Mary Blair—at that time Managing Director of the Rockefeller Foundation—picked up the article. Over the years the Rockefeller Foundation had developed a substantial footprint in humanitarian aid in Africa and was interested in deepening its ties with WFP. In the mid-2000s Joseet Sheeran, Executive Director of WFP, and Judith Rodin, President of the Rockefeller Foundation, had several meetings discussing a possible collaboration. Following these discussions, WFP made two proposals to the Rockefeller Foundation for a common project. The Rockefeller Foundation, however, rejected these proposals. Instead, in 2008 it came up with its own proposal, namely the visionary article written by Syroka and Wilcox.

Instrumental here was Mary Blair. She joined the Rockefeller Foundation as Managing Director in 2005 and led two of the foundation's new initiatives: The *Developing Climate Change Resilience* program and the *Advancing Innovation Processes to Solve Social Problems* program. The climate change initiative sought "to catalyze attention, funding, and action in building climate change resilience for poor and vulnerable people globally", the innovative solution program, in turn, sought "to test the use and scaling of new innovation tools and methods" (Rockefeller Foundation, 2015a). Both programs were very well resourced in financial terms as can be seen from the foundation's website.[11] In her search for an innovative proposal, she came across the Syroka/Wilcox article and made the case for sponsoring the idea. The Rockefeller Foundation was keen on financing something 'off-the-shelves'.

As the Rockefeller Foundation was keen on realizing the Syroka/Wilcox idea, Sheeran, Executive Director of WFP, called Wilcox in Belgrade to persuade him to develop and realize a project based on his vision. After the Ethiopian project both Syroka and Wilcox had abandoned WFP and embarked on different projects: Syroka moved from WFP to the World Bank to realize another sovereign index insurance pilot project and Wilcox became the Political Director for the United Nations in Iraq in 2007 and representative of the UN Secretary-General to Serbia in 2008, where he negotiated the Secretary-General's Kosovo package. Sheeran, Executive Director of WFP, convinced Wilcox to rejoin WFP for the project.

In October 2008 the Rockefeller Foundation agreed to collaborate with WFP. The Rockefeller Foundation provided the WFP-hosted project *Climate and Disaster Risk Solutions* with a grant of US$ 3,000,000 to develop ideas on how to restructure the "international natural disaster assistance from a largely ad-hoc, post-emergency aid system toward an optimized global risk management system with pilot efforts in selected developing countries" (Rockefeller Foundation, 2015b).

It is noteworthy that the Rockefeller Foundation funded a *vision* and not a worked out concept note: An interviewee emphasized that WFP received the grant solely on the basis of that visionary article and thus on *trust* (Personal Interview, ARC official, 2015, January 9). The *Climate and Disaster Risk Solutions* project team submitted a more detailed concept note only later when WFP had the funds already at its disposal. Moreover, it is noteworthy that it was the

Rockefeller Foundation and not WFP who came up with the proposal. It appears WFP did not want to (or feared to) depart from existing, established structures by promoting a vague but arguably more innovative proposal. Whatever the precise reason was, this rather unusual kick off story highlights the larger role of foundations—in this case of "grant-making foundations" (Anheier, 1999)—in fueling innovations and thus in securing social and political diversity.

The incubator: the Climate and Disaster Risk Solution project

Richard Wilcox led the newly constituted *Climate and Disaster Risk Solutions* project. He nominated three other teammates: Joanna Syroka, Fatima Kassam and Brownwyn Cousins. As Wilcox assembled the team, he relied on his existing social network by calling the persons he worked with in the 2004 Ethiopian project. In other words, he reassembled the old and apparently well-attuned team. It seems that trust and predictability and voice[12] (Aldrich & Ruef, 2006) gained in these previous social interactions in the context of this work experience along with the necessary expertise of these protagonists were the main rationale behind that choice.

The team first specified the goal of the project and agreed to develop a tool for a pan-African information system to enable humanitarian actors to plan a response to a food shortage (WFP, 2009); the tool they envisioned should give humanitarian actors some financial estimates of how much such an intervention to prevent a famine from happening would cost. The team agreed to development of a pan-African software that translates weather data into an estimation of crop failures and the probable costs thereof to replace the harvest. The development of this software took about one year, that is, until November 2009 and was labeled Risk View (later Africa Risk View).

ARC developed out of the *Climate and Disaster Risk Solutions* project and the play with ideas. Throughout the year of software development, the project team took care to spread the news by sensitizing the Red Cross, Oxfam and other humanitarian actors about the project and the possible usefulness of that software for their activities. As one interviewee put it: "We saw this as a funds-management tool not just for WFP but for the humanitarian aid community" (Personal Interview, ARC Official, 2015, January 9). The idea to use the software as basis for a regional insurance mechanism came up only in September 2009 and was inspired by the launch of CCRIF in 2007.

LEARNING FROM THE CARIBBEAN EXPERIENCE

After it had developed the software, the project team played around and experimented what it could do with it. In search for applications the team considered CCRIF and explored whether a regional insurance arrangement would work for African countries. The ARC Project Team was running simulations to find out if there are any cost-savings for African countries if they would pool their financial risks associated with the provision of food security in a common regional insurance facility. After the team discovered that there would be substantial cost-savings, the project team directly reached out to the CCRIF to obtain further information and knowledge. It held several informal conversations with stakeholders involved in the creation of CCRIF to learn more about CCRIF, its organizational form, its political genesis and the lessons that have been learned, as interviewees revealed.

Getting in touch with CCRIF staff was not a big issue since Simon Young, at that time CEO of CCRIF, was a former colleague of a project team member. Hence, personal networks were crucial to obtain insider knowledge, collect 'lessons learned' and 'good practice', and develop the idea of ARC accordingly. In these informal conversations, Young suggested the project team up 'to get something out there', as one of the interviewee put it. Young advised the team to use the opportunity of not being attached to large bureaucracies like the UN, WFP or the World Bank, which are characterized by long-lasting decision-making processes, and to create an insurance facility as quickly as possible and without great political involvement.

Following his very pragmatic expert advice ("just do it"), the project team drafted the idea of a pan-African insurance facility. It sought to imitate the political procedure in the Caribbean. In the Caribbean the regional insurance facility was kicked off through a letter exchange between the CARICOM and the World Bank, which mandated the latter to develop the idea of a regional catastrophe pool (see Chapter 3, *The beginnings of CCRIF*). Following the Caribbean example, the project team approached the African Union (AU), the political counterpart to the CARICOM, with the idea to mandate it for the development of a regional insurance arrangement for Africa.

Agenda-setting: the project team approaches the AU commissioner for a mandate

In November 2009 the project team reached out to the AU Commissioner for Rural Economy and Agriculture to set the idea on the agenda of the African Union. It approached the AU Commissioner for Rural Economy and Agriculture since food security fell primarily under his department. The project team presented the software Africa Risk View and made the case for the establishment of a pan-African insurance facility. The team informally asked the AU Commissioner for Rural Economy and Agriculture to officially mandate the ARC Project Team to realize that idea.

The feedback the project team received was basically positive: The AU Commissioner pointed out that the idea of a pan-African insurance pool for food security resonates well with the AU's implementation of the *Comprehensive African Agriculture Development Programme* (CAADP) and the efforts of some AU member states to reduce hunger and improve responses to food emergency crises. Yet, he also argued that the AU Commission cannot or does not want to simply mandate the ARC Team to develop a regional insurance instrument for AU Commission without its involvement. In a later letter to WFP, the AU Commissioner argued that a pan-African insurance facility, intended to benefit all AU Member States, requires pan-African ownership and thus the *active* involvement of the AU in key decisions. In other words, he demanded a *participatory approach* and wanted the project team to work with the AU policy organs. The result of the meeting was that the Commissioner announced its political support to the idea, promised to and also introduced the idea into the policy organs of the AU and other key stakeholders, including regional economic communities. In the meantime, the AU Commissioner for Rural Economy and Agriculture formally asked the project team, respectively WFP, to develop a more comprehensive proposal by providing the Commissioner with some more concrete technical ideas on the actual functioning of such a pan-African insurance facility.

Mobilizing political support

Following this meeting, the AU Commissioner for Rural Economy and Agriculture took the political lead and disseminated the idea of creating a pan-African insurance entity in AU's policy circles. The commissioner presented the idea at several interministerial conferences in 2010, notably the *Third Joint African Union Conference of African Ministers of Economy and Finance* and *UN Economic Commission for Africa ("ECA") Conference of African Minister of Finance, Planning and Economic Development* held in Lilongwe, Malawi (25–30 March 2010), and *the Second African Ministerial Conference on Disaster Risk Reduction* in Nairobi (14–16 April 2010). Both conferences discussed and finally supported the idea in the form of resolutions that called for enhanced efforts to mitigate exposure to disaster risk through risk sharing across regions (Malawi-conference) and to explore the feasibility of a pan-African disaster risk pool (Nairobi-conference); a support that further was endorsed by decisions of the AU Executive Council (EX.CL./Dec.564 XVII and EX.CL./Dec.604 XVIII) in its report to the conferences at the AU Summit in Kampala, Uganda (25–27 July 2010). The AU Assembly, the highest political organ of the AU, followed the recommendation of the AU Executive Council and adopted the resolution "Towards Realizing a Food Secure Africa" (EX.CL/596 XVII) at its XV Session. The resolution supported "efforts to enhance national and regional capacities to mitigate exposure to disaster risk by institutionalizing effective financial instruments such as strategic grain gains, budgeted contingency funds as well as through sharing risk across sub-regions". The AU Commissioner for Rural Economy and Agriculture also briefed the group of African Ambassador based in Rome about the idea to create a pan-African insurance pool. The AU Commissioner for Rural Economy and Agriculture also established contacts with USAID and the World Bank to secure further donor support. The AU Commissioner was thereby strongly supported by the project team, as evidenced by several exchanges of letters between the AU Commission and WFP.

The pan-African insurance idea quickly became a subject of 'high politics'. The AU Assembly, the Assembly of the Heads of State and Government of the African Union and thus the highest political entity of the AU, adopted a supportive resolution already in July 2010. This is only eight months after the idea was first introduced to an AU representative—albeit to a Commissioner and thus already to a 'political champion', that is, a top-level representative. This is remarkable insofar as the AU is a highly consensus-driven organization—decisions by the Executive Council and AU Assembly are taken by two-third majority—that lacks a pre-meeting consultative culture (Lung, 2014, p. 4): All negotiations within the AU tend to take place at the meeting and *not* before, in contrast to, for example, the negotiation culture within the UN or the EU. This characteristic typically extend negotiation processes within the AU (Lung, 2014, p. 4).

Instrument development

The division of labor between the AU and the ARC project team

While the AU Commissioner focused on securing political support, the project team focused on the technical development. In contrast to the development of CCRIF, the rebranded "ARC Project Team" did basically all the technical design work in-house. This does not imply that it did not seek advice. Quite in contrast, to further develop the insurance idea the ARC Project Team formed an informal advisory group. Moreover, informally it reached out to CCRIF several times (Personal Interview, CCRIF Consultant, 2015, May 18). The advisory group consisted of the African Development Bank, the World Bank and the United Nations International Strategy for Disaster Risk Reduction (UNISDR).

It might be noteworthy that the outreach to these 'high-profile' organizations was arguably not only motivated by gaining intellectual input but also by the need to address legitimacy concerns raised by (later) stakeholders such as African countries, donors and private sector entities. An interviewee from the ARC Project Team pointed out that one of the main challenges was the lack of credentials of the ARC Project Team when it presented the idea to AU representatives, donors and collaborators alike: "We did not just come in with ability or brand, like a World Bank" (Personal Interview, ARC Official, 2015, January 9). In other words, it did not enjoy the social recognition and thus social empowerment because it lacked a reputation.

The close and informal collaboration between the AU Commission and WFP finally ended up in a formal technical and administrative service provision agreement ("Agreement on the African Capacity Project Special Arrangement") between the AU Commission and WFP. It was concluded in June 2011, following a decision of the AU Executive Council. The Executive Council decision (EX.CL.Dec/607 XVIII) requested the AU Commission "to put in place the necessary administrative arrangement" for a cooperation to enable the creation of a pan-African contingency funding mechanism to address extreme weather events. The special agreement also established that WFP would *de facto* run such an arrangement—at least for limited time period. Based on that contract, WFP would provide ARC with services in its day-to-day operations. This implied also that ARC staff would be UN staff.

Although there was formally a clear division of labor with the AU Commissioner basically taking over the political lead and the ARC Project Team focusing on advancing and tackling technical issues, the subsequent instrument development process was, in fact, largely shared between the ARC Project Team and the AU policy organs. Throughout the instrument development and implementation process, the AU Organs were involved into technical key decisions rejecting and validating various ideas. Taken together, the whole adoption process from staging the insurance idea over selecting and implementing it with the subsequent 'customization' of insurance policies took about four years. Hence, this process is best captured as a complex negotiation process where different interests were played out and where various technical and political challenges had to be overcome.

Political tensions: rivalry within the AU commission

The close collaboration between the ARC Project Team and the AU Commission to devise ARC overplays, however, the great political challenges faced by the team to realize the insurance instrument. The process from getting a political mandate to the formal constitution of ARC took about two years, from January 2010 until November 2012, when the Establishment Treaty of the ARC was finally signed. It is only then when AU members started to negotiate—that is to customize—insurance policies. ARC went 'online' only in 2014. This long adoption process offers a stark contrast to the comparatively quick development and implementation process in the Caribbean. Complex and long-term negotiations between the ARC Project Team, AU Commission, AU Executive Council, AU member states, donors and WFP characterized the process.

Although the AU Commission and several AU member states had apparently a strong political interest in creating a regional catastrophe risk pool, as evidenced by the resolution-based endorsements, some actors challenged the development and implementation process. Political key discussions centered around where such an insurance pool should sit within the AU and which precise institutional form it should take. Interviewees highlighted that the project had several moments where the project was close to fail "because of politics" (Personal Interview, ARC official, 2015, January 9).

So how did the instrument development process play out? What where the main challenges? One of the major challenges in this process was the power dynamics and the political culture within the African Union and AU member states, as revealed by some interviewees. These challenges were amplified through the crosscutting character and goal of the ARC project.

In contrast to the adoption process in the Caribbean, where the CARICOM's Head of State Meeting was the only regional forum[13] involved in mandating the World Bank Team, the ARC Project Team needed to secure the support of *several* stakeholders of the AU to get the ARC Agency off the ground. They needed the support of the AU Commission, the AU Executive Council, the AU Assembly, AU member states and donors. This was the result of the very institutional structure of the AU and turned out to be challenging for the ARC Project Team right from the beginning.

As elaborated previously, the ARC Project Team approached the AU Commissioner for Rural Economy and Agriculture. Problematically, the idea of creating a pan-African insurance pool was based on a crosscutting vision, linking different thematic fields such as climate change, agriculture, food security, disaster emergency and finance in a novel manner. For all of these thematic fields the AU had (and has) different commissioners in place. While the ARC Team initially opted for the AU Commission for Rural Economy and Agriculture as leading department—basically on the rationale that the department is responsible for food security—other AU Commissioner soon claimed that the administration of a potential pan-African catastrophe insurance would fall in their field of competence and, thus, within their department, as an interviewee revealed.

In other words, as the AU Commissioner for Rural Economy and Agriculture took forward and pushed the initiative within the policy organs of the AU other commissioners—while, in principle, and collectively as the Commission endorsing the idea—were fighting to have the competences of the pan-Africa insurance facility within their department. This rivalry within the Commission is understandable from a theoretical, power-sensitive perspective: Having a regional insurance pool under the aegis of Commissioner A, for example, enhances its scope vis-à-vis the other Commissioners B and C to shape the overall course of the AU. Moreover, supervising an insurance facility signifies the access to and distribution of substantial financial resources. The rivalry within the Commission must, however, also be read in its historical context. During the time period in which the creation of a regional catastrophe pool idea was discussed, the AU itself underwent a reform process that involved the reallocation of competences. The intra-Commission struggle around were the ARC should sit institution-wise was arguably just part of a larger political power play that was already going on for a longer time.

These intra-Commission power struggles brought the ARC initiative close to an end. In January 2011, the AU Executive Council, based on the report of the second ministerial conference on disaster risk reduction, enhanced the political pressure on the AU Commission in the form of a resolution that demanded the AU Commission to accelerate action (EX.CL/DEC.607 XVIII). Also, the Fourth Joint AU Conference of African Ministers of Economy and Finance, Planning and Economic Development and ECA Conference of African Ministers of Finance, Planning and Economic Development in March 2011 exerted some pressure as it called the AU Member States "to support the Commission's efforts through contributions to the design phase of the Risk Capacity project" (AU Resolution L5).

The solution and the final result of these political struggles within the AU Commission was the decision to create a two-tiered institutional structure and establish the ARC Agency as a *Specialized Agency* of the AU. ARC Agency has been set up as being *independent* from the AU Commissioner. It received the immunities and privileges of the AU organizations[14] but its *own* Governing Board with its *own* member states. Yet, it is to note that the Chairperson of the AU Commission secured considerable influence on the actual composition of the Governing Board, which consists of five members and one alternate for each: The Chairperson of the AU Commission has the right to appoint one of the Governing Board members according to its preferences—with the restriction that the member must be an AU national and have experience in insurance-related matters—plus the right to appoint another member in consultation with the Executive Director of WFP with the restriction that the member must be an AU national and have experience in food security-related matters (cf. Establishment Treaty). The AU Commission—or more precisely speaking, the *Chairperson* of the AU Commission—secured for a privileged position.

It is worth emphasizing that this compromise, this solution came from *within* the AU Commission and not from the ARC team, WFP or another actor involved. This indicates, that the design of ARC is throughout *substantially* AU-owned, in particular AU-Commission-owned, although the whole process was initiated and guided by the ARC Project Team and thus ultimately by WFP.

Instrument selection

The ARC establishment treaty: the constitution of ARC in November 2012

At their annual conference in Addis Ababa on 27 March 2012 the African Ministers of Finance and Economic Planning unanimously adopted Resolution 16 and recommended to establish the ARC Agency as a *Specialized Agency* of the AU and requested an expert meeting to fledge out the precise structure. This call was endorsed and thus formally legitimized by the AU Assembly in Addis Ababa, Ethiopia in July 2012. The AU Assembly decided that the "the ARC shall be established as a Specialized Agency of the African Union and accorded the privileges and immunities specified in the OAU General Convention on Privileges and Immunities" (EX.CL/724 XXI) by the end of the year. It is this decision that mandated the implementation of ARC.

The innovative legal solution, namely to create a specialized entity of the AU, signified also a new challenge for the AU Commission and the ARC Project Team: It implied the negotiation of an establishment treaty between all 54 AU member states.

Following the call from the African finance ministers in March 2012, lawyers from the AU and the FAO drafted a treaty, which was then presented and discussed in a "Validation Workshop" formally convened by the AU Commission in Johannesburg, South Africa in May 2012 and a subsequent government expert meeting in September 2012. The draft establishment document entailed the latter two-tiered institutional structure (*ARC Specialized Agency*, the political entity, and *ARC Limited*, the financial affiliate). The two-tiered structure came up since "nobody wants to do business with anyone who has got immunities" as one interviewee put it (Personal Interview, ARC official, 2015, January 15). For the validation workshop the ARC Project Team invited AU lawyers, AU member states representatives, potential donors, development partners, re-insurers, representatives from the CCRIF and the Munich Climate Insurance Initiative (MCII) to comment on the draft. Following several minor changes in the treaty the ARC Project Team finally invited all AU 54 member states to a four-day conference in Johannesburg in November 2012 to establish the ARC. Empowered diplomatic representatives of 41 AU member states[15] attended of which 18 finally also signed the treaty to become founding members of the specialized agency. The treaty was written in a way that it came provisionally into force with the tenth signature, that is, without further ratification, approval or acceptance by another legitimizing entity, such as national parliaments.

The cost-benefit analysis: a (critical) case for ARC

In the meantime, that is in January 2012, the ARC Project Team commissioned a cost-benefit study "to examine the economic advantages and disadvantages of establishing a risk pooling facility as an early response mechanism to severe drought in sub-Saharan Africa" (ARC, 2012a, p. 3). The study was carried out by researchers from Oxford University and the International Food Policy Research Institute and completed in June 2012. The researchers compared ARC to the counterfactual "whereby donors pay what they would have contributed to ARC as annual budget support" (Clarke & Hill, 2012, p. 5). The study made a case for ARC; and yet, it offered also a fairly critical review. It suggested the greatest argument for ARC is not so much the cost-saving achieved through the regional pooling of resources (the main selling point of the Project Team in the beginning) but the faster provision of assistance that results when index-based payouts are explicitly linked to well-targeted social safety net programs. The researchers argued that the speed-benefit of ARC depends crucially on functioning of contingency schemes and aid distribution systems at national level. Indicating that some of the existing schemes might not be fit for purpose, the authors suggest that a smooth operation would require major efforts on the part of ARC member countries. The study suggested, in other words, that the African countries themselves would play the key role in making ARC more valuable than traditional instruments like budget-allocations and the reliance on external donor support and assistance. Economically speaking, the researchers suggested that ARC should keep the costs down by retaining as much risk possible in order to not spend too much on reinsurance and brokerage fees. Moreover, they recommended that ARC should focus on providing insurance only for extreme losses as "[i]nsurance is not the right financial mechanism for managing recurrent losses such as those that occur once every five years or less, on average" (Clarke & Hill, 2012, p. 5). Clarke and Hill suggested for losses below that level, countries should turn to 'traditional' budget reallocations.

The ARC Project Team published the study with a foreword 'to provide some background on ARC'. In this "ARC Response to the Cost-Benefit the African Risk Capacity" the project team acknowledges the more critical comments and promises to build on these study's insights when it comes to the design of ARC and the development of contingency plans (ARC, 2012a, p. 3). The reading of the study by the ARC Project Team is arguably a more optimistic reading than the study would permit. This indicates that ARC wanted to make sure that the study is read 'properly'; it appears the team sought originally further external—in this case epistemic—legitimation to set up ARC. In the end they succeeded: In November 2012 ARC was established. It was only then when the discussion on the precise form and structure of ARC as well as country negotiations on the contingency plans and insurance policies would take hold. The study provided the support to develop state-contingency schemes and, thus, may have aided in-country negotiations.

Instrument implementation

Shaping ARC and developing contingency plans

With the enforcement of the Establishment Agreement in November 2012, ARC formally came into being. The treaty established ARC as independent international organization and clarified the basic design of the ownership and governance structure of the ARC. It constituted the "Conference of the Parties of the ARC" (= ARC Member States) as the supreme authority to adopt rules, gave the Governing Board of the ARC the authority to enforce such rules and established the ARC Secretariat—the former ARC Project Team—as the authority to operate the facility on a day-by-day basis.

Yet, ARC was still in a state of transition and not a functioning organization. The new international organization and its constituting organs still needed to be filled with concrete goals, tasks, procedures and staff. A wide range of crucial issues, elements and puzzles were still not clarified and needed to be solved. These included institutional and financial questions, including the precise goals and functions of the ARC Agency, the composition of and the procedures for the election of the ARC Governing Board— the precise design, governance structure, location and capitalization of the ARC Ltd – but also operational questions such as how to convince re-insurers to provide reinsurance coverage for the facility; how to secure a critical mass of governments stick to their pledges buy the insurance coverage; what standards, practice and compliance mechanisms should guide the development of contingency plans.

Following the treaty, the founding ARC member states came together in two *fora*, called "Conference of the Parties of the African Risk Capacity" (CoP), to deal with these issues under the leadership of the ARC Secretariat. The first CoP was held in Dakar, Senegal in February 2013, the second one in Nairobi, Kenya in November 2013. The first conference in Dakar adopted a range of draft papers involving the precise goals and tasks of the facility (see later on), the Governing Board and ARC Secretariat, established the voting procedure for the Governing Board and elected its members and the criteria to be considered for selecting a suitable location for ARC Ltd (the financial affiliate), clarified the criteria to be considered for selecting a suitable location for ARC Ltd, created a work agenda for the ARC Secretariat and signed a service provision agreement with WFP through December 2014.[16]

Among the number of decisions that were taken at the first CoP, the most important one—next to the election of the Governing Board Member and the appointment of an interim Director General—was arguably the *Provisional Programme of Work and Budget for the ARC Agency*. It represented a sort of mission statement that outlined the task of ARC. The *Provisional Programme of Work and Budget for the ARC Agency* carved out that ARC's key goal would be to "provide participating African countries with predictable, quick-disbursing funds in the case of a drought to implement pre-defined, timely and effective responses" (ARC, 2013a, p. 19). Moreover, it clarified that ARC would be more than a purely financial tool; it would also be an intra-regional knowledge-sharing tool to build up knowledge and capacities in terms of how to adapt to disasters:

> However, ARC is more than a timely risk-financing tool. As a Specialized Agency of the AU it is also a powerful African platform that will promote dialogue at both the national and continental levels on how best to support vulnerable populations at-risk to natural disasters, setting the groundwork for continental standards in planning for early response to food security emergencies through its member-country peer review structure
>
> (ARC, 2013a, p. 19)

With the enforcement of the Establishment Agreement, the ARC Secretariat started to run its operations to achieve the objectives. One of the key tasks assigned to the ARC Secretariat was to lead the negotiations with those ARC member states that expressed an interest in buying an insurance coverage for the first pool.

Beginning in 2013 the ARC Secretariat was entering in parallel negotiations with nine countries to formalize the insurance contract: Burkina Faso, Ethiopia, Kenya, Niger, Rwanda, Senegal, Malawi, Mauritania and Mozambique. After they had expressed their interest in joining the first pool, the countries concluded a so-called "pre-participation Memoranda of Understanding" with the ARC Secretariat. These MoUs clarified the roles and responsibilities of the ARC Secretariat and the countries when they entered into an insurance contract. In 2014, Kenya, Niger, Malawi, Mauritania, Mozambique and Senegal became members of the first insurance pool (Insurance policy year 2014/15). The reasons for this fairly small number can be found in the political and practical challenges that accompanied these negotiations (see later on).

The ARC Secretariat created an ARC Government Coordinator for each country. Together with the in-country counterparts the ARC Government Coordinator had the task to

1 Customize Africa Risk View; that is, to feed the model with the necessary data to produce "a robust model that can accurately capture drought events in a country and, as closely as possible, predict the impact of these drought events" (ARC 2013a p. 20) in financial terms

2 Select the attachment and exhaustion point
3 Develop drought-response contingency plans that would outline how a potential ARC payout will be used

The leading in-country counterpart ("in-country Manager") tended to be either from the ministry of finance or the ministry of agriculture. Interviewees reported that the leading department depended very much on the country. The in-country manager was responsible for overseeing several in-country (sub-)working groups. The in-country (sub-)working groups were composed of experts (in finance, agriculture, vulnerability assessments, food security, disaster management), NGO representatives, government and technical staff from the appropriate entity at the different levels (e.g., high-/mid- and low-level representatives from the finance ministry, the agricultural ministry, the ministry of health and disaster management agencies). Interviewees pointed out that the precise composition and size of these working group members varied from country to country and ranged from only 2 people to about 15–20 people. Some countries such as Kenya, for example, had already working group in place ("Kenya Food Security Steering Group") that had the necessary expertise and whose work was compatible to or even matched the tasks the ARC Secretariat was demanding from the countries in these negotiations. Overall, the ARC Secretariat together with the in-country coordinator sought to engage and scale-up existing in-country structures.

Major challenges

These negotiations turned out to be more difficult than envisioned for a number of reasons. The impediments can be grouped as practical and political challenges. Yet, it is important to note that the challenges subsequently presented here cannot be generalized in a straightforward manner. They refer to in-country experiences and varied from country to country.

LACK OF EXPERTISE

The lack of expertise of in-country staff was one of the main practical problems. Proponents had difficulties in reading excel spreadsheets, doing regression analyses and writing contingency plans. Yet, interviewees emphasized that the lack of expertise was distributed unevenly among the countries. Some countries such as Kenya have a large administrative apparatus with substantial resources and substantial experiences in Disaster Risk Management as reflected in the sophisticated institutional mechanisms to deal with disasters while other countries simply do not have any structures in place. The lack of expertise was rooted in different structural problems. They ranged from the lack of education and experience to understaffing of some governmental agencies. Some countries had never written contingency plans, others had never had ownership of these plans. One interviewee pointed out, that contingency plans are often written by consultancy companies, hired by a country's development partner (Personal Interview, ARC official, 2015, January 19). As consultancy companies do not transmit the knowledge how to write these plans governments have simply little experience in how to draft such documents. Moreover, some departments in countries were entirely run by one person. Other departments were run by people that did not have the needed qualifications.

To address these structural problems the ARC Secretariat offered capacity building measures and held several workshops throughout the negotiation processes. Moreover, it worked more closely with the countries than it originally intended. It sometimes even engaged in one-to-one support. Some interviewees from the ARC pointed out that ARC might have lost a bit focus here and, yet, they were quick to emphasize one of the key tasks of the ARC is precisely this: Provide capacity building and strengthening by transferring skills and knowledge.

ACCESSIBILITY AND AVAILABILITY OF DATA

Another challenge was the political sensitivity of data. As any piece of information, the release of data might lead to advantages and disadvantages of some political groups in terms of their security, the legitimacy of their concerns, goals and political program, among others. Interviewees pointed out that some governments were reluctant in the disclosure of data. Often the release thereof needed to be approved and signed off at the highest political level. Apart from the reluctance to make data accessible, some administrations were reluctant to share their data internally, that is, among their own departments and thereby lose their traditional privileges. The political sensitivity of data added to the lack of data or the quality thereof in some countries. For most countries, the systematic collection of data started only around the millennium. On interviewee summarized the challenges as "Either it [data] is available but not accessible or it's just not available" (Personal interview, ARC official, 2015, January 19). To overcome these challenges the ART team spent substantial time on lobbying to get access to data, disaggregating available, reconstructing older and validating data where it was lost or contradicting.

LACK OF COLLABORATIVE CULTURE

Interviewees highlighted that the formation of cross-sectorial working groups around focal questions, comprising experts and public sector officials from different ministries, presented a challenge. Political departments were and are simply not used to integrative, cross-sectorial collaborations and did not want to engage in such collaborations. In most countries there were no established links between the different political departments. Sometimes the relationships were even characterized by a hostile attitude: A disaster management officer, for instance, did not want to engage with representatives from the finance ministry since he believed "these people think they are better", as one interviewee told (Personal Interview, ARC Official, 2015, January 9). The prevailing, *non-collaborative* political culture extended the negotiation process. It made it more difficult to align different understandings and interests to achieve a common agreement, respectively a decision with financial implications. The way this was overcome was mostly through collective decisions at the cabinet level in the form of cabinet papers. A related challenge to the lack of collaboration and communication was the lack of communicative infrastructure. In some countries there was simply no internet available and/or mobile phones were not working. This implied that ARC staff had to be on the ground or reach the appropriate persons via alternative means and messengers. These conditions did not help to disseminate ideas, get feedback and make a decision.

POLITICAL DISRUPTIONS

Another political challenge was the frequent changes of government. Elections, cabinet shuffles, political upheavals, *coups d'état*, have all affected the negotiation processes as outlined by interviewees. In The Gambia, for example, the Secretary General—equivalent to the Prime Minister—changed four times during the negotiations. Another country manager could not even remember the number of government changes in Mali that happened during the negotiation process. In some cases, political changes brought the whole negotiation process to a breakdown. As one interviewee explained:

> Just imagine you do start a conversation with a certain group of people and suddenly there is an election, government changes, everybody moves, you've a totally different government in place: new faces, new individuals. You have to start the conversation all over. That is a bit of a challenge . . . yeah, it is one of the biggest challenge we face.
>
> (Personal interview, ARC official, 2015, January 26)

The ARC Secretariat anticipated such political disruption and took care to involve mid-level and top-level bureaucrats (such as permanent secretaries), which tend to stay within the new administration. In addition it tried to bridge these gaps of action through cabinet papers: "that gives you the buy-in and is a very critical step", as another interviewee put it (Personal Interview, ARC Official, 2015, January 14). Yet, the same interviewee also clarified that "if someone hesitates and does not want to move forward we don't push it because we can't governments tell what to do". In one case one government wanted to conclude an insurance policy contract, while the successor government did not want to.

FUNDING THE PREMIUM

Another key challenge, arguably *the* key challenge, was the inherent lack of financial resources. ARC member states are hardly able to finance their expenditures via taxes or other revenue-generating means such as the issuance of sovereign bonds. The budgets are partly heavily subsidized through co-funding arrangements by donors, as pointed out by an interviewee. Given the apparent financing problems in the African context, adopting a new budget line was therefore a particular difficult undertaking for ARC members. One interviewee described the situation as this: "You have gone through all the technicalities and there might even be the want to pay but not the ability" (Personal interview, ARC official, 2015, January 9). However, several interviewees also pointed out that the situation and challenge varied from country to country. For some countries, like Kenya, the financing of the premium turned out to be not a big political issue; for others, such as Senegal, it was more difficult, as one interviewee reported.

Adopting a new budget line under financial constraints means usually cutting expenditures, reallocating budget items or finding alternative sources of funding (e.g., private capital markets, loans, external budget support). From a political perspective, cutting and reallocating budget items are not desirable strategies. No ministry wants to lose its distributive power or disappoint the potential beneficiaries of a financial allocation. Taking these considerations into account it is notable that despite the financial constraints *all* countries but Senegal financed the premiums for the first insurance season out of their recurrent budget by reallocating budget items. The premium of Senegal, by contrast, was financed through an external grant from the Japanese government. For the second season, Senegal followed the other ARC insurance policy holders, created a new budget item and paid the premium by itself.

The fact that nearly all countries of the first insurance pool paid their premiums out of their recurrent budget is throughout remarkable given their limited financial room for maneuver and the lack of experience in budgetary planning process in some countries. An interviewee explained this, with reference to political support that payouts buy: "Getting the funds and responding to the affected communities, that's a political plus for their side [the government; N.S.]. Because the community affected, they get response from the government. That definitely buys them a point" (Personal interview, ARC official, 2015, January 26). Another explanation might be that the ARC Secretariat simply knew how to convince the participating countries as it had the right expertise at hand. In that context, in January 2013 ARC hired Ekhosuehi Iyahen from CCRIF on a part-time basis and in January 2014 on a full-time basis as *Lead Client Service Manager for Risk Transfer*. Her task was to deal with the countries in all finance-related matters. With this link, the ARC could draw on firsthand experience from the Caribbean context regarding the conclusion of insurance contracts. As ARC member countries, some CCRIF members struggled to buy insurance coverage due to limited financial room for maneuver.

A second major problem during the negotiation for the first-year pool was that "premium payment times and budget development timelines were not in line" as one interviewee put it (Personal interview, ARC official, 2015, January 26). In other words, when countries decided to sign up for an insurance contract with ARC, the budgets were already approved. This asynchronous problem was overcome through budget reviews, which are usually conducted by the executive. A problem resulting from this time lag was, however, the subsequent time pressure: Given that the ARC Secretariat decided that the insurance policies should come into effect by 1 May 2014 in order to cover the dry Sahel zone, governments had only very little time to find funding to source the premium payments, usually just two to four months. The premiums were not necessarily financed out of the ministry of finance. In some cases, the funding came out of the ministry of agriculture or was pulled together from a range of ministries.

All interviewees from the ARC carefully pointed out that while they have been supportive in facilitating a premium payment (for example, through consultative work where to take the money from or reminding countries of upcoming deadlines), they did not see their task in convincing the members from buying an insurance coverage, emphasizing that some countries stepped back from the insurance program. They emphasized that the decision to enter into an insurance contract was and *must* be a government-owned decision. Countries must be the drivers behind ARC to ensure its longevity ('sustainability') and enable a real political change in term of getting more independency from external actors. Accordingly, all interviewees pointed out that the countries also should pay the premium from their *own* resources, that is, from their own budget because it is the financial allocation which gives them local ownership and allows them to set direction of the ARC. As one interviewee from the ARC explained:

> We don't want to be seen as standard-loan program and we don't want that sort of that they [the countries; N.S.] report on another partner . . . it is very important that governments *own* the premium payment and they say that it is their own.
>
> (Personal interview, ARC official, 2015, January 14)

Another interviewee pointed out:

> You don't want to have an AU-mechanism where the premiums are paid by non-AU members. The idea is for the countries, you know, be able to have their *own* response mechanism, so that they don't have to wait for money from, you know, international partners.
>
> (Personal interview, ARC official, 2015, January 19)

The launch of ARC

While the process of customization was ongoing, the CoP, under the leadership of the ARC Secretariat, transformed ARC into a proper functioning insurance mechanism (cf. ARC, 2013b). The CoP2 gave the ARC Secretariat a mandate to establish the ARC Ltd, the financial affiliate, provisionally in Bermuda and clarified its relation to the parent (ARC Agency). Moreover, it developed standards and guidelines for the peer review mechanism to assess the contingency plans and designed a compliance policy (in other words, it developed criteria and established consequences if an ARC member country would deviate from the pre-approved contingency plans and misuse the payout). Finally, the conference endorsed that the contingency operations plans should be applied in a flexible manner and 'encouraged' the ARC Agency to develop new insurance products for flood and cyclones.

Following a 14-month long customization process, Kenya, Mauritania, Mozambique, Niger and Senegal were granted a "Certificate of Good Standing" by the ARC Governing Board in March 2014. The certificate established that the countries had appropriate contingency plans in place that detail how they would use an insurance payout. The "Certificate of Good Standing" was issued by the ARC Secretariat and allowed its carrier to enter into an insurance contract with the ARC Ltd.

On 14 May 2014 the ARC Ltd was finally launched in Bermuda. The ARC Ltd is a self-insurance vehicle and functions akin to a commercial insurer. ARC Ltd is a letterbox company without physical infrastructure. It is composed of several private service providers, which include *Stroock & Stroock & Lavan LLC*, *Appleby Bermuda*, *Marsh Ltd* as insurance manager (the same the CCRIF uses) and *Willis Group* as reinsurance broker (the same as the CCRIF uses). Simon Young, the former CEO of CCRIF (until April 2012), was appointed to become CEO of ARC Ltd.

ARC Ltd was capitalized with a first tranche of about US$ 100 million and an additional US$ 100 million commitment by the Germany (via KfW) and the UK (through DFID) (ARC, 2015b). Moreover, it collected about US$ 17,5 million in premiums from Kenya, Mauritania, Mozambique, Niger and Senegal. According to the report of the CoP3 conference in Addis Ababa in January 2015, "ARC Ltd's first insurance portfolio attracted great interest in the reinsurance market—the final reinsurance price was below that anticipated, and subscription at that price were three times more than required" (ARC, 2015b p. 20). It secured around US $ 55 million in reinsurance coverage and was able to provide Kenya, Mauritania, Mozambique, Niger and Senegal with a total drought insurance coverage of $135 million. The first insurance policy year began on 1 May 2014 and ended on 30 April 2015.

Instrument modification: the evolutionary phase

The first year of operation (2014/15)

In its first year of operation ARC made three payouts to countries in the Sahel following a poor agricultural season. ARC paid US$ 26,3 million to Mauritania, Niger and Senegal, which had paid combined premiums of about US$ 8 million. Senegal got the largest payout with just over US$ 16,5 million, followed by Mauritania (US$ 6,3 million) and Niger (US$ 3,5 million). The funds arrived ahead of international humanitarian funds and each country used the payouts for different purposes and as outlined in their respective contingency plans (see Table 4.3) According to ARC, the payouts benefited an estimated 1,3 million food insecure people and 500,000 units of livestock (ARC, 2016b, p. 23). The season showed that ARC works—in principle.

Table 4.3 ARC payouts in 2014/15

Country	*Payout*	*Activities*	*Beneficiaries*
Senegal	16,5 Mio	1 Targeted food distribution	700,000 people
		2 Subsidized sales of livestock fodder	90,000 people (855,000 livestock)
Niger	3,5 Mio	1 Cash for work	227,048 people
		2 Targeted food distribution	42,000 people
Mauritania	6,3 Mio	Targeted food distribution	250,000 people
Total	**26,3 Mio**		**1,309,048 people**

Source: adapted from ARC, 2017d, pp. 14–15

PAYOUT TO SENEGAL

Already in July 2014, about six months before the actual drought materialized, Africa Risk View predicted that the registered shortfall in rain and the uneven distribution thereof across the country would lead to a considerable decline in crop yields in many parts of the country by the end of season. It estimated that crops will not be able to complete their full life cycle of 90 days by mid-October 2014 leading to lower amounts of harvest (ARC, 2014d, p.4). In September, the software predicted that the number of people affected by the decline in the overall agro-pastoral production—the agricultural and fodder production—would be substantial in some areas by the end of post-harvest season so that related disaster response costs for Senegal to prevent a broader food shortage would breach the threshold as fixed in the insurance contract. In other words, the model established that Senegal would get at least the minimum payout. After verification of the decline in agro-pastoral production in September and further needs assessment in October 2014 through Senegal's early warning platform, instituted by the Executive Secretariat of the CNSA (SE/CNSA) and compromising cross-sectorial stakeholders such as the National Food Security Council (CNSA) or the Civil Aviation and Meteorological Agency, Senegal prepared for early intervention measures to prevent a greater food shortage in March 2015 (ARC, 2014d, p. 11). An interministerial task force selected the support measures from the draft contingency plan submitted to the ARC Agency, clarified roles and responsibilities as well as the timeline for procurement and delivery. In November 2014, the task force submitted the Final Implementation Plan to the ARC Secretariat. In January 2015, ARC'S Governing Board approved the Final Implementation Plan, providing ARC Ltd with the notice to trigger a payout.

Senegal decided on unconditional food distribution and the subsidized sale of animal food at the national level for the time period between March and August 2015. The decision to provide the aid on a national scale followed from the assessment that 37 out of 45 administrative departments would be affected by a shortage on food. The FIP established that the Executive Secretariat of the CNSA (SE/CNSA), an office under the Prime Minister, would coordinate the intervention and target beneficiaries while the buying and delivery of food would be realized by the Food Security Commission (CSA). The intervention plan foresaw that each targeted person would be provided monthly with 15 kilograms of rice throughout the period by the respective regional administrative departments (ARC, 2014d, p. 11). In January, 2015 Senegal's Ministry of the Economy, Finance and Planning received a payout of US$ 16,5 million from ARC Ltd. ARC claims that around 790,000 people and 855,000 cattle benefited through subsidized sales of animal fodder (ARC, 2015a, p. 2).

PAYOUT TO MAURITANIA AND NIGER

Similar to Senegal, Mauritania and Niger also started to develop early intervention measures once trigger threshold was breached. Their preparations were, however, considerably slower and primarily based on the estimates provided by the software. Only in October 2014 did Mauritania's specialized interministerial working group GTS, compromising the national Food Security Commission (CSA), the Ministry of Agriculture and the Ministry of Livestock but also external actors (WFP, FAO, FewsNet), confirm the massive rainfall deficit and the lower or nonexistent prospects for harvest in the southern areas of the country. Following the report, the GTS engaged in a preliminary qualitative needs assessment which was then substantiated with quantitative data in a survey conducted by CSA and WFP in December 2014 (ARC, 2014c, p. 4). The government of Mauritania finally decided to opt for food distribution to the families most in need in the eight drought-affected regions with weak agricultural production. The measures consisted of providing households with an average of five members with a monthly ratio of 50 kilograms of wheat and four liters of oil for four months (ARC, 2014c, p. 8).

Niger, in turn, had no early warning platform in operation and relied entirely on Africa Risk View to establish the impact of rainfall shortfalls. No verification was done on the ground. Following the prediction of ARV, the "[m]ain structures of humanitarian aid as well as representatives of technical ministries met in October to discuss developments in agricultural season and developed a preliminary support plan" (ARC & DNPGCCA, 2014, p. 4).

Several "field missions assigned by the partners and other state services" (ibid., p. 4) were created to assess the situation on the ground. Based on these field trips and ARV data, the cabinet decided to use conditional cash transfers in drought-affected regions to targeted households for two months (CFA 32,500/~$ 53 per month). In addition, it decided to support (state-controlled) school canteens with the purchase of food on local markets over five months (ibid., pp. 8–14).

In February 2015 the situation in the Sahel had, indeed, turned out to be catastrophic so that the UN launched an international appeal to fund humanitarian support. At this stage ARC had already made payouts to two of the three most-affected countries of the Sahel. In January 2015, Senegal launched the planned social interventionist measures. In Mauritania, the distribution of aid to beneficiaries started in April 2015 and was completed shortly after

Ramadan. The cash transfer program in Niger, in turn, started in June 2015 (ARC, 2015a, p. 2).

EFFECTIVENESS OF PAYOUTS

To what extent ARC really contributed to a timely and effective response in practice is not entirely clear from outset. Limited information available forecloses a definitive assessment. From what is publicly available it appears that ARC works as intended: It enabled to speed up the mobilization of funds and distribution of aid.

As a follow-up on the payouts, ARC contracted independent organizations to conduct post-payout evaluations. The process audit on the first round of payouts is, however, not publicly available. Instead, ARC published a summary report that is available on its website. The summary report of a first processual evaluation demonstrated that ARC's mechanisms worked in principle. The countries received funds well ahead of international donor funds and the interventions carried out by the countries "largely matched the interventions outlined in the FIP" (ARC, 2017d). Yet, the report also noted some room for improvement and pointed to some political and procedural barriers that hindered an effective implementation. For instance, at the beginning some countries did want to declare an emergency as in their recent history this has been associated with a government failure to act. This non-declaration prohibited governments to use their fast-track procurement processes (ARC, 2017d), leading to some delays. The report also noted a lack of robust coordination processes and transparency on the technical side at the country level. Some countries relied on more informal communication structures when designing the Final Implementation Plan, preventing scrutiny (ARC, 2017d).

A recent and more comprehensive evaluation report from a consultancy (Oxford Policy Management, OPM), which had access to the full audit, also highlighted difficulties in disbursing the funds on the national level for two (Niger and Senegal) of the three countries. According to its report, Niger received its payout in February 2015 (OPM, 2017, pp. 35–36). Yet, due to a change in national policy, the payout was firstly transferred to the National Treasury where it remained until the Prime Minister intervened and it was forwarded to Niger's National Mechanism for the Prevention and Management of Disasters and Food Crises. Also, in Senegal ARC funds were retained within the Senegalese National Treasury rather than being immediately transferred to government departments responsible for implementation. The late allocation of funds led to substantial delays in program implementation. In sum, although ARC made timely payout it is not clear if for domestic reasons they were as effective.

Unfortunately, there exists no documentation on how the aid distribution worked out in practice. No documentation is available if the aid reached those it should reach. In other words it is not clear if the payouts were ultimately effective in mitigating food insecurity. According to Oxford Policy Management, "[s]econdary data on payouts in Mauritania and Senegal suggests that payouts did reduce sale of assets for some households and protect their livelihoods" (OPM, 2017, p. vii).

Key developments 2014–2018

DEVELOPMENT OF PAYOUTS

The following years ARC made another two payouts so that between 2014 and (December) 2018 ARC has made five payouts (see Table 4.4) to four countries in total. The payouts summarizes thereby to US$ 36,8 million.

In 2017, Malawi—after a major issue between ARC and Malawi (see later on) received a (belated) payout of US$ 8,1 million. According to ARC, Malawi used the payout to support about 800,000 Malawians by scaling-up a cash transfer program with humanitarian partners and replenishing strategic grain reserves. In February 2018, Mauritania received a payout of around US$ 2,4 million, which it used to subsidize livestock feed for pastoralist. It procured 11,000 metric tonnes of rice from local suppliers to support around 250,000 affected people (ARC, 2017a). According to ARC, Mauritania received the funds well ahead of other international humanitarian assistance. In sum, ARC claims that by now around 2,1 million food-insecure people and 900,000 livestock have benefited from its early payouts (ARC, 2017a).

Table 4.4 Payouts made by ARC until December 2018

Policy year	*Country*	*Payout*	*Hazard*
2014/15	Senegal	US$ 16,500,000	Drought
	Mauritania	US$ 6,300,000	Drought
	Niger	US$ 3,500,000	Drought
2015/16	Malawi	US$ 8,100,000	Drought
2017/18	Mauritania	US$ 2,400,000	Drought
Total		**US$ 36,800,000**	

Source: data taken from ARC, 2019b

MALAWI—FLAW IN THE MODEL?

A contentious issue was the belated payout to Malawi. In 2015 Malawi bought insurance for the 2015/16 season and did not get an immediate payout from ARC despite the fact that it was affected by a severe drought. ARC's payout structure is based on the output of Africa Risk View software. As highlighted previously, Africa Risk View is based on a number of assumptions. It estimates crop losses and translates these losses into the costs of an assistance program. Africa Risk View is customized for and by each country. While the software predicated a drought, its estimation was that only 20,594 people would be affected—a far lower number compared to what would turned out to be as the actual situation on the ground. Given the low number, the threshold for a payout was not reached and no payout was made. The drought on the ground was, however, quite severe. In April 2016 Malawi declared a national emergency and with support from WFP estimated that about 6,5 million people in 24 of its 28 districts were in need of food assistance. The government estimated that its Food Insecurity Response plan cost would cost around US$ 395 million. As Malawi complained about the massive mismatch and contested the usefulness insurance policy, ARC initiated a technical review process (involving field missions and meetings with the Malawi technical working group, and household surveys) and employed a consultant from the Centre for Agricultural Research and Development to identify the causes why the model had failed to predict the drought.

ARC found out that farmer had been growing "a different variety of maize in significant and increasing amounts in the recent past than initially selected as the reference crop in Malawi's customisation of *Africa Risk View*" (ARC, 2017f). The model took a maize variety with a longer maturation period (120–140 days) than the average maturation period for the varieties actually planted by farmers (90–100 day). Using "incorrect critical information in the model resulted in inaccurate output from *Africa Risk View*"(ARC, 2017f). Based on these findings, ARC revised the input of the model and rerun it. The results then matched the situation on the ground. As a consequence, ARC decided to pay out based on the magnitude and severity of the drought event and the amount of insurance coverage purchased by the government. ARC claims that the Africa Risk View model performed as it was designed to but that it was informed by inaccurate data to correctly capture the situation on the ground. At the CoP5 in March 2017, the ARC Secretariat announced it would invest more research and development to improve the model and verify its validity (ARC, 2017e). It has started, among others, to work with the European Commission's Joint Research Council to review Africa Risk View and the country engagement process (ARC, 2017e).

The political fallout of the incident was significant and created substantial negative press (see for instance The Economist, 2016). Both the lack of an immediate payout but also ARC's reaction, especially its decision to make a belated payout, was subject to heated discussions. The UK-based NGO ActionAid investigated the case in-depth and, in May 2017, issued a highly critical report on the Malawi case, blaming not only the model but ARC's customization process. Regarding the eventual payment, ActionAid concluded that it was "too little and too late" and politically motived. ActionAid used the Malawi incident to make a case against marked-based insurance products urging ARC, the G7, the World Bank and others to halt their work of expanding the market for climate risk insurances (ActionAid, 2017). In a press statement, ARC argued against the allegations of ActionAid and underscored that its insurance contracts

> do not include a negotiation component with respect to payouts as the policies are based on an objective index that captures the severity of an event . . . as is the case with all ARC payouts, the payout to Malawi was not the result of negotiations, nor was it done on an ex gratia basis.
>
> (ARC, 2017f)

The delay in the eventual payout led to frustrations and made Malawi hugely skeptical of ARC. It has not bought further coverage since.

EXPANDING THE PRODUCT LINE

From the beginning, ARC has been keen on expanding its product line. In January 2015, ARC announced that the company would develop an insurance product for floods, cyclones and for epidemics and outbreaks such as Ebola (ARC, 2015b, p. 4). While the flood insurance product is still under development, ARC has made some progress with developing a product for tropical cyclones as well as for epidemics and outbreaks.

In 2016/17 ARC launched a *tropical cyclone forecast and early warning tool* with regard to tropical cyclone activity in the South-West Indian Ocean. Based on 70 years of historical cyclone records and a simulated track set for 1,500 years containing almost 10,000 cyclones, ARC developed a tropical cyclone model that covers the hazards of wind, storm surge and wave damage and converts these hazard maps into US$ damage estimates per storm and per country. The model generates modeled loss calculations for these events—and for events that are being forecasted—creating information for early warning, disaster preparedness and planning purposes. It is informed by CCRIF's model (OPM, 2017, p. 22) and currently being tested. Among ARC Member States, Madagascar, Mauritius, the Comoros and Mozambique are exposed to tropical cyclones. So far, ARC uses the model to issue real-time warnings on tropical cyclones on its website. ARC is still in the process of integrating the model into Africa Risk View, the software that underpins the insurance contracts.

In April 2018, after about three years of modeling, ARC announced to pilot a parametric insurance product to allow for rapid payouts to finance public health programs. ARC's *Outbreak and Epidemic Pilot Programme* will be piloted in Uganda and Guinea and covers Ebola, Marburg, Meningitis and Lassa Fever. Analog to the drought insurance, ARC will work with countries to determine their epidemic risks and select indices to be covered, improve early warning systems, create contingency plans for rapid response and provide access to swift disbursement of financing through parametric insurance. The product is financed by the Rockefeller Foundation and realized with support from the Swiss Development Cooperation, Metabiota Inc, Health Systems Consult Ltd and public health economists affiliated with Columbia University are collaborating with ARC on the development of its pilot product (ARC, 2018b).

The demand for these products was country-driven, as interviewees from the ARC Secretariat emphasized. In fact, the interviewees revealed that they have a rather critical stance towards the development of new products and emphasized the amount of research that is needed and the complexity therein to develop models and products; interviewees highlighted that they view refining existing models as more important.

REPLICA COVERAGE

Since 2017/18 ARC offers its insurance policies also to humanitarian organizations and NGOs such as WFP or Oxfam. Since the share of coverage bought by countries ranges, according to ARC, from less than 10% to 30%, with the remainder largely remaining with the UN appeals process, ARC launched a "replica coverage" (ARC, 2017c). The replica coverage may be understood as a top-up to the "normal" ARC insurance. It seeks to scale-up interventions as detailed in the contingency plan. While an international organization or NGO plan must not go through the full technical process in-country, the plans for use of funds flowing from replica coverage must still be approved by the government. For the 2018/19 insurance policy year, the START Network, a UK-based Consortium of NGOs, bought replica coverage for Senegal. The policy replicated an equivalent one signed by the Government of Senegal, and allows the network members "to roll out operations alongside and in complementarity with the Government" (START Network, 2019).

ARC EXTREME CLIMATE FACILITY

In addition to developing new insurance products, ARC has started to work on developing a new pan-African investment vehicle to finance climate adaption measures. In September 2014 ARC announced to develop the *ARC Extreme Climate Facility* (XFC). The goal of the ARC XFC would be to address "key market failures" by providing the ARC members "with direct access to finance for adaption in the event that the frequency and magnitude of severe events increases" (ARC, 2015b, p. 25). To that end, ARC wants to develop a multi-hazard Extreme Climate Index for each African climatic region. The index will track changes in the frequency and magnitude of extreme climate events over and above a 30-year baseline climatology (ARC, 2016c). When the index exceeds predetermined thresholds, countries will automatically receive payments from the XCF to support their pre-approved climate adaptation plan. To fund these measures, the XCF plans to issue catastrophe bonds of an targeted amount of roughly US$ 1 billion. "At bond maturity, if the XCF index does not trigger payments during a financing window then the capital provided would be returned to investors in addition to the yield collected" (ARC, 2014a, p. 4). The basic idea behind developing a third organization (next to the ARC Agency and ARC Ltd) is to leverage funds from the private sector for climate adaption measures and increase the amount of international funding available for climate adaptation in Africa. Only in 2017 the UN reported a massive financing gap with regard to climate change adaptation (UNEP, 2017). In May 2018, ARC XFC was still in its research and development phase.

NEW COOPERATIONS

Analog to CCRIF, the ARC Agency has also started to build up strategic partnerships to improve its products and services. It has signed or is seeking to sign MoUs with the African Development Bank (AfDB) for premium support; weather and food security information services (e.g., AgriMET) to improve the ARV model; regional trade groups such as West Africa's Inter-African Conference on Insurance Markets (CIMA), Southern African Development Community (SADC) and East Africa's Intergovernmental Authority on Development (IGAD), to help promote ARC. Yet, "the scope and reach of such agreements is not yet fully defined" (OPM, 2017, p. 19).

Policy renewal

While ARC started well, more recently it seems to be at a critical juncture. Between 2014 and 2018 eight different African countries have bought insurance coverage. The annual number of countries purchasing insurance policies is, however, shrinking (see Table 4.5). In this context it is also notable that the most food insecure countries are not even engaged with ARC (OPM, 2017, p. 29). As of 2018, ARC's original plan to provide insurance coverage for up to 24 countries by 2020 seems to be out of reach.

The shrinking number of countries taking out insurance does not necessarily imply a decreasing legitimation of the instrument, but it could. Various reasons are thinkable as to why countries decided to not buy insurance. They include all the challenges that the ARC Secretariat managed to overcome during the implementation phase (e.g., political disruption, staff turnover, lack of funding for premium financing, inadequate data) and are most likely to vary from country to country. While uncovering the reasons is beyond the scope of this study, a recent study by OPM gives, a clear hint. According to a perception survey, country-level officers in Kenya, Malawi and Mauritania find the premiums too high in comparison to other available options (donor support and CAT DDO). Countries appear to struggle with the premium payments. According to OPM, in some cases countries missed deadlines to make their premium payments. Also, in at least one case, a country signed a policy but did not make the premium payments in full (OPM, 2017, p. 29). This implies that in the long run the ARC team failed to build up momentum. There might also be a limited understanding of what insurance is. In its evaluation OPM reports that "[i]n some

Table 4.5 Insured ARC countries by season

Country	*Insurance season*			
	2014/15	*2015/16*	*2016/17*	*2017/18*
Burkina Faso			✔	✔
Kenya	✔	✔		
Malawi		✔		
Mali		✔	✔	✔
Mauritania	✔	✔	✔	✔
Niger	✔	✔	✔	
Senegal	✔	✔	✔	✔
The Gambia		✔	✔	✔
Number of insured countries	4	7	6	5

Source: ARC, 2017a,b, 2019

countries, insurance premiums seemed to be viewed more as a savings account, with an expectation that you should get out at least what you put in" (OPM, 2017, p. 30).

An exception seems to be West Africa. It is also notable that ARC seems to be more attractive in West Africa, where two countries (Senegal and Mauritania) have bought coverage for four consecutive years. This might have to do with a more pressing need, a more sustained engagement of the ARC team or with a 'demonstration effect'. Overall, three of the five payouts were made to West Africa. This might have created a more favorable stance and helped ARC to secure the political momentum.

By contrast, in East Africa the belated payout to Malawi seems to have induced a more unfavorable stance where both Malawi and Kenya decided to not further buy insurance coverage. Both countries seem to have lost trust in Africa Risk View as the OPM report highlights:

> Many stakeholders and political-decision makers in both Malawi and Kenya continue to question whether ARV can be adapted accurately and extensively enough to their unique country contexts. Specific concerns include the accuracy and resolution of satellite rainfall data, use of the Water Satisfaction Requirement Index (WRSI), accuracy in predicting vulnerability, lack of regional sensitivity to show localised droughts, inability to capture dry spells that occur during critical times within the growing season, and inability to capture evapotranspiration, water-logging or vegetation cover.
>
> (OPM, 2017, p. 21)

Challenges and doubts seem to be growing. So far, ARC has shown a high level of responsiveness to country demands and addressing challenges. It has started to research and develop a range of new products (e.g., flood insurance, tropical cyclone insurance, epidemic and outbreaks insurance, ARC Extreme XCF) and concluded a number of MoUs to better serve its member countries, including one with the African Development Bank to facilitate premium sponsorship. This responsiveness to country needs conflicts increasingly with complexity. What ARC is and what it might become seem to be increasingly unclear. ARC continues to evolve.

Part III: summary

The African Risk Capacity is a **pan-African public-private self-insurance entity** that provides **six African countries** (insurance policy year 2016/17) with **insurance coverage against the financial costs of a drought-related social interventionist program** aimed at the provision and distribution of food. It does so in exchange for an annual premium and the submission of contingency plans that detail how the payout will be used. ARC members are comparatively free in how they design their contingency plans, as long as they develop some plans that outline a drought-related emergency response. ARC makes payouts *before* a drought has even occurred. It does so via a newly developed software labeled Africa Risk View that estimates the occurrence of a drought, the number of people in need of assistance and the financial costs of a public interventionist program to assist those in need. The software uses real-time hazard data (rainfall/crop monitoring data) and customized exposure and response cost data (target population data/cost of response program). Insurance payouts are capped at certain level (US$ 30 million per season) and are tied to the submission of contingency plans.

The generic goal of ARC is **to prevent a drought-related famine**, by ensuring food security—that is, the access to food—**of the most vulnerable segments of a population** through a timely provision funds and aid. With this early provision and distributions of aid, ARC seeks **to prevent the most vulnerable segments of society from engaging in typical survival and coping activities** such as the down-shifting of food habits or the selling of assets and thus **secure their achieved socioeconomic status**. Through an early intervention ARC hopes to **enhance the effectiveness and leverage of (inter)national aid and assistance.**

ARC is an organizational hybrid made up a **political entity** (ARC Agency) and a **financial entity** (ARC Ltd). While the inter-governmental ARC Agency provides general oversight and strategic guidance, the ARC Ltd is the insurance vehicle that carries out the parent's insurance functions on a day-to-day basis. To be eligible for the conclusion of an insurance contract a country must first of all become a member of the ARC Agency. A *specialized entity* of the AU, membership is in principle open to all AU members. ARC members must then develop and submit contingency plans to the ARC Agency. The contingency plans must detail how countries will use insurance payouts. They must be *targeted at the most vulnerable* segments of society and will be reviewed by the ARC Agency for approval. Once approved, the ARC Agency offers its members insurance coverage and the possibility to tailor the insurance contract according to financial needs and possibilities.

Between 2014 and 2018 ARC has made **five payouts** to four different governments (Senegal, Niger, Mauritania and Malawi) totaling to about **US$ 36,8 million** to finance early intervention measures and prevent famine. ARC claims that at least 2.1 million people and 900,000 livestock could be assisted with the payouts. Apart from one notable exception (Malawi in 2016/17), countries received the payout *before* an official international appeal was even launched. Despite some more recent problems, it appears that **ARC works largely as intended**.

Historically seen, **ARC developed a series of *experimentations*, a *scientific article, financial sponsorship*, the *play with ideas, policy learning* and a *participatory design process*.** The idea to develop a pan-African insurance scheme was carried by a **small WFP-affiliated project team**. Endowed by the Rockefeller Foundation, it was keen on realizing ARC and thus may be interpreted as ***instrument constituency***. Since the 1990s, various ideas and projects had been discussed and tried out within the UN system to restructure the financing of humanitarian aid. WFP and the World Bank, for instance, tested both the viability of index-based drought insurance in Africa, yet, with rather mixed results. Trigger for the development of ARC was, however, a scientific article that proposed restructuring the funding mechanisms of drought-related aid in Africa and the financial sponsorship of the Rockefeller Foundation. The foundation was keen on financing something 'off-the-shelves' and sought a collaboration with WFP. Inspired by the scientific article it endowed a small team hosted by WFP to develop the idea of the article further and come up with some projects. The team developed the software Africa Risk View. After it had developed the software, the team searched for an application and played around with it. It reached out to CCRIF, the newly established public-private regional insurance facility in the Caribbean, to obtain knowledge about its workings and learn lessons. The team decided to transfer the model to the African continent and replicate the political genesis of CCRIF. The team approached the AU to get a political mandate for the development a pan-African regional insurance scheme.

In contrast to CCRIF, the **WFP-affiliated project team was not 'simply' mandated by the AU** to develop a regional insurance arrangement, as the African Union insisted on local ownership. It demanded to be involved in the development of the idea and thus in technical key decisions. As a consequence, **ARC was developed in a participatory manner**. While the WFP-affiliated project team took over the technical lead, the AU Commission secured political approval by various AU policy organs. The **leading involvement of the AU Commission** and other AU policy organs in technical key decisions and subsequent struggles over where ARC should sit institutions-wise substantially extended the instrument design and development process. And, yet, it **provided the instrument with substantial political input legitimacy**. The result of a complex two-year negotiation process was a two-tiered institutional structure. The ARC Agency saw the light as a specialized agency of the AU. Only when the organization was set up, precise goals of the ARC (provide insurance and capacity building) were specified. The newly established ARC Secretariat, run by WFP through a service provision agreement, provided support in customizing the insurance policies and helped the ARC member countries to secure financial donor support to capitalize ARC.

Notes

1 I conceive 'food security' here in the way the term has been defined by the 1996 World Food Summit, held in Rome: "Food security exists when all people, at all times, have physical and economic access to sufficient, safe and nutritious food that meets their dietary needs and food preferences for an active and healthy life" (FAO, 1996).

2 All figures refer to the insurance policy year 2016/17, if not stated otherwise.
3 Eighteen member states signed the Establishment Agreement, creating the ARC Agency as a Specialized Agency of the African Union. Fifteen additional countries signed the treaty at a later stage.
4 A mutual insurance company is an insurance company that is fully owned by its policyholders.
5 ARC member states are: Burkina Faso, Burundi, Central African Republic, Chad, Comoros, Republic of Congo, Côte d'Ivoire, Djibouti, Gabon, The Gambia, Ghana, Guinea, Guinea Bissau Kenya, Liberia, Libya, Madagascar, Malawi, Mauritania, Mozambique, Niger, Rwanda, Sahrawi Arab Democratic Republic, Sao Tome and Principe, Senegal, Sierra Leone, Sudan, Togo, Zambia, Zimbabwe.
6 In the first year all premiums but Senegal's were paid by the countries themselves and not by donors.
7 I ordered the countries chronologically, according to the date they signed the MoU (cf ARC, 2014g).
8 UK's DFID provided about GDP 90 million and Germany's KfW EUR 50 million. The sum is equivalent to about US$ 200 million (in 2014 average exchange rates). About US$ 100 million have already been paid out. For further information, see www.africanriskcapacity.org/arcltd/ and ARC's financial audit (ARC, 2017b).
9 Stakeholders included the Malawi Department of Climate Change and Meteorological Services, the National Smallholder Farmers' Association of Malawi (NASFAM), the Malawi Rural Finance Company (MRFC), the Opportunity International Banking Malawi (OIBM), Malawi Meteorological Service, the International Crop Research Institute for the Semiarid Tropics, (ICRISAT) and the International Research Institute for Climate and Society (IRI) in New York.
10 The early history of ARC is heavily based on one key informant. Further cross-validation through additional interviews with donors, WFP and internal documents remains a desideratum to substantiate claims.
11 For a list and the amounts of grants, see the website of the Rockefeller Foundation (Rockefeller Foundation, 2015b).
12 "*Trust* tells founders whom they can count on in difficult situations, and it substantially enhances predictability. *Predictability* refers to how the other party will behave if situations change. And, finally, using *voice* in a relation means the persons involved will make their complaints known and negotiate over them, rather than sneaking away" (Aldrich & Ruef, 2006).
13 In the Caribbean, the design team received its mandate from a sole forum, notably CARICOM's Head of State meeting.
14 The privileges and immunities specified in the OAU General Convention of Privileges and Immunities.
15 The following countries participated in the conference: Benin, Botswana, Burkina Faso, Burundi, Cameroon, Central African Republic, Chad, Comoros, Congo, Côte d'Ivoire, Djibouti, Egypt, Ethiopia, Gabon, The Gambia, Guinea, Kenya, Lesotho, Liberia, Libya, Malawi, Mali, Mauritania, Mozambique, Namibia, Niger, Rwanda, Sahrawi Arab Democratic Republic, Senegal, Seychelles, Sierra Leone, South Africa, South Sudan, Sudan, Swaziland, Tanzania, Togo, Tunisia, Uganda, Zambia, Zimbabwe.
16 The agreement was subsequently extended to 31 August 2015 (ARC, 2015, p. 18).

References

ActionAid. (2017). *The wrong model for resilience: How G7-backed drought insurance failed Malawi, and what we must learn from it.* Retrieved September 19, 2019, from https://actionaid.org/publications/2017/wrong-model-resilience

Aldrich, H., & Ruef, M. (2006). *Organizations evolving*. London: Sage.

Anheier, H. K. (1999). *Foundations in Europe: A comparative perspective* (Civil Society Working Paper No. 18). London. Retrieved from https://core.ac.uk/download/pdf/94098.pdf

ARC. (2012a). *ARC response to the cost—Benefit analysis of the African risk capacity*. Retrieved September 19, 2019, from www.africanriskcapacity.org/wp-content/uploads/2016/11/ARC_CBA_and_Response.pdf

ARC. (2012b). *Report of the ACR contingency planning peer review meeting September 2012*. Johannesburg. Retrieved from January 15, 2015, from www.africanriskcapacity.org/c/-document_library/get_file?uuid=a7174e64-f92a-4702-b41d-ff6209e3376c&groupId=350251

ARC. (2013a). *Report and decisions of the first meeting of the conference of the parties of the African Risk Capacity (ARC) Agency*. Retrieved from https://www.africanriskcapacity.org/wp-content/uploads/2016/11/Conference-of-Parties-Final-Report-EN.pdf.

ARC. (2013b). *Report and decisions of the second meeting of the conference of the parties of the African Risk Capacity (ARC) Agency*. Retrieved from https://www.africanriskcapacity.org/wp-content/uploads/2017/02/COP2_Report_EN_20140526.pdf.

ARC. (2014a). ARC Ltd will pay out $25 million to countries in the Sahel. *ARC Focus*. Johannesburg. Retrieved January 15, 2015, from www.africanriskcapacity.org/newsletters;jsessionid=4C0D6FAC4BA819B13478D8FC04487579

ARC. (2014b). *ARV methodology*. Retrieved February 22, 2016, from www.africanriskcapacity.org/de/africa-risk-view/methodology

ARC. (2014c). *Final implementation plan Mauritania November 2014*. Retrieved December 3, 2014, from www.africanriskcapacity.org/documents/350251/840546/MR_Formatted+FIP_EN_20140128_v02_CR.pdf

ARC. (2014d). *Final implementation plan Senegal November 2014*. Johannesburg. Retrieved November 25, 2019, from https://www.africanriskcapacity.org/wp-content/uploads/2018/10/SE_Formatted-FIP_EN_20150128_v01_CR.pdf

ARC. (2014e). *How ARC works*. Retrieved November 3, 2014, from www.africanriskcapacity.org/about/how-arc-works

ARC. (2014f). *Memorandum of understanding*. Retrieved November 3, 2014, from www.africanriskcapacity.org/countries/mou-countries

ARC. (2014g). *Sovereign disaster risk solutions—A project of the African Union*. Retrieved November 3, 2014, from www.africanriskcapacity.org/de/documents/general

ARC. (2015a). *African risk capacity's 2015 payouts: At least 1.3 million people set to benefit*. Johannesburg. Retrieved September 29, 2019, from https://unfccc.int/sites/default/files/alaskary_additional_resourses_payouts_and_successes.pdf

ARC. (2015b). *Report of the third meeting of the conference of the parties of the African Risk Capacity (ARC) agency*. Retrieved May 13, 2016, from www.africanriskcapacity.org/documents/350251/847459/CoP3_+FinalReport+_EN_20150216.pdf

ARC. (2016a). *About us*. Retrieved May 13, 2016, from www.africanriskcapacity.org/about/vision-and-mission;jsessionid=45BC93BFD74FE6093375F1675E85DC87

ARC. (2016b). *African risk capacity strategic framework 2016–2020*. Retrieved May 13, 2016, from www.africanriskcapacity.org/wp-content/uploads/2017/01/PI_Strategic-Framework-2016-2020_20161207_EN_TA.pdf

ARC. (2016c). *Factsheet: Extreme Climate Facility (XCF)*. Johannesburg. Retrieved May 13, 2016, from www.africanriskcapacity.org/wp-content/uploads/2016/11/XCF-flyer-EN.pdf

ARC. (2017a). *African risk capacity funds assisted at least 2.1 million people affected by drought*. Johannesburg, South Africa. Retrieved November 11, 2017, from www.africanriskcapacity.org/arcdocuments/

ARC. (2017b). *African risk capacity insurance company limited: Audited financial statements.* Retrieved June 8, 2018, from www.africanriskcapacity.org/wp-content/uploads/2018/03/ARC-Q3-2017-Audited-IFRS-Financial-Statements-Electronic-Release.pdf

ARC. (2017c). *Fact sheet replica coverage.* Johannesburg. Retrieved June 8, 2018, from www.africanriskcapacity.org/wp-content/uploads/2017/07/Replica-Coverage-flyer-EN.pdf

ARC. (2017d). *Lesson learned summary report: 2014/15 ARC payouts Senegal, Niger, Mauritania.* Johannesburg. Retrieved June 8, 2018, from www.africanriskcapacity.org/wp-content/uploads/2017/10/ARC-2015-Payout-Lessons-Learned-Summary-Report.pdf

ARC. (2017e). *Report of the fifth session of the conference of the parties of the African Risk Capacity (ARC) agency.* Johannesburg. Retrieved June 8, 2018, from www.africanriskcapacity.org/wp-content/uploads/2017/05/CoP5_Report-of-the-Fifth-Session-of-ARC-Conference-of-the-Parties_EN_20170411.pdf

ARC. (2017f). *Responses to actionaid's flawed claims.* Retrieved June 2, 2018, from www.africanriskcapacity.org/2017/07/10/african-risk-capacity-response-to-actionaids-flawed-claims/

ARC. (2018a). *African risk capacity: Transforming disaster risk management & financing in Africa.* Retrieved August 15, 2019, from www.africanriskcapacity.org/

ARC. (2018b). *Press release: Ebola, Meningitis, Marburg and Lassa fever to be piloted in African risk capacity outbreaks & epidemics insurance programme.* Johannesburg. Retrieved June 8, 2018, from www.africanriskcapacity.org/2018/04/09/press-release-ebola-meningitis-marburg-and-lassa-fever-to-be-piloted-in-african-risk-capacity-outbreaks-epidemics-insurance-programme/

ARC. (2019a). *ARC food security & risk management.* Retrieved August 22, 2019, from www.africanriskcapacity.org/food-security-risk-management/

ARC. (2019b). *ARC impact—Since 2014 and over four risk pools, ARC Ltd has paid out over US $36 million to countries affected by drought.* Retrieved August 22, 2019, from www.africanriskcapacity.org/impact/

ARC & DNPGCCA. (2014). *Republic of Niger final implementation plan, November 2014.* African Risk Capacity and Republic of Niger Office of the Prime Minister National Mechanism for Disaster and Food Crisis Prevention (DNPGCCA).

Assa, J. (2012). Financialization and its consequences: The OECD experience. *Finance Research, 1*(1), 35–39. Retrieved from http://researchpub.org/journal/fr/number/vol1-no1/vol1-no1-4.pdf

AU. (2012). *Agreement for the establishment of the African Risk Capacity (ARC) agency.* Pretoria, South Africa. African Union. Retrieved November 9, 2017, from www.au.int/en/content/agreement-establishment-african-risk-capacity-arc-agency

AU & WFP. (2014). *Briefing book: Sovereign disaster risk solutions—A project of the African Union.* Conference, African Union & World Food Program.

Clarke, D., & Hill, R. (2012). *Cost-benefit analysis of the African risk capacity facility.* Retrieved from www.ifpri.org/publication/cost-benefit-analysis-african-risk-capacity-facility

DFID. (2012). *Annual review: African risk capacity.* Retrieved January 15, 2016, from http://iati.dfid.gov.uk/iati_documents/3763353.docx

The Economist. (2016). *ARC's covenant: A worthy insurance scheme goes awry.* Retrieved June 8, 2019, from www.economist.com/finance-and-economics/2016/08/25/arcs-covenant

FAO. (1996). *Rome declaration on world food security and world food summit of action.* World of Food Summit 13–17 November, Rome. Retrieved from http://www.fao.org/3/w3613e/w3613e00.htm

Hellmuth, M. E., Osgood, D. E., Hess, U., Moorhead, A., & Bhojwani, H. (2009). *Index insurance and climate risk: Prospects for development and disaster management.* New York: International Research Institute for Climate and Society. Retrieved from https://cgspace.cgiar.org/handle/10568/932

Kimetrica. (2014). *Independent review of the African Risk Capacity (ARC) contingency planning*. Nairobi.

Lung, F. (2014). The African risk capacity agency in the context of growing drought resilience in Sub-Saharan Africa. *Bologna Center Journal of International Affairs*, *16*, 1–26. Retrieved from http://bcjournal.org/volume-16/2013/6/6/the-african-risk-capacity-agency-in-the-context-of-growing-d.html

Michanek, E. (1981). An international disaster relief insurance. *Development Dialogue*, *1*, 30–37.

The Nordic UN Project. (1991). *The United Nations in development—Reform issues in the economic and social fields: A nordic perspective*. Stockholm: The Nordic UN Project.

OPM. (2017). *Independent evaluation of the African risk capacity annex 6—Organisational review*. Oxford: Oxford Policy Management. Retrieved from www.opml.co.uk/projects/independent-evaluation-african-risk-capacity

Rockefeller Foundation. (2015a). *About us*. Retrieved May 13, 2015, from www.rockefellerfoundation.org/about-us/our-team/maria-blair

Rockefeller Foundation. (2015b). *Grants and grantees*. Retrieved May 13, 2015, from www.rockefellerfoundation.org/grants/grants-and-grantees/c9ce8edf-db82-45cc-91af-05fd4eeae0af

Sen, A. (1982). The food problem: Theory and policy. *Third World Quarterly*, *4*(3), 447–459.

START Network. (2019). *Statement on ARC replica*. Retrieved August 15, 2019, from https://startnetwork.org/news-and-blogs/statement-arc-replica-start-network

Syroka, J. (2014, June 21–22). *African risk capacity: An African led strategy for managing extreme weather risks African risk capacity*. Presentation at the Second Forum of the Standing Committee on Finance, UNFCCC, Montego Bay, Jamaica.

Syroka, J., & Nucifora, A. (2010). *National drought insurance for Malawi* (Policy Research Working Paper No. 5169). Washington, DC: World Bank. Retrieved from https://openknowledge.worldbank.org/handle/10986/19950

Syroka, J., & Wilcox, R. (2006). Rethinking international disaster aid finance. *Journal of International Affairs*, *59*(2), 197–214.

UNEP. (2017). *Adaptation financing gap report 2016*. Bonn: United Nations Environment Program. Retrieved from http://web.unep.org/adaptationgapreport/2016

WFP. (2009). *NWP synthesis publication*. Rome: World Food Program. Retrieved from https://unfccc.int/files/adaptation/application/pdf/wfp_ap_update_sep_09_drr_3_sp.pdf

WFP. (2012a). *Fact sheet: Productive safety net programme*. Rome: World Food Program. Retrieved from www.wfp.org/sites/default/files/PSNP Factsheet.pdf

WFP. (2012b, June 13–15). *Risk management & transfer disasters—The Ethiopian experience*. Presentation at Addressing Loss & Damage Associated with Climate Change, Addis Abeba.

WFP. (2017). *Livelihoods, early assessment and protection*. Rome: World Food Program. Retrieved March 12, 2017, from www.wfp.org/climate-change/initiatives/livelihoods-early-assessment-protection

Wilcox, R. (2014, August 24–27). *ARC Extreme Climate Facility (XCF)*. Presentation at 5th International Disaster & Risk Conference IDRC Davos, Davos. Retrieved from www.africanriskcapacity.org/wp-content/uploads/2016/10/ARC-5th-IDCR-27-August-2014-Richard-Wilcox-v3-XCF-Presentation.pdf

Wilcox, R., & Arthurs, H. (2007, February 19–21). *Final report on the Ethiopia drought insurance pilot project*. Presentation at Executive Board First Regular Session Rome, Agenda Item 10, Rome. Retrieved from www.alnap.org/help-library/wfp-final-report-on-the-ethiopia-drought-insurance-pilot-project

5 The Pacific Catastrophe Risk Insurance Pilot (PCRAFI Pilot)

I will now turn to the third and last case of this book: the *Pacific Catastrophe Risk Insurance Pilot*. The Pacific Catastrophe Risk Insurance Pilot was part of the larger Pacific Catastrophe Risk Assessment and Financing Initiative (PCRAFI), a regional development program set up by the World Bank, and sought to test the waters for a regional insurance pool. Between 2012 and 2015 the Pacific Catastrophe Risk Insurance Pilot (PCRAFI Pilot) provided up to six small Pacific Island States (PICs) with parametric insurance coverage against tropical cyclones, tsunamis and earthquakes. In 2016, the PCRAFI Pilot was transformed into a stand-alone regional insurance company, sometimes called PCRAFI Facility, at other times Pacific Catastrophe Risk Insurance Company (PCRIC). Due to resource constraints and the lack of publicly accessible documentation, this chapter considers only the pilot.

In the first part of this chapter, I will center stage the goal and functioning of the PCRAFI Pilot. I will explore the goals of the PCRAFI Pilot and the way it worked on a practical and technical level to fulfill these objectives. Analogous to the other case studies, I will first provide a short overview of the PCRAFI Pilot to forestall key takeaway points. I will then elaborate on these points zooming onto (1) the goals (2) the operational structure (3) the technical basis and (4) the reinsurance structure of the arrangement.

In the second part of this chapter I will explore how the PCRAFI Pilot has come about and illuminate in particular the making of the insurance instrument (0–6), that is, the activities leading up to the creation of the PCRAFI Pilot. Finally, I will briefly review the first year of operation and point to more recent developments: The establishment of a stand-alone insurance facility (7).

I will finish this chapter with a concluding summary (III).

Part I: understanding the PCRAFI pilot

The *Pacific Catastrophe Risk Insurance Pilot* was a **pilot program** by the World Bank that provided up to **six Pacific Island Countries** (PICs) between 2013 and 2015 with an index-based **insurance coverage against the *fiscal* consequences of *major* tropical cyclones, earthquakes and tsunamis** in exchange for an annual premium. The premiums were fully paid by the World Bank. The goal of the program was to test the viability of a regional insurance facility for the small Pacific Island Countries.

The pilot was an application of the broader Pacific Catastrophe Risk Assessment and Financing Initiative (PCRAFI), a joint initiative by the Secretariat of the Pacific Community (SPC/SOPAC), the Asian Development Bank (ADB) and the World Bank, co-financed by Japan, the Global Facility for Disaster Risk Reduction and Recovery (GFDDR) and the ACP/EU Natural Disaster Risk Reduction Program. PCRAFI is a long-term development program that aims to enhance the 'financial resilience' of Pacific islands with regard to disasters; in other words, their ability to withstand disasters *financially*.

PCRAFI aimed at providing 15 Pacific Island Countries[1] with disaster risk and assessment tools. It developed detailed hazard maps, collected information on assets, population and sought to promote understanding of the potential costs of disasters to national fiscal budgets (PIFS, 2014, p. 5). In addition, PCRAFI—through regional conferences, workshops, studies—offered 'technical assistance' and 'capacity building' to the PICs. It took stock of and evaluated the PICs financial strategies and fiscal instruments to deal with the economic consequences of disasters. Based on these results, PCRAFI explored, proposed and supported the implementation of new financial tools—such as contingent credit or insurance—for the small Pacific Island Countries to reduce their exposure to the financial consequences of tropical cyclones, storm surge, rain, earthquakes and tsunamis. In this context, the World Bank set up the *Pacific Catastrophe Risk Insurance Pilot Program*.

The **generic goal** of the PCRAFI Pilot was **to secure the liquidity of the small Pacific island states** by providing them *immediately* with *limited* funds when they are hit by *major* tropical cyclone, earthquake or tsunami. *Major* events refer here to weather events that were statistically modeled to happen historically 1:10, 1:15 or 1:20 years.[2] Being modeled on the generic goal of CCRIF, the insurance sought **to prevent short-term cash-flow problems and secure additional funding for early emergency relief measures.**

The level of the payout was calculated via a model that used real-time weather data and combined it with geophysical and economic data to estimate the costs of a peril ('modeled loss approach'). The payouts reflected a certain percentage of the modeled losses and were capped at a certain level. Analogous to CCRIF, the PCRAFI Pilot may be understood akin to a **budget-support tool**. The payout was not tied to a rule-based mechanism. The PICs were free to decide what they wanted to do with the insurance payout.

In contrast to CCRIF and ARC, the PCRAFI Pilot did *not* comprise its own stand-alone regional catastrophe insurance facility. Instead, the insurer role was performed by the World Bank. It acted as 'intermediary' between the PICs and private re-insurers. The countries bought insurance coverage from the World Bank, which immediately ceded—that is, transferred—this coverage to private reinsurance companies in the form of a cat swap agreement. That means the World Bank was fully reimbursed by commercial re-insurers in case of a payout obligation.

The PCRAFI Pilot was launched as a three-year program in 2012/13 and ended in 2015/16, providing six countries (the Cook Islands, Marshall Islands, Samoa, Solomon Islands, Tonga and Vanuatu) with insurance coverage. In 2016, the World Bank restructured the PCRAFI program and thus the insurance pilot. In June 2016, PCRIC was founded. Unfortunately, resource constraints and lack of publicly available data prohibited an in-depth investigation into this new but important development. This case study focuses therefore only on the insurance pilot. Yet, from what can be inferred from the few available documents, it appears that the key principles underpinning the new facility remain largely the same.

Between 2013 and 2018 the PCRAFI Pilot and newly created stand-alone facility have made **four** payouts to two countries, totaling to about **US$ 6,7 million**. Only in January 2014, in the first year of operation of the pilot, Tonga received a payout of US$ 1,27 million following Tropical Cyclone Ian, a Category 5 cyclone. In April 2015, Vanuatu received a payout of US$ 1,9 million following Tropical Cyclone Pam, a Category 5 cyclone. In February 2018, Tonga received a payout of US$ 3,5 million from the newly created PCRIC.

The generic goals of the PCRAFI Pilot

The generic goal of the PCRAFI Pilot was "*to ensure immediate and effective post-disaster response*" (World Bank, 2011, p. 1, my emphasis) by providing "limited but immediate *budget support* following a major disaster (earthquakes including tsunamis, and tropical cyclones)" (PIFS, 2014, p. 5 my emphasis).

The generic goal of the PCRAFI Pilot related back to CCRIF upon which it was largely modeled. Similar to the insurance arrangement in the Caribbean, the PCRAFI Pilot sought *to prevent short-term cash-flow problems*—the problem when states can fund only a part of the required amount that is necessary to finance *immediate* post-disaster responses—that is, emergency and early relief and recovery measures deemed to be necessary by the state authorities to restore 'normality', such as the clearance of debris and the provision of shelter, medicine and food.

Similar to the Caribbean arrangement, the PCRAFI Pilot had an emphasis on flexibility: it did *not* come with strings attached, namely a contingency scheme that channels a payout into certain activities. Countries were free to allocate an insurance payout wherever they want. The insurance arrangement did neither prescribe nor specify these emergency response and early relief measures in terms of a pre-agreed and recommended expenditure list. However, what the World

Bank had in mind when it designed the pilot were the costs for fuel to distribute goods and aid in kind, as the islands are widely dispersed, the purchase of satellite phones and subscriptions to service providers to collect information and coordinate the response, the payroll of public servants, the purchase of tools and material to restore key public infrastructure (Personal interview, World Bank official, 2015, May 7). From the beginning, the insurance was *not* designed to finance the replacement of greater disaster-related physical losses of assets and infrastructure, tax shortfalls, greater social intervention programs or reconstruction efforts; instead it was designed to secure an *additional* budget for *major* catastrophes "such as those which can disrupt core public services" (PIFS, 2014, p. 5) as the World Bank put it. In this sense the arrangement was—and still is—much more limited than the Caribbean and the African arrangement.

Given its focus on providing additional funds for core public services (and thus its limited maximum payout) the PCRAFI Pilot is best understood as a *budget-support tool* that *supplemented other financing tools* used to fund post-disaster policies. According to PCRAFI's country studies (Bailey, Cook, & Mahul, 2015a, 2015b, 2015c; World Bank, 2015b, 2015c, 2015d, 2015e), the region is not only heavily exposed to natural disasters (in particular tropical cyclones, earthquakes and tsunamis) but has only limited financial tools in place to deal with the consequences of a disaster. The Pacific region relies heavily on international aid to finance post-disaster expenses (World Bank, 2012b, p. 12). It is only very recently and in the context of more global efforts to deal with disasters that the PICs have started to develop a tool kit to finance disaster-related policies. The available existing instruments include annual allocations for the unforeseen ('contingency budgets'), budget reallocations, national reserves funds and in some cases traditional public property catastrophe insurance. The Cook Islands, the Marshall Islands and the Solomon Island, for instance, insured key public assets against perils such as cyclones and earthquakes, although in a decentralized and uncoordinated way: The ministries decide if and, in case, which of its assets will be insured against what. Yet, again it is worth emphasizing that instruments such as national reserves funds or property insurance are all comparatively new; they were created only over the last decade. Pacific DRFIP adds to these existing financial instruments.

The operational structure

The PCRAFI Pilot was a development program offered by the World Bank. In exchange for an annual premium, the World Bank offered the small Pacific islands individually tailored insurance policies against the financial consequences of the region's key disaster risks: tropical cyclones, earthquakes and (earthquake-triggered) tsunamis. Countries could choose (within certain parameters) how much coverage they wanted to buy. Yet, for each hazard they must conclude a separate insurance contract.

In contrast to CCRIF or ARC, the PCRAFI Pilot did initially not comprise a stand-alone regional insurance facility. Instead, this role was played by the World

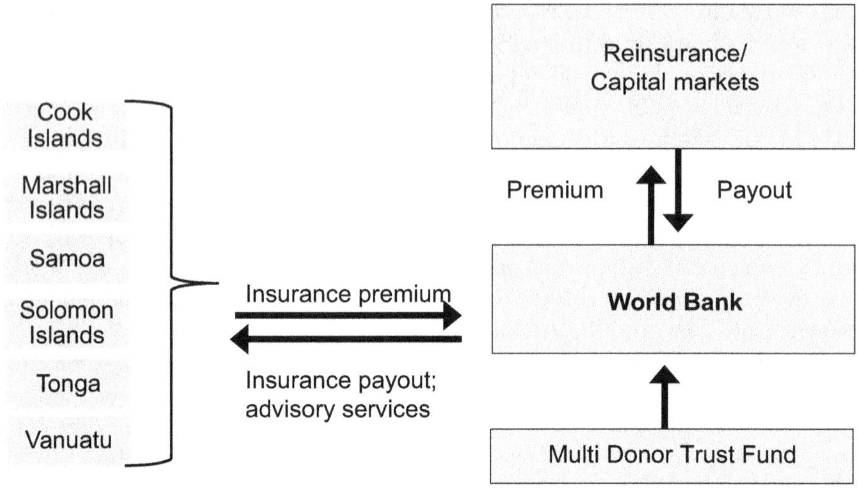

Figure 5.1 Operational structure of the PCRAFI Pilot

Bank. It acted as *de facto insurer* (see Figure 5.1). The World Bank acted as *de facto* insurer insofar as it transferred the cost of payout immediately onto private reinsurance companies via a **swap agreement**. Simply put, the World Bank put an agreement in place that ensured it would get fully reimbursed by private (re-)insurers in the case of a payout. The World Bank did not hold the financial risk of payout at any time (for more details see later on). With the aid of a swap agreement, the World Bank enabled the Pacific countries to gain access to private (re-)insurance capital.[3]

As *de facto* insurer, the World Bank also *coordinated* all insurance-relevant activities (as a 'real' insurer). Similar to CCRIF and ARC, the World Bank outsourced all insurance-related activities—from risk mapping and modeling to post-disaster loss calculations—to various public actors and private actors (modeling firms, insurance brokers, data providers) via service-provision agreements.

The insurance contract and premium level

The PCRAFI Pilot offered the insurance contract for one year for three consecutive seasons. The insurance policy year ran from 1 November each year until 31 October. The insurance policy holders paid an annual premium, which reflected their 'risk profile', respectively the financial risk they wanted to transfer. The 'risk profile' was based on actuarial analyses describing a country's probabilistic losses based on historical hazard data, exposure data and vulnerability assumptions.

In order to obtain insurance coverage, countries had to select the trigger level. Countries could choose between three layers of coverage: low, medium and high. The lower layer stood for events with a statistically estimated return period of

1:10 years. The medium layer covered events with a 1:15-years return period, while the higher layer events with a return period of 1:20 years and thus more severe events. The World Bank allowed the countries to request also a more customized option (World Bank, 2015d, p. 10).

The selection was in fact, however, largely circumscribed by premium support. All premiums but the Cook Islands' were fully funded by the World Bank. The reason was that all countries but the Cook Islands are low-income countries, according to the statutes of the World Bank and thus can apply for ODA funds. In the second year (2013/14), participating PICs had to contribute a rather symbolic lump sum of US$ 20,000 from their own budget, the third year lump sum increased to US$ 40,000. The following table illustrates the premium and coverage limits for the second insurance season[4] (2014/15):

Table 5.1 Coverage and premium level PCRAFI Pilot in the insurance policy year 2014/15

Country	*Return period*	*Premium (in million US$)*	*Coverage limit (in million US$)*
Marshall Islands	1:15	US$ 0,38	US$ 15,0
Samoa	1:20	US$ 0,38	US$ 16,0
Solomon Islands	1:10	US$ 0,38	US$ 10,0
Tonga	1:10	US$ 0,38	US$ 12,0
Vanuatu	1:20	US$ 0,38	US$ 12,0
Cook Islands	1:10	US$ 0,1	US$ 3,0
Total		**US$ 2,1 million**	**US$ 68 million**

Source: Bailey et al., 2015a, 2015b, 2015c; World Bank, 2015b, 2015c, 2015d, 2015e

A modeled-loss approach

Akin to CCRIF, the PCARFI Pilot was based on a *modeled-loss approach* to calculate losses and determine if a payout is due. The model used various real-time hazard data derived from a network of weather stations that measures the intensity of the peril in question (tropical cyclone, earthquake, earthquake-related tsunami) and combines it with geophysical and economic data in a model to estimate the losses *as the event happened.* The losses were derived from the exposure of various assets to the hazard. Participating countries received a payment if a certain (modeled) loss level was breached.

The technical basis

To estimate disaster-related losses the World Bank used an industry-wide standardized catastrophe model. It contained four basic components (AIR, 2012): (1) hazard module (2) exposure module (3) vulnerability module (4) insurance module. Each of those modules was built 'from scratch' (World Bank, 2015a, p. 28).

1 *Hazard module*

The hazard module measured the intensity of the perils in real time. The perils considered in the model were *tropical cyclones*, *earthquakes* and earthquake-related *tsunamis*. For tropical cyclones the model drew on the wind-speed, precipitation and coastal flooding (due to storm surge of the sea level), for earthquakes on the magnitude and intensity[5] of the ground shaking, for tsunamis on the sea wave heights and velocity. The hazard data was obtained from the Hawaii-based Joint Typhoon Warning Center[6] (JTWC) and the US Geological Service, based in the US federal state of Virginia[7] to prevent political fraud. Hazard conditions at more detached places were based on established modeling techniques "but from a relatively limited set of input data" (World Bank, 2015a, p. 28).

2 *Exposure module*

The exposure module contained information on the exposure and distribution of assets and people. It compiled the replacement costs of three types of assets: buildings (residential, commercial, industrial and public buildings), major infrastructure (e.g., airports, ports, power plants, bridges, dams) and major cash crops[8] (e.g., sugarcane, taro, rice, banana).
The data was based on variety of sources such as interviews with local experts, engineers, specialists, government reports, crop production reports, industrial data. Different replacement costs were assigned according to quality, construction class and occupancy class for each country (AIR, 2012, p. 43). The compilation of these comprehensive datasets was new. It was developed and processed with state-of-the-art modeling technologies and reflects largely the output of PCRAFI (see later on).

3 *Vulnerability module*

The vulnerability module combined the data from the hazard and exposure module to estimate the degree of damage to specific assets due to a peril in financial terms. This 'hazard footprint' was based on several assumptions (captured in 'damage functions') such as geophysical characteristics (e.g., land surface, soil properties, topographic slopes), asset characteristics and/or event experiences (e.g., historical/simulated earthquakes) (AIR, 2012). These assumptions were largely theory-driven as there existed no empirical data on the impact of disaster on Pacific islands building types (World Bank, 2015a, p. 28).

In addition to calculating the replacement costs, the module also calculated "emergency losses". Emergency losses "reflect[ed] the cost that local government may sustain as a result of providing necessary relief and recovery efforts" (AIR, 2012, p. 52), such as debris removal or demolition of damaged buildings. Emergency losses are calculated as a percentage of the modeled total replacement costs. The percentage was fixed at 23% for tsunamis and tropical cyclones (including floods), and at 16% for earthquakes, following 'best practice' standards in loss estimations.

4 Insurance module

The insurance module tied all information together and computed the level of the payout based on the insurance contract details (see later on). The insurance contract detailed the conditions when a payment was due; that is, when the loss level is breached and when the payout is capped ('excess-of-loss arrangement').

The model and loss analyses were compiled by AIR Worldwide. Given the limited number of data collection points (weather stations), the primarily satellite-based mapping of physical assets, the lack of (historical) data on destruction and claims, the model was largely theory-driven and had, therefore, its limitations. The output provided, at best, only a rough estimation of the real loss, as one interviewee from the World Bank highlighted (Personal interview, World Bank official, 2015, May 7).

The (re-)insurance strategy

How did the (re-)insurance arrangement work? What was the relationship between the World Bank and the capital markets? As indicated previously, the World Bank acted as *de facto* insurer. It provided the countries with an insurance coverage. Yet, it did not hold the financial risk of a potential payout on its balance sheet. The financial securitization did the trick (for details see Chapter 2, *What is reinsurance?*). The World Bank re-packaged the various risks and sold them via a swap agreement to (re-)insurers. In other words, it did not resell single or slices of the insurance policies but various insurance portfolios. Each of those insurance portfolios bundled payment obligations to various countries and, thus, comprised various level of risks. The World Bank bundled the insurance policies to accommodate the varying 'risk appetite' of (re-)insurers and create "a critical mass of business" (PIFS, 2014, p. 8).

As mentioned previously, the World Bank and Japan sponsored the premiums. The necessary preparatory work involving the risk modeling were financed through the World Bank, Asian Development Bank, Australia and other donors.

Part II: the genesis of the PCRAFI Pilot

Ideational origin

The Forum Economic Ministers Meeting (FEMM)

The idea to develop a Pacific insurance instrument dates back at least to 1999. At the Global Conference on Small Island States in February 1999, the World Bank presented the idea to set up a (sub-)regional insurance scheme for the Eastern Caribbean to a high-level audience, comprising representatives from the Pacific Forum—the Pacific islands' main regional political forum. The World Bank's presentation of the Pollner Report followed a long-term discussion within the Caribbean on how to deal with dramatic increase in reinsurance prices following Hurricane Andrew (for more details on the Pollner Report, see Chapter 3, *The proposal by Pollner*, pp. 123). The idea was to set up a regional re-insurer to provide a hedge the volatility of commercial (re-)insurance prices. Inspired by the Global Conference on Small Island States, the Pacific island's economic and finance ministers considered the issue of a regional catastrophe insurance scheme at its annual Forum Economic Ministers Meeting (FEMM) in Samoa the same year. The idea came up in the context of discussions on the development of a comprehensive regional disaster management strategy (PIFS, 2003, p. 1).

The PICs economic ministers discussed the World Bank's work on a regional catastrophe insurance scheme for the Caribbean and concluded that a similar concept note should be developed for the Pacific. In a resolution, the FEMM mandated the Pacific Forum Secretariat to establish a task force to assist the Pacific island states in identifying "appropriate regional and international financial organisations" (PIFS, 2003, p. 1) to pursue a regional catastrophe insurance initiative similar to the one in the Caribbean. Australia, the region's largest political and economic actor (World Bank, 2000, p. 21), offered the lead in advancing the initiative by coordinating the tasks and funding some activities through its development agency AusAID.

In the follow-up, Australia, the Pacific Secretariat, the South Pacific Applied Geoscience Commission (SOPAC), the PICs intergovernmental organization for regional geoscience and the World Bank collaboratively started to conduct some exploratory work in the form of preliminary analyses and discussions to assess the general interest and feasibility of a regional insurance scheme for the Pacific region (PIFS, 2003, p. 2). At the FEMM 2000 in Niue, the economic ministers discussed the results of this first exploratory undertaking and mandated SOPAC, the World Bank and Australia to progress with the preparation of a pilot study. They were tasked to prepare a pilot study for a country for which the sufficient data for an insurance scheme would be available and to report the study results the subsequent year (PIFS, 2003, p. 3). It is important to note that the FEMM continued to discuss a regional insurance scheme just as *one* instrument in a broader tool kit that can be used to limit fiscal and economic volatility following natural disasters (ADB, 2009; PIFS, 2003; Shorten et al., 2003). The FEMM was not fixed on the creation of a *regional* insurance instrument; it also considered the creation of *national* catastrophe insurance instruments. Hence, in contrast to discussions in the regional fora of the Caribbean and Africa, the annual discussions might be understood as 'open explorations'.

At the FEMM 2001 the development of a regional insurance scheme received a backslash: In their progress report, the consortium informed the Pacific ministers about a substantial lack of information pertinent to the requirements of the insurance industry (PIFS, 2003, p. 3). It reported that in contrast to the Caribbean, systematic data on hazard impacts and knowledge of existing physical assets such as existing buildings, infrastructure etc. in the Pacific region are largely nonexistent[9] (PIFS, 2003, p. 3; Shorten et al., 2003, p. 4). It would not be able to realize a pilot study on a country. In response to the insufficient information basis, the FEMM 2001 mandated SOPAC to conduct a more limited pilot study on Port Vila, the urban center of Vanuatu and a major economic center of the Pacific islands to learn more about the feasibility of an insurance scheme. Australia provided the necessary funding to collect and analyze the data. The FEMM 2001 asked SOPAC and the World Bank to report back "on the applicability of catastrophe insurance as a possible component of a broader, more comprehensive disaster management" (PIFS, 2003, p. 3).

The Vanuatu Pilot Project 2001–2002

Port Vila was chosen as some hazard impact and physical asset data—data on existing infrastructure, buildings, etc.—was already available. From 1991–1994, Vanuatu had collected some data on hazards-related impacts and existing physical assets in the context of the Vanuatu Government Risk Management Scheme (GRMS). The GRMS at that time envisaged the creation of a national catastrophe insurance scheme. This idea was, however, not developed further. Moreover, Port Vila was part of another regional data collection exercise, the risk assessment program *Pacific Cities*, conducted by SOPAC. The program started in 1997 and collected geographic and geotechnical data on Port Vila, Suva, Nuku'alofa, Honiara and Apia to assess hazards and the stability of building and infrastructure assets on some Pacific islands cities to model and predict earthquake effects (PIFS, 2003, p. 8; Shorten et al., 2003, p. 5).

The Vanuatu Pilot Project (Port Vila) was conducted by SOPAC and financed by Australia. It should provide a glimpse of what needs to be done to develop an insurance scheme and provide the underpinning for a regional insurance scheme. The scope of the Vanuatu pilot project included (1) to classify zones and buildings at risk (e.g., well-engineered structures made of concrete vs shacks and sheds of poor quality) to establish a rough model that captures likely damage and associated costs for cyclonic winds, storm surges and waves, earthquakes and tsunamis and (2) to identify and evaluate an appropriate insurance model.

At the FEMM in 2002, SOPAC presented the findings from the pilot study. The study estimated hazard pathways and the replacement costs of the assets 'at risk'; in addition, it roughly scaled-up related replacement and repair costs for a more comprehensive sub-regional scheme compromising Fiji, Solomon Islands, Samoa, Tonga and Vanuatu. The reason for the selection of these islands was that, according to SOPAC, they all face comparable hazards and similar (financial) risks (Shorten et al., 2003, p. x). Finally, the pilot study roughly estimated the funds necessary to insure earthquake-, tsunami-, cyclone- and storm surge-related events for such a sub-regional scheme. It suggested that it would require a fund of about AS$ 100 million "generated from a combination of donors, governments, premiums and possibly contingent credit from donors" (PIFS, 2003, p. 3); in any case, it would need the support of a multilateral development bank (ibid.).

In response to the results of the pilot study, the FEMM 2002 once again "noted that catastrophe insurance is only one possible component of broader, more comprehensive disaster mitigation and management strategies" (ibid., p. 4) and called the Pacific Secretariat to conduct a comparative cost-benefit analysis on *national* and *regional* insurance arrangement and report back at the next FEMM.

The cost-benefit analysis

The Pacific Secretariat's 2003 cost-benefit analysis made a clear case for a *regional* or at least a *sub-regional* scheme. First, it highlighted the *financial* cost-benefit ratio: "The small size and limited internal capacity of a number of countries in the region limit what can be undertaken at the national level and strongly argue for greater emphasis on collaboration at the sub-regional and regional level" (ibid., p. 5). Moreover, the cost-benefit analysis indicated that the Australian insurance industry might be interested in a *regional* scheme but not in a national scheme due to the small business volume of a national arrangement (ibid., p. 6). Second, the cost-benefit analysis underscored the empowering political potential of a regional insurance instrument: A (sub-)regional scheme could

> reduce the reliance on donors in times of disasters, thus freeing that financial assistance for other development purposes; and provide incentive (in the form of lowered premiums) to engage in disaster mitigation and hazard analysis, including hazards mapping, implementation of building codes and standards and the retrofitting of existing structures to mitigate loss.
>
> (ibid., p. 5)

The arguments were not so different from the conclusion drawn by the World Bank's Pollner Report from 1999 (see Chapter 3, *The proposal by Pollner*). In fact, they are largely framed in World Bank language. Yet, while making a case for a (sub-)regional insurance scheme, the report emphasized the tremendous hazard and vulnerability information that is necessary to realize such a scheme. It would require a comprehensive (sub-)regional data collection program and the *a priori* input of catastrophe modeling agencies and the insurance industry. Such a comprehensive data collection exercise would be complex, costly and lengthy and hardly feasible without the aid of donors. It argues that this 'price tag' must be factored into a decision pro insurance.

Despite the tremendous costs, the report made a case for such comprehensive hazard and financial risk data collection program for insurance purposes. It would not only neatly fit into existing, long-term SOPAC data collection programs such as the *Pacific Cities* program (PIFS, 2003, p. 6) but also allow the development of *alternative* instruments to deal with disasters. The report pointed out that the data necessary for an insurance scheme would also provide "the information necessary for mitigation, building standards, retro-fitting and simply for future planning" (ibid., p. 8) and thus for the development of more regulatory instruments to deal with disasters. Hence, the data obtained would be of multiple usages and not foreclose any specific policy instrument.

The report also noted the lack of regulations and enforcement thereof across the Pacific Islands and recommended giving way to developing and strengthening national regulatory approaches. The cost-benefit analysis highlighted that insurance mechanisms are "appropriate for risks that cannot be mitigated through structural or *ex ante* measures, and against events that have the potential to cause large economic loss" (ibid., p. 9). It concluded that while countries should focus on preventive measures, they should "urge donors to support the development of the hazard and vulnerability information so as to create the basis for the further consideration of risk financing for disasters" (ibid., p. 10).

Following this cautious case for a (sub-)regional insurance scheme, the idea seemed to have lost high-level political momentum. It is striking that over the following years the PICs did not consider catastrophe insurance as a separate talking point at a high-level meeting such as the FEMM. This, however, does not mean that considerations to develop such a scheme stopped, quite in contrast. The PICs largely embraced the recommendations made by the report. In 2006, they approached the World Bank for a comprehensive regional risk and financing assessment program, a program that became then known as the Pacific Catastrophe Risk Insurance Initiative (PCRAFI). This multi-year data collection program would form the basis for the Pacific's regional insurance scheme.

The Pacific Catastrophe Risk Assessment and Financing Initiative (PCRAFI)

The *Pacific Catastrophe Risk Assessment and Financing Initiative* was set up in 2007 following an official request by the PICs at the annual World Bank/IMF/Pacific Island Forum Countries Meeting in Washington DC in 2006 (PIFS, 2013). PCRAFI was basically a massive data collection program on the Pacific region's exposure to natural hazards. Purpose of the program was to provide all 15 PICs[10] with country-specific 'risk maps' to allow them to identify areas where people and assets 'at risk' from the region's major threats: tropical cyclones, earthquakes and tsunamis.

Given the lack of available geo-referenced data, PCRAFI created the risk maps and compiled the underpinning datasets largely from scratch. It did so by leveraging state-of-the-art technologies such as 'remote sensing analysis'—the analysis of high-resolution satellite images—field visits or 'crowd-sourced mapping'.[11] The exposure database, for instance, includes a comprehensive inventory of buildings (residential, commercial or industrial), infrastructure (roads, bridges, airports, ports and utility assets) and major cash crops (e.g., coconut, palm oil, coffee, pepper, etc.). It contains information on the location, their characteristics that affect their vulnerability (or types in the case of cash crops) and their replacement costs.

To compile the database on buildings, PCRAFI digitized about 500,000 buildings from high resolution satellite images and visited about 80,000 buildings to characterize these buildings and create the classification methodology (World Bank, 2013, p. 7). About 3 million buildings, mostly located in rural areas, were grouped and interpreted using moderate- or low-resolution imagery. To establish subsequent unit repair and replacement costs for buildings and infrastructure, PCRAFI consulted a variety of sources, including reports from construction firms, government reports, historical disaster reports and local experts (ibid., p. 20). The replacement costs for cash crops were collected from crop production budgets issued by local governments.

This undertaking was throughout remarkable: According to the World Bank "[m]illions of building and infrastructure assets and hundreds of thousands of hectares of cash crops with their characteristics and location" (World Bank, 2013, p. 30) were included in a geo-referenced database. It reported that the database would constitute "the most comprehensive exposure dataset for the pacific" (ibid., p. 30). The total replacement cost for all assets in the 15 PICs was estimated to be in the range of US$ 113 billion (ibid.). Next to developing an exposure database, PCRAFI produced also a comprehensive historical loss database. PCRAFI reconstructed and modeled the pathway of more than 2,400 tropical cyclones and 32,000 earthquakes in the Pacific region and their impacts. For the first time ever, the impact of tsunamis was modeled.

Taken together, according to the initiative's website PCRAFI "assembled, processed, developed, and organized the *largest* collection of geo-referenced data for hazard modeling in the region" (SOPAC, 2014). AIR Worldwide, the catastrophe modeling agency involved in PCRAFI, claimed that "the resulting database is of a level of accuracy and detail that is seldom, if ever, achieved (AIR, 2011)". Partly about 32 local, national and regional actors have been involved in this complex and doubtless remarkable data-creation exercise. With SOPAC, GNS Science and the Pacific Disaster Center (PDC), for example, three major regional scientific organizations were involved just in field visits to photograph and classify buildings (AIR, 2011). The involvement of so many actors made the coordination very difficult, as an interviewee from SPC/SOPAC revealed.

Without doubt, PCRAFI was astonishing in terms of coordination but even more so in terms of output: It produced two new bodies of knowledge: Geoscientific knowledge and knowledge of the economic costs of disasters. For the first time, the PICs exposure to hazards had been systematically modeled. Moreover, for the first time the region was economically quantified. The total replacement cost for all assets in the 15 PICs was estimated to be in the range of US$ 113 billion (World Bank, 2013, p. 30).

PCRAFI had an initial total budget of US$ 3,3 million through a multi-donor trust fund. US$ 1,3 million therefore came from Japan under the World Bank-administered Japan Policy and Human Resources Developed (PHRD) Technical Assistance Program; the Japanese Government provided an additional US$ 1 million to cover the costs of the field verification of data. The UN-mandated Global Facility for Disaster Reduction and Recovery (GFDRR) contributed US$ 1 million. More recently, in 2012, the ACP-EU Natural Disaster Risk Reduction Program funded by the European Commission provided another layer of funding, namely US$ 1,4 million to further strengthen the Pacific risk information system and its use by the PICs (World Bank, 2012c).

The comprehensive data collection started in 2007 and took until 2011. Once the data was compiled and catastrophic risk profiles created, SPC/SOPAC and the World Bank started to take stock of the existing strategies and instruments used by the PICs to deal with the economic and fiscal impact of disasters. The country centered on how the PICs finance post-disaster relief, recovery and reconstruction measures (Bailey et al., 2015b, 2015c; Cook, 2011a, 2011b; World Bank, 2015b, 2015c, 2015e). They build a case for "an integrated risk financing and insurance strategy" (World Bank, 2014), an instrument mix that includes national reserves, budget allocations and insurance instrument. Why then the focus on insurance? This development must be seen in the context of the successful launch of CCRIF.

The beginnings of the PCRAFI Pilot

The World Bank as ideational sponsor: a policy transfer

In 2007, following an official request by the FEMM and in parallel to the data collection under PCRAFI, the World Bank started to work out a concrete technical concept for the design of a regional insurance scheme under the umbrella of PCRAFI. The initiative to develop a regional insurance instrument was labeled *Pacific Disaster Risk Financing and Insurance Program*. The program was inspired by the successful launch of CCRIF in the Caribbean the same year.

The basic goal of the newly set up Pacific Disaster Risk Financing and Insurance Program was to develop a regional catastrophe scheme for the Pacific islands to finance their immediate, short-term liquidity needs following a natural disaster. In other words, the World Bank basically sought to replicate the Caribbean instrument (see Chapter 4, *Generic goals of CCRIF*) for a different context, that is, for different perils and income levels. The adoption of the Pacific insurance program is thus a 'classic' case of a policy (instrument) transfer, "the process by which knowledge about policies, administrative arrangements, institutions and ideas in one political system (past or present) is used in the development of policies, administrative arrangements, institutions and ideas in another political system" (Dolowitz & Marsh, 2000, p. 5). Key was apparently the 'demonstration effect' of CCRIF, the example that a regional index-based insurance arrangement works.

Unfortunately, due to the limited and partly contradictory information available, the precise *nature* of the policy transfer remains a research desideratum. It remains unclear if the adoption of insurance program was tendentially more exogenously or endogenously driven, that is, primarily driven by the PICs, the World Bank or other actors involved in PCRAFI (such as the Asian Development Bank or scientific organizations such as SPC/SOPAC). Interviewees from the World Bank and SOPAC told different stories: One interviewee highlighted that at the annual World Bank/IMF/PIFS Meeting, the PICs asked the World Bank to develop such a catastrophe insurance scheme; another interview, in turn, indicated that the World Bank had a more active role. Asked about the role of the World Bank, the latter interviewee replied that the World Bank "created the demand for them [the Pacific Islands; N.S.] to ask someone from the Bank to go to the Forum Economic Ministers Meeting in order to be able to develop the product" (Personal Interview, World Bank official, 2015, June 22).

Given the fact that from 2002 to 2007 a regional catastrophe insurance scheme did not appear anymore as an issue of 'high politics' in the Pacific region—it was not a separate talking point anymore at the annual FEMMs—it seems that the demand to develop a regional insurance instrument and model it upon the Caribbean was rather created by the World Bank. Only Fritz Koshiba, Minister of Resources, Development and Environment of Palau raised the issue of a Pacific insurance scheme in a speech at the FEMM 2007. In that speech he noted that the Caribbean launched a regional catastrophe insurance facility and pointed out that the PICs had once considered such a catastrophe insurance scheme and might be interested to follow up on this development (Koshiba, 2007, p. 6).

Whatever the precise nature of policy transfer was, the World Bank started to develop a concept note although the demand from the region was at least not overwhelming. This does not necessarily imply that the PICs were not interested anymore in the advancement of an insurance scheme or that they were somehow 'forced' to adopt an insurance instrument by the World Bank. Overall, they formally *did* ask the World Bank, that is, they *mandated* the World Bank to develop a regional insurance instrument.

And, yet, it appears that the World Bank together with the Asian Development Bank (ADB) had a strong(er) interest in creating a regional insurance instrument for the Pacific. The ADB conference held in November 2008 serves as a case-in-point. It is here where the ADB and the World Bank sought to mobilize public and private actors alike.

Stimulating interest: the ADB conference in Tokyo

The ADB considered the development of (catastrophe) insurance instruments at least since the adoption of the Hyogo Framework in 2005. Following the adoption of this framework, the ADB initiated a technical assistance program, the "Development of Catastrophe Risk Insurance". The purpose of this two-year program (2006–2008) was to explore the possibility of a regional catastrophe insurance mechanism for Asia and the Pacific (ADB, 2009, p. vi). The ADB considered ways in which it "can play a catalyst role to jump-start ex ante initiatives without impeding or competing with possible private sector interest in the catastrophe risk-insurance area" (ibid., p. 2). As a result of this two-year program, the ADB organized a two-day, high-profile conference in Tokyo in November 2008. The workshop "Natural Catastrophe Risk Insurance Mechanisms for Asia and the Pacific" was jointly financed by the ADB, the World Bank and Japan's Ministry of Finance and inspired by the foundation of CCRIF in 2007 as outlined in the conference report (ibid., p. vii).

The ADB brought together a range of public actors from the region and elsewhere (e.g., regional and national disaster management representatives, representatives from financial institutions, development actors, multilateral development bank representatives) and private actors (insurers and re-insurers, modeling agencies, insurance brokers, rating agencies). The over 200 participants included high-profile representatives from ADB, AIR Worldwide, AON Benfield, Bloomberg, CCRIF, Jardine Lloyd Thompson, General Insurance Association of Japan, International Finance Corporation, government representatives of Cayman Islands, India, Indonesia, Japan and the Philippines, Guy Carpenter, International Institute for Applied Systems Analysis, Mitsui Sumitomo Insurance, Munich Re, Risk Management Solutions, Pacific Island Applied Geoscience Commission, Swiss Re and the World Bank. All participants were invited on a by-invitation-only basis (ibid., p. xii)—which demonstrates not only the clear goal-orientated character of this networking event but also the convening power of the ADB.

The conference had the explicit goal to "(i) alert public actors and private sector actors to new opportunities to transfer natural catastrophe risk using both traditional insurance and recent capital market mechanisms; (ii) assess the opportunities for catastrophe risk transfer in the Pacific island countries" (ibid., p. vi). In other words, the conference allowed public and private actors to interact and thus to exchange information, experiences, views and explore common interests. In different panels participants discussed, among others, the building blocks of catastrophe risk modeling and challenges thereof, the pros and cons of different catastrophe insurance triggers, possible applications of risk transfer mechanism in Asia and the Pacific, the creation of the CCRIF and subsequent 'lessons learned'.

One panel explicitly explored the technical feasibility of a regional catastrophe insurance mechanism for the Pacific region and how international and regional financial institutions could assist in the creation thereof. Based on its experience in the Caribbean, the World Bank presented three possible options for regional insurance mechanism for the Pacific: "a risk pool, supported by reinsurance; a combination of risk retention, backed by a donor-based reserve fund and complemented by reinsurance; and finally, complete reliance on a donor-funded reserve pool" (ibid., p. vii). Doing so, the World Bank sought not only feedback but also arguably the support from the private sector (ibid., p. 28).

Overall, the workshop demonstrated that the ADB (together with the World Bank) had a significant facilitating role in transmitting knowledge from the insurance sector (e.g., disaster risk finance experts, insurance industry representatives, catastrophe modeling agencies) to policymakers (e.g., ADB representatives, government representatives) and vice versa. They provided a forum for regional policymakers and insurance representatives to share new developments, 'best practices' and learn about each other's interests. It could be argued that the workshop, not to say the conference, was a market-making event, as it revolved to a great degree around identifying and fueling supply and demand for catastrophe-related insurance products. As the conference report highlights:

> The key finding from the conference was that market interest in developing risk transfer mechanisms for Asia and the Pacific is substantial. This interest derives not only from traditional insurance and reinsurance sources, which have and continue to be active in developing risk transfer mechanisms for the region, but also from global capital markets, which have a more recent interest in becoming engaged with the region. That interest derives principally from a desire to develop a portfolio asset class uncorrelated to traditional capital market asset risk parameters, such as interest rates and economic cycles. This interest in developing risk transfer solutions also stems in part from the gradual merging of the traditional insurance market and the deeper more liquid capital markets.
>
> (ibid., p. vi)

The conference revealed in particular the determination of the ADB to advance the idea of disaster risk financing in the region. In its conference report the ADB recommended a range of initiatives that could promote catastrophe risk financing in Asia and the Pacific, such as the proposal to establish a regional public-private partnership for catastrophic risk finance—a common program that would link national governments, ADB programs and commercial interests in the region—or the suggestion of providing funding for a regional catastrophe model development.

Apparently, the Asian Development Bank (with support from the World Bank) sought to (re-)politicize the issue of a regional catastrophe insurance instrument by putting it on the table of a high-level meeting, involving key representative from public and private sector. Through the dissemination of knowledge on the requirements of functioning insurance markets and cutting-edge insurance products, the banks engaged in capacity building and facilitated the creation of a public-private network.

Systematizing instrument development efforts

THE WORLD BANK'S DISASTER RISK FINANCING INSURANCE PROGRAM (DRFI)

In 2010, while preparing the Pacific arrangement, the World Bank systematized its efforts in devising disaster risk financing by establishing the *Disaster Risk Financing and Insurance Program* (DRFI). The purpose of DRFI has been to "increase governments' financial response capacity to natural disaster" (World Bank, 2012a), that is, a government's financial power to fund disaster-related services and expenses. DRFI is a joint partnership between the World Bank and the 2006-founded Global Facility for Disaster Risk Reduction and Recovery (GFDDR). GFDDR is a World Bank-administered agency to implement the UN's Hyogo Framework for Action, respectively the new Sendai Framework for Action (2015–2030).

DRFI promotes to integrate of disaster risk reduction and climate change adaption measures into the World Bank's economic development agenda. The idea is to institutionalize "sustainable and cost-effective risk financing strategies" in countries which are highly exposed to natural disasters and which have low/middle income basis. The political purpose is to "increase their financial and fiscal resilience" (World Bank, 2012a). The term 'resilience' serves hereby as a metaphor for 'bouncing back'—to being able 'to cope with' the financial impacts of disasters, that is, to have some money available to restore the status *ex ante*.

The larger goal of DFRI is to foster systemic shift in disaster risk financing, namely to "help government shift from post-disaster assistance to proactive ex ante budget planning" (World Bank, 2012a). It does so through the provision of 'technical assistance' (e.g., policy advice on national strategies, assessment studies, workshops to tie together knowledge, establishment of an *ex ante* budget planning system and assessment tools, etc.), the promotion of a private catastrophe risk market (e.g., data collection, advice on necessary legal and regulatory frameworks to promote private sector insurance penetration to offset public sector costs, etc.) and arguably most importantly the concrete design—that means, the explicit ideational sponsoring—of "affordable and cost-effective" policy instruments (World Bank, 2012a).

The policy instruments the World Bank lists exemplarily are the Catastrophe Deferred Drawdown Option (CAT-DDO), a contingent credit with low interest rates that can be used following a disaster, and a range of insurance instruments (insurance of public assets, property catastrophe risk insurance, agricultural insurance, disaster-linked micro insurance, regional disaster insurance schemes) for different beneficiaries (governments, homeowners, farmers, poorer segments of society). The World Bank understands all of its tools as a form of "bridge financing"—as a form of financing which is limited but which "provide *immediate* liquidity in the aftermath of a natural disaster" (World Bank, 2012a my emphasis). The foundational assumption is that the timely provision of funding allows to accelerating the process of economic recovery. This foundational assumption appears to be driven by the development of CCRIF.

Overall, DFRI seeks to build on and leverage existing collaborations and programs. Accordingly, the World Bank incorporated and initiated a range of projects under its umbrella—among them the *Pacific Disaster Risk Financing and Insurance Program*, which was then labeled *Pacific Catastrophe Risk Insurance Pilot.*

Instrument development and implementation

The pacific catastrophe risk insurance pilot

In the second half of 2011, following the conclusion of PCRAFI's massive data collection program and the availability of 'risk maps', the World Bank started to design and implement the *Pacific Catastrophe Risk Insurance Pilot*. The process was made up by six key activities: (1) securing approval: 'Expression of Interest' letters (2) learning from success and failures: Transferring the model of CCRIF (3) fostering understanding and commitment: Providing risk management education (4) overcoming institutional challenges: Realizing the World Bank's role as *de facto* insurer (5) stimulating the interest of re-insurers.

In September 2011, at the World Bank/IMF Annual Meeting in Washington DC, the World Bank offered all countries that were covered by the data collection program PCRAFI—and thus all countries for which risk maps and catastrophe models were available—to participate in a regional insurance pilot project to secure *ex ante* funds for disaster response by submitting a formal letter to the World Bank. According to World Bank reports, the Marshall Islands, Samoa, the Solomon Islands, Tonga and Vanuatu expressed their interests first orally and then confirmed their participation by submitting a formal "Expression of Interest"- letter, signed by their respective ministers of finance, to the World Bank (World Bank, 2015a). A small but interesting detail is that the countries also had to communicate officially that they would like to have the World Bank be *actively* involved in the insurance pilot as they lack experience and financial market expertise. This suggests that the World Bank was keen on preventing a public image of being an instrument 'promoter'.

Adapting CCRIF: learning from success and failures

After having secured sufficient participants, the World Bank moved on to design the Pacific insurance pilot. This time, and in contrast to the creation of CCRIF (see Chapter 3, *Instrument development*), the World Bank designed the insurance pilot largely in-house,[12] drawing on its experience and expertise gained with the establishment of CCRIF in 2007. The design team was led by Olivier Mahul who, along with Francis Ghesquiere, had been a key figure in the context of the creation of CCRIF. Apparently, they are the connecting dots between the Caribbean and the Pacific arrangement. The design team considered the "ability to implement the product in near time" and the "speed of the payout" (World Bank, 2015a, p. 19) as guiding key factors in the design process.

The team basically copied and adapted the Caribbean model. The World Bank first established that PCRAFI would have the same goal as CCRIF: To provide timely *limited* cash. Analog to CCRIF it conceived the insurance pilot as a *budget support tool*: A financial tool that would not reimburse 100% of the losses derived from a natural catastrophe but only a part of it. The idea was no different from CCRIF's, namely to empower countries to concentrate on emergency and early relief measures and so to speed up the macroeconomic recovery process. This implies that the Pacific design is basically informed by the impact of Hurricane Ivan and the experiences made in Grenada—that is, on the problems and 'lessons learned' in a different political context.

When it came to devising the precise institutional structure of the Pacific pilot, the internal discussions revolved not so much around the legal structure of the

Pacific arrangement—a major challenge in CCRIF's design process—but rather around the best trigger and how to integrate government emergency response costs into the equation. Largely copying the Caribbean model, the World Bank decided to base the Pacific insurance arrangement in principle on a parametric trigger. A 'classic' indemnity trigger where insurers dispatch loss adjuster to establish claims was ruled out because of pragmatic reasons. There was simply no appropriate insurance market infrastructure in the Pacific region. Consequently, the establishment of a third-party system to estimate and check the losses was deemed to be unfeasible (World Bank, 2015a, p. 18).

The World Bank considered a rather simple, purely parametric trigger and a more complex modeled-loss trigger. A purely parametric trigger uses only real-time weather data (e.g., wind-speed, ground-shaking, wave height) as trigger for a payout. A modeled-loss trigger uses 'variable' weather data alongside other 'stable' data such as economic and population data to estimate losses and establish claims. The payout is not tied to the breach of a certain hazard value but to the breach of a certain loss level. The modeled-loss trigger is more complex than a purely parametric trigger as it is based on more data; the advantage is, however, that a modeled-loss trigger is also more precise than a purely parametric trigger. It reduces *basis risk*—the situation where the payout does not meet the actual costs. Although a modeled-loss approach still entails per definition basis risk—a modeled-loss approach pays out against the *representation* of losses and not real losses—the estimation tends to be much closer to reality. Yet, the loss calculation is also slower given the tremendous information that needs to be computed.

Ambitious to reduce basis risk the World Bank opted for a modeled-loss trigger. The decision for this more complex trigger was arguably the consequence of a 'lesson learned' from CCRIF: In its first years of operation CCRIF was based on a purely parametric trigger. Yet, the model underpinning CCRIF in the first year hardly fitted its purpose. A non-payout to Jamaica following the passage and destruction due to Hurricane Mitch in CCRIF's first year of operation (2007) triggered some debate within Jamaica about the usefulness of the insurance instrument so that CCRIF decided to revise its model and develop a more precise second generation model.

The decision to develop a modeled-loss trigger demonstrates that the World Bank was ambitious to further minimize 'basis risk'. This is further evidenced by the fact that the World Bank even sought to explicitly model secondary, government emergency response cost. Yet, this turned out to be a greater problem: Governments emergency response costs (such as the provision of shelter, food, gas, material, etc.) depend first of all on political decisions whom to support and what to rebuild; yet, they vary also according to "the location and size of the event and the type of damage caused" (World Bank, 2015a, p. 20). Government response costs are tendentially higher when a catastrophic event such as an earthquake happens in a remote area with difficult access, few residents and bad building standards because of the comparatively higher inter-island transport costs and the greater damage.

The World Bank tried to find a formula to account for these cost variations but, given the complexity involved, finally opted to use a portion of the total "ground-up losses" (the damage to buildings, infrastructure and cash crops): 23% of the total ground-up loss for tropical cyclones and 16% for earthquakes (including tsunamis) (ibid., p. 19). Hence, the problem was fixed very roughly—and pragmatically. After the World Bank had developed the basic structure, the World Bank team offered four options from which countries could choose. The four options are illustrated by Table 5.2:

Table 5.2 Options offered for PICs

Option	*Coverage*
A	1:10 years attachment 1:150 years exhaustion point for tropical cyclones and earthquake
B	1:15 years attachment 1:150 years exhaustion point for tropical cyclones and earthquake
C	1:20 years attachment 1:150 years exhaustion point for tropical cyclones and earthquake
D	Bespoke coverage

Source: adapted from World Bank, 2015a, p. 14

Fostering understanding and securing political commitment

In parallel to the instrument development and design process, the World Bank, in collaboration with SOPAC, set up a capacity building-program (World Bank, 2015a, p. 12) to educate and train country representatives on the key features and operation of parametric insurance schemes in general and the Pacific scheme in particular. It organized a number of workshops and provided a number of support services targeted at high-level and upper-level decision-makers—mainly from the finance ministries and other relevant departments—to foster commitment, secure a political buy-in and allow for an 'informed' choice when it finally came to selecting a country's insurance coverage. In various workshops the World Bank explained to high-level and upper-level decision-makers how different levels of coverage would affect the premium level. The World Bank finally asked the PICs to send a formal Commitment Letter in which they detail their selected insurance coverage. The World Bank demanded that the formal Commitment Letter was signed by the minister of finance and thus at the cabinet member level (World Bank, 2015a, pp. 12–14). Doing so, the World Bank secured political buy-in at the highest political level.

Despite the capacity's building activities, it is striking that the scope of country involvement in the instrument design and development process was arguably much more limited, as in the case of CCRIF (see Chapter 3, *Instrument development*). In contrast to the Caribbean arrangement, the World Bank did not involve the Pacific countries in the instrument design process. This time the process, it appears, was merely World Bank-driven.

Major challenges

THE WORLD BANK AS *DE FACTO* INSURER

While the World Bank engaged in capacity building measures it still had to find a way around how it could perform the role it had envisioned for itself, namely the role as the PICs' *de facto* insurer (or as "intermediary" between the countries and capital market actors, as the World Bank puts it). The problem was that the organizational mandate of the World Bank prohibited the bank from underwriting its own insurance policies (World Bank, 2015a, p. 18). Hence, the challenge was to find a way around how the World Bank could pass on a potential payout obligation to the capital markets so that it does not hold financial risk at *any* time. To solve this problem the World Bank created a catastrophe swap. The catastrophe swap allowed the World Bank to shift a potential payment obligation to a third party, namely to re-insurers, in the very moment it was obliged to make a payment. This was not the first time the World Bank developed a catastrophe swap. It had a few precedents to draw on, such as the Malawi weather derivative in 2005 (see Chapter 4, *The genesis of ARC*).

As elaborated in more detail in Chapter 2, a catastrophe swap is basically a *derivative contract*. A derivative contract is a financial obligation that *derives* from the performance of some criteria as fixed in another contract. In this case, this other contract (Contract A) was the contract between the World Bank and the participating PICs. Simply put, *Contract A* stated that 'if certain parameters are breached the World Bank has to make a payment to a country'. *Contract B*—the derivative contract— in turn stated that 'if the World Bank has to make a payment to a country it gets reimbursed by re-insurers'; in other words, the contract arrangement foresaw that the re-insurer will pay instead of the World Bank. As both transactions have to happen simultaneously, concluding catastrophe swap agreements are best understood as a financial practice that involves two 'mirror transactions'. Re-insurers usually take over payment obligations only for a premium. Through the investment of this stream of premiums, re-insurers generate their profit and accrue capital reserves to pay potential claims (Jarzabkowski, Bednarek, & Spee, 2015, p. 31).

Playing the role of a *de facto* insurer of sovereign entities, the World Bank broke new ground. This new role as *de facto* insurer needed approval from the top-level, the World Bank's Board of Executive Directors. The Board demanded that the World Bank needed to mitigate the unpredictable credit risk involved the insurance arrangement; in other words, it demanded that the World Bank would undertake preventive measures to reduce the financial risk that the re-insurer(s) might not be able to finance its payment obligation in case there is a valid claim from one of the PICs. The World Bank team eventually decided to reduce this financial risk by diversifying its insurance portfolio across *several* re-insurers so that not only one re-insurer would be assuming the financial risk for the Pacific region. In addition, all the re-insurers had to have a good reputation and a credit rating of at least A+. Moreover, the World Bank decided to pay the premiums to the re-insurers only 'in arrears' (every quarter) and not up front. Such a delayed payment practice is unconventional on the catastrophe insurance market and "required flexibility on part of re-insurers, since it involved working according to a nonstandard premium payment schedule driven by the nature of the institution involved" (World Bank, 2015a, p. 22), as the World Bank Implementation Report points out.

Why did the World Bank decide at all to perform the role of a *de facto* insurer given all these complexities involved? According to the World Bank's own Implementation Report, the Pacific countries themselves asked the World Bank to be part of an insurance arrangement. The World Bank reports "they felt they themselves lacked experience accessing international reinsurance markets and dealing with sovereign disaster-risk transfer more generally" (ibid., p. 18). While this explanation might be one reason, another one might be simply re-insurers' *mistrust* towards small developing islands states (SIDS). As the same World Bank Implementation Report highlights on a different note, re-insurers have little business experience with SIDS. Yet, for re-insurers to become part of the arrangement "previous experience with sovereign transactions was considered a critical perquisite" (ibid., p. 27). That would mean the World Bank used its reputation, the ascription as a *trustworthy organization* to overcome this mistrust between countries and the private sector. Doing so, it constituted itself as *the* market maker. Against this backdrop it seems plausible that the regional insurance arrangement would not have been possible with the World Bank performing 'just' the role of a promoter as it, for example, in the case of CCRIF.

STIMULATING THE INTEREST OF RE-INSURERS

Having secured the participation of the countries, the World Bank approached market participants for reinsurance coverage respectively for the conclusion of a swap agreement. Securing the interest of (re-)insurers to conclude a deal with SIDS respectively with the World Bank was not easy. The World Bank used a two-step process to fuel the interest of (re-)insurers and filter appropriate candidates.

As a first step, they briefed (re-)insurers about the insurance arrangement and the underlying catastrophe model (ibid., pp. 28–29). In several sessions they clarified technical questions regarding the underpinning catastrophe risk model. The World Bank undertook these briefing sessions together with AIR Worldwide, the catastrophe modeler. This first step was already challenging for all actors involved: The insurance pilot used a new catastrophe model for which interested (re-)insurers had no equivalent to benchmark. Normally there is a standard catastrophe model, the "vendor model" which is used by all catastrophe risk modelers and catastrophe re-insurer to compare and evaluate risks and, thus, negotiate a price upon these. These 'vendor models' are essential to attribute economic value to uncertain events and thus for the overall marketization of catastrophe insurance deals (Jarzabkowski et al., 2015, pp. 68–71). Yet, since the pilot was based on a new model (which included, for instance, earthquake-triggered losses from tsunamis and secondary response costs), "re-insurers were required to build a ground-up understanding of the modeling processes, the input data and assumptions used for the model, and the uncertainties in outputs from the model when deployed for real-time loss estimation" (World Bank, 2015a, p. 29). In addition, as the Pacific is poorly covered by insurance, (re-)insurers had little if any data for the perils covered under the insurance pilot to verify the relations built in the equation. Re-insurers had to invest significant time and resources to validate the model and the risk analysis that underpinned the insurance contract.

As a second step and following model validation, the World Bank asked (re-) insurers to formally express their interest (ibid., p. 27). After they received a formal confirmation, the World Bank screened the applicants and selected them based on pre-defined criteria involving a (re-)insurer's reputation and credit rating. It then entered into the negotiation process about the insurance terms and premium level. In this negotiation process the World Bank had to fuel the 'risk appetite' of (re-)insurers; in other words, it had to promote the attractiveness of the reinsurance deal and, thus, the willingness of re-insurers to allocate some capital reserves to a catastrophe swap.

As indicated in Chapter 2, there is no magical formula that determines the 'risk appetite' of re-insurers. Depending on a variety of dynamic and interrelated factors (such as available liquidity, strategic goals of the firm, general or geographic risk exposure, etc.), some re-insurers prefer high-risk, others low-risk deals. A re-insurer's interest in high-risk or low-risk deals is highly variable throughout the year (Jarzabkowski et al., 2015, p. 18; Power, 2009) and depends, among others, on overall market conditions. Generally speaking, re-insurers aim to hold a diversified package of deals as they seek to secure the profitability of the sum of deals (Jarzabkowski et al., 2015, p. 136).

The negotiation processes in the catastrophe insurance markets are typically a closed and interactive process that follows some informal rules and social expectations (Jarzabkowski et al., 2015). In this interactive negotiation process, the World Bank needed to fuel the interest of (re-)insurers by convincing them not only of the 'soundness' of their catastrophe model but also of the long-term attractiveness of the deal offered.

As elaborated in Chapter 2 (*The specificities of the reinsurance market*), in catastrophe markets re-insurers typically do *not* compete with each other by offering a lower price than a potential competitor to conclude a reinsurance deal; in contrast to other financial markets, there exists no price competition but only a consensus price among re-insurers: All re-insurers offer their reinsurance, that is their capital reserves, at the same price (Jarzabkowski et al., 2015, p. 30) to primary insurers. Re-insurers compete in a different way, namely for the *share* of a deal; they bid to take "greater shares of the deal that—for various reasons—they consider most valuable" (ibid., p. 127) and usually receive them in the long-term from the insurer, namely in the context of insurance renewals. Insurers are expected to give those re-insurers who paid a loss a greater share as a kind of payback the following year (ibid., p. 40).

PRACTICAL CHALLENGES

The negotiation process between the World Bank and the re-insurers involved several challenges. First of all, it was an organizational challenge: It required that the World Bank had to coordinate with multiple public and private actors (across a very large geographical area), multiple re-insurers (across multiple time zones) in a limited time span (World Bank, 2015a, p. 31). The bank had to organize the input and signatures of senior officials from multiple actors in a timely and organized manner. A particular timing challenge in the finalization of the catastrophe swap was posed through the very nature of catastrophe reinsurance market. In contrast to other financial markets, the catastrophe reinsurance market is of a punctuated nature (Jarzabkowski et al., 2015, pp. 40–42). The convention is that reinsurance deals are usually concluded on specific dates. These dates vary according to where the re-insurers are registered. For European re-insurers, for example, the main renewal date is 1 January, while for Japanese re-insurers it is 1 April. Re-insurers are usually very busy prior to these dates with analytical work to create and cross-check models and with the subsequent price negotiations (World Bank, 2015a, p. 32). The World Bank identified 1 November as the best renewal date for the three-years lasting pilot program and approached the re-insurer early to avoid greater timing challenges.

Although the World Bank managed to negotiate a premium price by late October 2012 the planned inception date was pushed back due to delays of "several sources" (ibid., p. 32). The insurance pilot was launched on 18 January 2013 and did not coincide with major renewal dates. However, this delayed launch created additional complexities and challenges. An insurance contract is usually negotiated for one year, meaning the price is established for an insurance coverage from 1 November—31 October. Consequently, the premium price had to be adjusted for the new inception date. This required a renegotiation (ibid., p. 33). While a shorter duration of an insurance contract would normally translate into a price reduction this is not necessarily the case with global re-insurers: Individual re-insurers would usually try to recoup the costs they incurred to them across their portfolio. Consequently, as it became clear that the insurance policy would be launched after 1 January—the main renewal date of European re-insurer—the World Bank faced the risk that some re-insurers would have changed their 'risk appetite' by then and thus their pricing strategy. Yet, since no major event was happening it managed to obtain a reduced pricing (ibid.).

Second, the 'quoting process'—the process of establishing the premium price—was a particular evaluative challenge. The World Bank had to come up with a price suggestion without knowing if the price would be too high/too low and thus be accepted by the re-insurers they intended to attract. The World Bank had to pitch their evaluation more or less blind. The initial price pitching is, however, a crucial bet: A 'right' evaluation (from a re-insurer perspective) signals technical expertise, it demonstrates that the calculative practices of re-insurers and the (throughout the year highly variable) price have been well-researched and, thus, creates confidence in the soundness of the deal (Jarzabkowski et al., 2015, p. 128) in the sense of 'the people behind the deal, they know what they are doing'. The World Bank's price pitching was therefore a crucial move. Moreover, the price had to express the particular and partly unconventional terms of the deal, such as the premium payment in-arrears (instead of the conventional up-front payment), as described previously. In order to get a rough idea of 'reasonable' premium price, the World Bank looked at deals with similar terms and conditions (World Bank, 2015a, p. 32) but again, there was no clear reference model; it came up with a conservative price, according to the Implementation Report (ibid., p. 31).

In retrospect, World Bank interviewees concluded that they got a "competitive price", a price that "had the benefit of maximizing value for money to the client [the PICs; N.S.], but it also set a precedent that can be used in negotiating future pricing" (ibid., p. 32), as the Implementation Reports explains. The portfolio was thereby particularly attractive to Japanese re-insurers, "partly due to their history of development cooperation with the Pacific region, and the support from the Japanese Ministry of Finance for the project" (ibid., p. 27).

A third challenge was to secure the participation of re-insurers: There was little financial incentive for re-insurers to conclude such a deal. The catastrophe swap provided only limited business volume for re-insurers. Although the precise premiums are not known, the grants provided by Japan to the PICs give a rough approximation. The aggregated volume thereof was US$ 2 million. If the grant was only used for premium payment, then the four re-insurers' profit calculation was US$ 2 million divided by four minus the re-insurers' costs for administration and the opportunity costs of holding back enough capital. The income revolved likely around US$ 400,000. This implies that the insurance pilot offered rather peanuts relative to the billion dollar business volumes of the companies involved.

In addition, the way the catastrophe swap was structured by the World Bank was not very appealing to re-insurers. In contrast to a conventional catastrophe swap, the package was not 'layered', that is, divided into different high-risk and low-risk tranches. This made the catastrophe swap economically unattractive. As explained previously, normally re-insurers have a different 'risk appetite': Some want to underwrite more risky products, others, by contrast, are attracted by low risk products. So, usually some re-insurers pick the high-risk layers and others low-risk layers. Yet in this case, the catastrophe swap was structured in a way that all re-insurer had to take the same risk, that is, the same slice compromising both high- and low-risk. By high- and low-risk layers, I refer to probability of insurance claims and thus the level of premiums the re-insurers receive for holding back capital.

Despite the 'unattractiveness' of the deal, the World Bank managed to secure reinsurance. The question arises why re-insurers still signed up for the product despite this unattractiveness. A consultant who was involved in CCRIF sees the attractiveness of the Pacific deal for re-insurers more in the innovative character and the publicity involved in supporting such an innovative product than in the marginal profit it offers: "It's not a great risk they are insuring relative to their business and it's innovative, it's new; and finally, it's business to them. The premiums were reasonable" (Personal interview, former World Bank consultant, 2015, May 18). The interviewee also pointed out that the World Bank did all the work for the re-insurers, they offered them a business possibility at zero costs:

> if it [the reinsurance market; N.S.] has somebody who could do that arrangement for it and just put a bit of a margin on top for packaging that risk, that is, doing all the hard work, they are pretty happy to do that. That enables them to give us a much better rate . . . just because there is no cost to the reinsurers to get this business, right? They are not paying us a commission to bring them business. So, you know, it's good for them.
>
> (Personal interview, former World Bank consultant, 2015, May 18)

Several experts pointed out that re-insurers are facing more and more direct competition from the capital markets as investors put up their own disaster-linked securities. The available liquidity and, thus, the competition by capital market investors would drive down prices (see Chapter 2, *Alternative risk transfer today*). Practitioners claim that the 'the market has softened out'; one underwriter even claimed that "effectively it is a soft market forever now" (Personal interview, former World Bank consultant, 2015, May 18). Moreover, re-insurers are interested in getting off demand and thus in developing new markets. Some pointed out that for re-insurers, these insurance-linked securities are not only business but they help them to present themselves as technological innovators at the forefront of fighting climate change. One interviewee from one of the regional catastrophe insurance facilities pointed out that the participation of re-insurers is largely driven by Corporate Social Responsibility (CRS) considerations.

A fourth challenge concerned legal issues: Many re-insurers faced internal difficulties in closing a contract with an international political entity as the World Bank—just as the World Bank had some difficulties in closing a contract with re-insurers. The World Bank reports, for example, that there was one re-insurer that wanted to underwrite the pool but could not do it as "it writes its global catastrophe business out of a jurisdiction that the World Bank deemed unacceptable from a counterparty risk perspective" criteria (World Bank, 2015a, p. 28).

Instrument modifications: the evolutionary phase

The first year of operation (2012/13)

On 18 January 2013 and with about 2,5 months in delay the PCRAFI Pilot was launched. In contrast to CCRIF or ARC, it did not make a payment in its first insurance policy year 2013—despite the occurrence of a heavy earthquake on the Solomon Islands.

THE NON-PAYMENT TO THE SOLOMON ISLANDS

In February 2013 and thus only one month after the launch of the Pacific insurance arrangement, a heavy earthquake struck Santa Cruz (Solomon Islands). The earthquake with a magnitude of 8.0 on the Richter-scale caused substantial loss of life and damage and even triggered a minor tsunami (measuring wave heights of about meter in the Lata wharf in the Solomon Islands and about 3,5 meters in some areas of the Nendoe Islands) (World Bank, 2015a, p. 53). The Solomon Island government reported 480 destroyed houses, 4500 affected people by the earthquake and the tsunami and 10 fatalities. The water supply infrastructure and Lata airport and wharf were significantly damaged (ibid., p. 56).

Despite the considerable impact, the insurance did not pay out. The reason can be explained with flaws in the modeling. The center of the earthquake was located in the Temotu province and, thus, far from the inhabited and economic center of the Solomon Islands. The Temotu province accounts for only about 5% of the population (approximately 20,000) and about 2% of the asset value as modeled by PCRAFI. In other words, the damage was not severe enough to breach the necessary threshold for a payout. The non-payment did, however, not match the need on the ground as the remote location meant also relatively high transportation costs associated with government's relief efforts.

The non-payment clearly demonstrated the limits of a parametric-based trigger. As any parametric trigger, it suffers from 'basis risk', a situation where the model losses do not match the actual losses. Similar to CCRIF, the insurance model valued in particular concentrated areas and expensive assets. That means, it downplayed the costs of geographic remoteness or dispersion; and, yet, paradoxically it is precisely the remoteness the Pacific islands or dispersion of its inhabitants which drives up public service costs (such as the provision of shelter, medical support) due to the higher transportation costs, lack of available infrastructure and adequate human resources. In other words, it demonstrated that the model was not sufficiently tailored to the Pacific context. The non-payment caused substantial frustration among the Solomon Islands and nurtured doubts on the usefulness of the insurance arrangement.

Key developments 2014–2018

DEVELOPMENT OF PAYOUTS

While no payout was made during the first insurance policy year of the pilot, three payouts were made between 2014 and 2018 (see Table 5.3). Tonga received payouts in January 2014 and February 2018 following Tropical Cyclones Ian and Rita, respectively. Vanuatu received a payout following Tropical Cyclone Pam in March 2015. The total payouts summarized to US$ 6,7 million.

Two of the payouts were still made under the pilot, the third payout under the pilot's successor the *Pacific Catastrophe Insurance Company* which was established in June 2016 (see what follows).

Table 5.3 Payouts made by PCRAFI until December 2018

Policy year	*Country*	*Payout*	*Peril*
2013/14	Tonga	US$ 1,270,000	Tropical Cyclone
2014/15	Vanuatu	US$ 1,900,000	Tropical Cyclone
2017/18	Tonga	US$ 3,500,000*	Tropical Cyclone
2013–2018		**US$ 6,670,000**	

Source: data taken from Artemis, 2015; PCRIC, 2018; World Bank, 2015a

THE PAYOUT TO TOGA IN THE 2013/14 INSURANCE POLICY SEASON

Tropical cyclone Ian approached Tonga as a Category 5 cyclone of the Saffir-Simpson Hurricane Wind Scale. It passed close to Vava'u and made direct landfall on Ha'apai islands with a sustained wind-speed of 205 km/h. At least one person was killed, 14 injured and about 1094 houses have been destroyed or damaged. Significant damage was also done to cash crops. Several islands were unreachable by phone; moreover, the electrical grid collapsed (World Bank, 2015d, p. 3). Losses were modeled to be US$ 49,3 million with associated US$ 11,3 million emergency response costs. The highest level of modeled lost was attributed to residential buildings (66%), followed by crop losses (20%), public buildings and infrastructure (9%), commercial, industrial and other buildings (5%). Most modeled damage resulted from wind damage, losses from precipitation and storm surge were modeled—depending on the area—to make up less than 1–13% (World Bank, 2015a, p. 46). As stated previously, Toga received a payout of US$ 1,3 million within three weeks. The payout accounted for only about 2% of the modeled losses (modeled ground-up losses and modeled emergency losses). And, yet, the amount was equivalent to the country's contingency budget and more than half of the reserves of the Toga National Reserve Fund (World Bank, 2016, p. 5).

THE PAYOUT TO VANUATU IN THE 2014/15 INSURANCE POLICY SEASON

Tropical cyclone Pam was a Category 5 cyclone on the Saffir-Simpson Hurricane Wind Scale. The cyclone achieved a (one-minute) sustained wind-speed of about 269 km/h and is listed as the strongest cyclone ever registered. The eye of cyclone Pam passed east over Efate Island (population 66,000) and thus about 45 km from Port Vila, Vanuatu's capital and economic center. According to the UN's humanitarian office, about 24 people were killed, 15,000 buildings damaged or destroyed across the islands (OCHA, 2015a). The most damaging effects of the cyclone were largely concentrated in Southern part of the country and not in Port Vila as one might think, as Port Vila "has a higher percentage of masonry/concrete buildings and a lower percentage of traditional-style dwellings when compared to the country wide average" (World Bank, 2015a, p. 68), according to PCRAFI's Event Report developed by AIR Worldwide. In other words, the buildings there are more stable and provide more protection.

Root crops, which make up about 80% of the local food source, were "significantly damaged across all affected islands" (ibid.). The loss was modeled to be US$ 98,9 million. Most of the *modeled* economic losses, namely 60%, were made up by resident buildings, followed by commercial, industrial and other buildings (18%) and public buildings and infrastructure (18%). Cash crops accounted for about 4% of the modeled economic losses. It is notable, once again, that these are *loss estimates* are based on *representations* (e.g., resident building, commercial building etc.); as such, these figures should be taken with caution. The United Nations Office for the Coordination of Humanitarian Affairs (OCHA) estimated that about 166,000 people out of a population of 270,000 have been affected in Vanuatu; 75,000 were in need of an emergency shelter. It launched an international appeal calling for US$ 29,9 million in aid (which it did, indeed, collect by September 2015) (OCHA, 2015b).

As with Tropical Cyclone Ian, given the extent of destruction the insurance payout of US$ 1,9 million to Vanuatu was, indeed, limited. It accounted only for about 1,9% of the modeled losses and about 6,3% of the amount Toga called the UN for. And, yet, by putting the sum into the international humanitarian aid context, it becomes clear that insurance payout was, in fact, still the highest single 'external' financial contribution dedicated to relief and recovery (OCHA, 2015a).

Policy renewal

SOLOMON ISLANDS LEAVE THE PILOT AND THE COOK ISLANDS JOIN

The pilot has seen two changes with regard to insurance policyholders. In the second policy year (2013/14), the Cook Island joined the pool and bought insurance coverage, paying even the full premium. This brought the pilot's aggregated insurance coverage up to US$ 67 million. In the third policy year (2014/15), however, the Solomon Islands withdraw from the project reducing the overall coverage again to US$ 42 million.

The reason for the Solomon Island's decision to leave the pilot can be found in another non-payment: In April 2014 excessive rainfalls due to a slow moving tropical depression affected the Solomon Islands and created damages of about US$ 107,8 million, equivalent to 9,2% of GDP. Major infrastructure, including 675 houses and food gardens, were destroyed. The flooding caused 22 fatalities and displaced about 10,000 people, depriving them of their livelihood. And, yet, no payout was made. This time the reason was the specific insurance terms. Under the conditions of the contract, the payout was linked to the occurrence of a tropical cyclone as opposed to a tropical depression. As result, perceiving little value in the insurance, the Solomon Islands withdrew from the pilot.

This nurtured doubt, also within the World Bank. While the pilot demonstrated that a regional insurance scheme is attractive to the private market, it also showed that it comes with its own problems. As a senior official from the World

Bank who was involved in PCARFI pointed out, parametric-based insurances are "very complicated products" and the challenging part in educating others about the product is "the degree you can simplify them down". Given the complexity of the Pacific insurance, the interviewee concluded "I am not sure in retrospect if this is the right thing" (Personal interview, World Bank official, 2015, May 7). Insurance is still a largely alien concept.

Despite these experiences, with the exception of Vanuatu the Pacific countries did not change their attachment point for the third season. Vanuatu did increase the threshold in both renewal processes, that is, from a 1:10 year attachment (insurance policy year 2013), to 1:15 (insurance policy year 2013/14) and then to 1:20 (insurance policy year 2014/15) (World Bank, 2015a,17).

CHANGES IN PREMIUM FINANCING

While the countries' premiums were fully financed in the first insurance policy year, the countries had to contribute US$ 20,000 in the second insurance policy year and US$ 40,000 in the third insurance policy year. Although the World Bank demanded from countries to pay only a small share of the actual premium costs, this requirement complicated the renewal process. In contrast to the first year, the PIFs had to carry out a budget process. The World Bank's implementation report described the execution of this process as a political and administrative challenge (World Bank, 2015a, 16).

EXPANSION TO A CAPACITY BUILDING PROGRAM

Throughout the first years the World Bank extended the scope and depth of the capacity building workshops. During the first year of operation, for instance, the World Bank hosted a range of additional country workshops to review their selection. Moreover, it convened a regional workshop in March 2014 to allow for a "peer-to-peer exchange" on the PICs' existing disaster financing instruments. Moreover, the World Bank presented other disaster risk financing instruments (such as an insurance of public assets) (World Bank, 2015a, 12). According to the World Bank, the Bank wanted to give the countries "the opportunity to discuss past experiences, lessons learned, and ways to optimize post-disaster financial tools to improve post-disaster budget execution" (World Bank, 2015a, p. 12). It wanted to facilitate a dialogue so that countries could imitate and adopt each other's financial policy instruments. While this goal of assembling and engaging different PICs reveals indirectly a lack of regional integration in the Pacific, it highlights the role of the World Bank as regional 'carrier of ideas' and thus as a regional policymaker. It also demonstrates that the World Bank has slowly started to use the instrument as an entry point for the development of regional disaster policies and politics.

THE LAUNCH OF THE PACIFIC CATASTROPHE RISK INSURANCE COMPANY

With the end of the insurance season 2014/2015, the PCRAFI Pilot formally ended. In its final evaluation report, the World Bank concluded that the pilot's achievements were "satisfactory" (World Bank, 2016a). This assessment was apparently shared by the PICs' annual Forum Economic Ministers Meeting. During the 2015 meeting, 14 out of 15 PICs agreed to establish a common and stand-alone insurance facility. On 10 June 2016, the "PCRAFI Facility" (later labeled *Pacific Catastrophe Risk Insurance Company, PCRIC*) was established by legal statute in the Cook Islands as an insurance captive. Via a Multi-Donor Trust Fund, Germany, Japan, the United Kingdom and the United States provided US$ 40 million in grant funding.

While the countries started to develop a governance structure, the World Bank restructured its PCRAFI program and set up a comprehensive TA program. The goal of the PCRAFI TA Program is to provide " the PICs, regional organizations, and the PCRAFI Facility with technical assistance and capacity building on disaster risk finance and insurance and on public financial management of natural disasters" (World Bank, 2016b).

Unfortunately, no publicly available information exists on the precise governance structure of the new insurance facility. Since September 2016, the facility has offered insurance coverage to the five PICs that participated in the pilot. In February 2018, it made a payout of US$ 3,5 million to Tonga. To date, this represents the highest single payout to a PIC. Funds were transferred after seven days of the cyclone event and hailed as highly relevant. In April 2018, at the FEMM, Tuvalu proposed to establish a new climate insurance facility to support the PCRAFI Facility. Whatever form it has taken or may take in the nearer future, the Pacific insurance instrument, it appears, is here to stay.

Part III: summary

The **PCRAFI Pilot** was a World Bank-owned **pilot program that between 2013 and 2016** that provided up to six countries from the Pacific with insurance **coverage** against the financial consequences of ***major* tropical cyclones, earthquakes and tsunamis.** 'Major' refers to events that were statistically modeled to happen at least 1:10 years. Following the breach of a certain loss threshold, insured countries received rapidly ***limited funds* so that they can kick off emergency and early relief measures** immediately. The level of the payout was modeled on (real-time) hazard and exposure data and capped at a certain, pre-defined level. In 2016, the insurance pilot has been transformed into a regional stand-alone facility, PCRIC.

The **generic goal** of the PCRAFI Pilot was the **securitization of public liquidity to cover** explicit and implicit public service liabilities. The payouts flowed into national treasuries and were not tied to specific expenditures. Countries were free to decide how they wanted to use a payout. The PCRAFI Pilot was largely modeled on CCRIF. The secondary goals were therefore to enable the continuous functioning key public services, accelerate the rebuilding process and mitigate long-term financial impact of a disaster.

Between 2013 and 2018, the instrument has made three payouts totaling to about **US$ 6,6 million**. Tonga received payouts in January 2014 and February 2018 following Tropical Cyclones Ian and Rita. Vanuatu received a payout following Tropical Cyclone Pam in March 2015. While the first two payouts were still made under the pilot, the most recent payout was made under the new facility.

Since the **PCRAFI Pilot** was, empirically speaking, a pilot project within a larger development program it was much more limited in scope and differed substantially with respect to CCRIF and ARC in terms of its operational structure. The PCRAFI Pilot **did not comprise a stand-alone insurance facility**. Instead, it was based on a complex institutional arrangement in which the **World Bank** performs the role of a ***de facto* insurer.** In case of a payout the World Bank made a payment to the country and was fully reimbursed by the re-insurers the very moment it had to make a payment. This mirror transaction was possible through an innovative swap agreement. Moreover, in contrast to CCRIF and ARC, there was **no retention by the insurer** (that is, the World Bank) and the **reinsurance capacity was not layered**. All financial risks were transferred straight to re-insurers via the swap agreement. The larger purpose of the pilot was to test if there is demand from the PICs and re-insurers to support a regional insurance arrangement.

The history of the PCRAFI Pilot represents *largely* a **textbook case of a policy transfer.** It developed out of **intra-regional discussions, financial sponsorship, a demonstration effect and an instrumental learning process**. The idea to create a Pacific insurance scheme was discussed within the Pacific region since the late 1990s, when on a global conference the World Bank presented its idea to set up an insurance pilot project in the Eastern Caribbean to finance post-disaster expenses. This presentation triggered discussions within the Pacific Forum, the Pacific's main regional political forum, to consider the development of its own regional insurance scheme. A commissioned feasibility study concluded, however, that the Pacific region lacked adequate data. As a consequence, the idea lost its high-level political momentum in 2003. The study recommended the PICs to implement a massive data collection program and then consider various policy options. In 2007 the idea was then taken up by the World Bank, which during that time started to increase its efforts to develop instruments for disaster financing. Together with other regional actors the World Bank embarked on a massive data collection program by launching the Pacific Disaster Risk Financing Initiative (PCRAFI). For the first time, the PICs' exposure to hazard was systematically mapped and the economic costs thereof quantified. During that time the World Bank systematized its efforts to devise disaster financing instruments and launched, among others, the global orientated Disaster Risk Financing and Insurance Program. Together with the Asian Development Bank it organized a large, regional, high profile conference on the "Development Catastrophe Risk Insurance" in Tokyo to sensitize public and private representatives on the benefits of insurance. This created a favorable context for the realization of the idea. In 2011, finally, inspired by the 'success story' of CCRIF and in search for a new application, the World Bank offered the small Pacific states a fully sponsored pilot program on a regional insurance scheme. The World Bank basically intended to transfer the Caribbean model to test demand from PICs and appetite from re-insurers. In other words, **the emergence of the Pacific arrangement was ultimately rather supply-driven**; there was neither an overwhelming demand on the part of the PICs (not to speak of the insurance sector) nor a proper instrument selection process in the sense that the pacific countries considered various instruments and then mandated the World Bank for its assistance. Simply put, **the World Bank offered the PICs an insurance instrument at largely zero costs**.

Once a number of countries signed up for the pilot, the World Bank **designed** the instrument with **fairly limited local involvement**. With regard to design, the World Bank largely copied the Caribbean design and applied instrumentally the lesson it had learned in the Caribbean context. It created a more accurate (and sophisticated) modeled-loss trigger to account for damages and costs from a variety of sources, including secondary emergency response costs. With regard to the **implementation** process, the World Bank set up a **capacity building program**. It worked closely with political representatives, offering trainings and workings. The goal of these educational measures was to create an 'enabling environment' for the customization of the insurance policies. The main challenge for the World Bank was to find a way to perform the role of an insurer without breaking its organizational mandate. The fact that the World Bank was keen on finding a way around demonstrates its determination to realize an insurance solution. It is also against this backdrop that the **World Bank** must be understood as an **instrument constituency**.

Since its creation in 2013, the PCRAFI Pilot has seen significant changes. After the instrument's first year of operation (2013), the **World Bank expanded the capacity building component of the pilot**. It increasingly offered regional workshops to allow for peer-to-peer learning and enable countries to critically review and improve their disaster risk financing strategies and tools. The workshops were not restricted to the PICs participating in the insurance pilot but offered to all countries under the PCRAFI umbrella. Overall, it seems the World Bank used the pilot increasingly as an entry point to facilitate regional disaster management. The shift to more capacity building measures was possibly also related to two 'non-payouts' to the Solomon Islands. A heavy earthquake on a remote island of the Solomon Islands and a tropical depression with heavy rainfall that crossed the Solomon Islands did not trigger a payout. As a result the Solomon Island withdrew from the insurance pilot. In 2016, after three years of operating, the PCRAFI Pilot was transformed into a stand-alone insurance facility, the Pacific Catastrophe Risk Insurance Company.

Notes

1 The Pacific Island Countries involved in the PCRAFI are: Cook Islands, Federated States of Micronesia, Fiji, Kiriba, Marshall Islands, Nauru, Niue, Palau, Papua New Guinea, Samoa, Solomon Islands, Timor-Leste, Tonga, Tuvalu and Vanuatu.
2 All figures refer to the insurance year 2014/15, if not stated otherwise.
3 In this sense, the World Bank could also be understood as *facilitator* (or 'intermediary' in World Bank language) between the countries and private (re-)insurers.
4 This selection of the second season is related to data availability.
5 The intensity is dependent on the epicenter and the exact geographic characteristics.
6 The JTWC is a US government agency (jointly led by US Navy/US Air Force) located in Pearl Harbor, Hawaii and responsible for typhoon warnings in the North West Pacific Ocean, South Pacific Ocean, Indian Ocean and Micronesia.
7 The US Geological Service is a scientific agency of the US government and a bureau of the US Department of the Interior. It is headquartered in Reston, Virginia and provides the US with hazard maps.
8 Cash crops are crops that are planted for sale to return for profit and not for the use of a grower.
9 This can be contrasted with reference to the Caribbean. As the pilot project summary notes: "The World Bank has previously promoted work on Catastrophe Insurance in the Caribbean Region. The Caribbean study had the advantages of starting with a much more established, well-defined and coordinated insurance market, and a situation where the hazard climate had been more closely studied and modelled for a number of years and where certain risk-loss analyses had been conducted. Some of these jump-off points had yet to be established for the Pacific region and the first steps to do so are taken in the current project" (Shorten et al., 2003, p. 4).
10 Cook Islands, Federated States of Micronesia, Fiji, Kiribati, Marshall Islands, Nauru, Niue (New Zealand), Palau, Papua New Guinea, Samoa, Solomon Islands, Timor-Leste, Tonga, Tuvalu and Vanuatu.
11 Crowdsourced mapping allows registered users—in this case local organizations—to comment on and verify the produced maps. The method seeks to ensure that the maps are constantly updated.
12 The model underpinning the payout was developed by AIR.

References

ADB. (2009). *Natural catastrophe risk insurance mechanisms for Asia and the Pacific—Main report*. Manila: Asian Development Bank.

AIR. (2011). *AIR currents: The Pacific catastrophe risk insurance initiative*. Retrieved from http://air-worldwide.com/Publications/AIR-Currents/attachments/The-Pacific-Catastrophe-Risk-Assessment-and-Financing-Initiative/

AIR. (2012, September 12–13). *Pacific disaster risk financing and insurance initiative*. Presentation of Pacific Catastrophe Risk Models, Training for Pacific Disasters Managers, Nadi, Fiji.

Artemis. (2015). *Vanuatu paid $ 1.9m for cyclone pam from Pacific catastrophe facility*. Retrieved March 30, 2015, from www.artemis.bm/blog/2015/03/30/vanuatu-paid-1-9m-for-cyclone-pam-from-pacific-catastrophe-facility/

Bailey, B., Cook, S., & Mahul, O. (2015a). *PCRAFI country note—Fiji*. Washington, DC. Retrieved from http://documents.worldbank.org/curated/en/993781468029722952/Fiji-country-note

Bailey, B., Cook, S., & Mahul, O. (2015b). *PCRAFI country note—Marshall Islands*. Washington, DC. Retrieved from http://documents.worldbank.org/curated/en/209571468088772601/The-Marshall-Islands-country-note

Bailey, B., Cook, S., & Mahul, O. (2015c). *PCRAFI country note—The Cook Islands*. Washington, DC. Retrieved from http://documents.worldbank.org/curated/en/405171468244771591/The-Cook-Islands-country-note

Cook, S. (2011a). *Cook Islands investment in disaster risk management*. Suva, Fiji Islands.

Cook, S. (2011b). *Fiji investment in disaster risk management*. Suva, Fiji Islands. Retrieved from www.spc.int/DigitalLibrary/Doc/GSD/Bulletins/bulletin_d_information_de_la_SOPAC_april_september_2011.pdf

Dolowitz, D. P., & Marsh, D. (2000). Learning from abroad: The role of policy transfer in contemporary policy-making. *Governance*, *13*(1), 5–23. http://doi.org/10.1111/0952-1895.00121

Jarzabkowski, P., Bednarek, R., & Spee, P. (2015). *Making a market for acts of God—The practice of risk-trading in the global reinsurance industry*. Oxford: Oxford University Press.

Koshiba, F. (2007). *Statement on the economic impacts of climate change in the Pacific region*. SPREP Environment Ministers Meeting at the Forum Economic Ministers Meeting, Palua.

OCHA. (2015a). *Vanuatu: Tropical cyclone pam—Humanitarian snapshot as for 25 March 2015*. Retrieved August 20, 2019, from www.humanitarianresponse.info/operations/vanuatu/infographic/cyclone-pam-snapshot-19-march-2015

OCHA. (2015b). *Table A: Total funding and outstanding pledges as of 10 September 2015*. Retrieved from https://fts.unocha.org/pageloader.aspx?page=emerg-emergencyDetails&emergID=16569

PCRIC. (2018). *Press release—Tonga receives record insurance following cyclone gita*. Avarua, Cook Islands. Retrieved from https://reliefweb.int/sites/reliefweb.int/files/resources/TCGitaPCRICFinalPressRelease20feb.pdf

PIFS. (2003). *Risk financing for disaster*. Majuro, Marshall Islands.

PIFS. (2013). *Climate change financing: Pacific disaster risk financing and insurance program—Pacific catastrophe risk insurance pilot*. Nuku'alofa, Tonga.

PIFS. (2014). *Strengthening economic linkages: Climate change financing and disaster risk reduction—Pilot country experiences with the Pacific catastrophe risk insurance pilot*. Honiara, Solomon Islands. Retrieved from www.forumsec.org/resources/uploads/attachments/documents/2014FEMM.08_Pacific_Catastrophe_Risk_Insurance_Pilot.pdf

Power, M. (2009). The risk management of nothing. *Accounting, Organizations and Society, 34*(6), 849–855. http://doi.org/10.1016/j.aos.2009.06.001

Shorten, G., Goosby, S., Granger, K., Lindsay, K., Naidu, P., Oliver, S., . . . Walker, G. (2003). *Catastrophe insurance pilot study, Port Vila, Vanuatu: Developing risk management options for disasters in the Pacific region* (SOPAC Joint Contribution Report No. 147). Prepared for World Bank Office, Sydney, and AusAID, Canberra, Australia. Retrieved from http://siteresources.worldbank.org/INTPACIFICISLANDS/Resources/Catastrophe+Insurance+Volume+1_Report+Final.pdf

SOPAC. (2014). *About PCRAFI*. Retrieved March 16, 2017, from www.pcrafi.sopac.org/about/

World Bank. (2000). *Pacific regional strategy*. Washington, DC. Retrieved from http://documents.worldbank.org/curated/en/797421468774275109/Pacific-regional-strategy

World Bank. (2011). *Pacific disaster risk financing program*. Washington, DC. Retrieved from http://siteresources.worldbank.org/INTLACREGTOPURBDEV/Resources/840343-1319570618921/PDRFIS_pager.pdf

World Bank. (2012a). *Disaster risk financing and insurance program—Program pager*. Washington, DC. Retrieved from http://siteresources.worldbank.org/EXTDISASTER/Resources/8308420-1353538006746/4Pager_GFDRR-DRFI-program_final_Nov21_2012.pdf

World Bank. (2012b). *Pacific catastrophe risk insurance project—Project information document appraisal stage*. Washington, DC. Retrieved from http://documents.worldbank.org/curated/en/2012/10/16931813/pacific-islands-pacific-catastrophe-risk-insurance-project

World Bank. (2012c). *Pacific Islands: Disaster risk reduction and financing in the Pacific*. Retrieved August 1, 2016, from www.worldbank.org/en/results/2012/04/01/pacific-islands-disaster-risk-reduction-and-financing-in-the-pacific

World Bank. (2013). *PCRAFI—Catastrophe risk assessment methodology*. Washington, DC. Retrieved from http://siteresources.worldbank.org/EXTDISASTER/Resources/8308420-1342531265657/PCRAFI_REPORT_WEB_Final.pdf

World Bank. (2014). *The Pacific disaster risk financing and insurance program—Program pager*. Washington, DC. Retrieved from www.gfdrr.org/sites/gfdrr/files/publication/PCRAFI_Program Pager_FINALVERSION.pdf

World Bank. (2015a). *Pacific catastrophe risk insurance pilot—From design to implementation*. Washington, DC. Retrieved from www.gfdrr.org/sites/default/files/publication/Pacific_Catastrophe_Risk_Insurance-Pilot_Report_140715%281%29.pdf

World Bank. (2015b). *PCRAFI country note—Samoa*. Washington, DC. Retrieved from http://documents.worldbank.org/curated/en/443131468085732609/Samoa-country-note

World Bank. (2015c). *PCRAFI country note—Solomon Islands*. Washington, DC. Retrieved from http://documents.worldbank.org/curated/en/640541468100735571/Solomon-Islands-country-note

World Bank. (2015d). *PCRAFI country note—Tonga*. Washington, DC. Retrieved from http://documents.worldbank.org/curated/en/251921468291332622/Tonga-country-note 526

World Bank. (2015e). *PCRAFI country note—Vanuatu*. Washington, DC.

World Bank. (2016a). *Pacific catastrophe risk insurance pilot program: Implementation completion and results report*. Washington, DC. Retrieved from http://documents.worldbank.org/curated/en/655341475523018949/pdf/PIC-CatastropheRiskInsurancePilotSmallGrant-ICR-FINAL-09302016.pdf

World Bank. (2016b). *PCRAFI facility: Phase II*. Washington, DC. Retrieved from http://pubdocs.worldbank.org/en/178911475802966585/PCRAFI-4-pager-web.pdf

6 CCRIF, ARC and the PCRAFI Pilot in comparative perspective

This chapter approaches CCRIF, ARC and the PCRAFI Pilot from a comparative perspective. **The first part of this chapter ties the empirical insights together, elucidating commonalities and differences between the characteristics and histories of CCRIF, ARC and the PCRAFI Pilot.** It will provide seven tables (Table 6.1–6.7) to make the commonalities and differences between the goal, operational structure, (re-)insurance strategy, governance structure, additional features and histories of CCRIF, ARC and the PCRAFI Pilot visible.

The second part of this chapter abstracts away the differences and reflects on the commonalities between the characteristics and histories of CCRIF, ARC and the PCRAFI Pilot. It analyzes the attractiveness of CCRIF, ARC and the PCRAFI from a political and state-centric perspective. The purpose is to develop a more general and interpretative explanation as to why states from various regions across the Southern globe have come to make use of these instruments.

Part I: contrasting CCRIF, ARC and the PCRAFI Pilot: united in diversity

So what are commonalities and differences in characteristics and histories between CCRF, ARC and the PCRAFI Pilot?

Goals and payout modality

Objectives: economic vs social goal

It is notable that both CCRIF and the PCRAFI Pilot **have been designed to mitigate public cash-flow problems in the aftermath of a natural disaster.** The idea of the arrangements is to provide governments *readily* with limited funds so that they can intervene *immediately* and circumvent typical budget (re-)allocation processes and/or an international appeal. Accordingly, the instruments are best understood as *budget support tools*; they add to the existing tool kit to finance post-disaster emergency, relief and reconstruction measures and are not intended to replace them. CCRIF and the PCRAFI Pilot seek to provide a hedge against volatilities in public finances in the aftermath of a disaster and have a fiscal and macroeconomic goal. Designed akin to a private business interruption insurance, CCRIF and the PCRAFI Pilot provide governments ***quickly***—that is, within 14 days—**with limited funds** to overcome (potential) tax revenue shortfalls due the destruction of income-generating key centers of economic activity and enable them to secure the provision of public services.

ARC, by contrast, has been **designed to mitigate drought-related famine for the most vulnerable segments of society**; it is generically intended **to provide the poorest segments of society with cash or in-kind (e.g., food)** to prevent them from engaging in typical survival or coping strategies such as the selling of assets and stocks. The idea is to limit the direct and indirect, long-term impact of a drought on the poorest. ARC is best understood as a *social interventionist tool* that adds to (or scales-up) existing social safety net programs. It has a largely socioeconomic goal. The ARC is designed to enable insurance payouts *before* the event has even occurred. The idea is to empower governments to prepare interventionist measures *readily* in order to prevent the loss of life and/or prevent drought-affected people from engaging survival and coping strategies.

Payout modality: laissez-faire vs conditionality

Related to their diverging goals, CCRIF and the PCRAFI Pilot on the one hand and ARC on the other can also be distinguished in terms of their payout modality. **CCRIF and the PCRAFI Pilot are both insurance instruments characterized by a *laissez-faire* approach.** In case of a payout, the governments or, more precisely speaking, the finance ministries can spend the payout for whatever purpose. In contrast to international (development) aid, the funds are not 'earmarked', that is, tied to specific expenditures. A government may adopt any measures and distribute the money to whatever agency or target group—in line with needs, social expectations or domestic political pressure.

In contrast to CCRIF and the PCRAFI Pilot, **ARC is based on the conditionality principle.** To ensure the provision of cash, in-kind or services are received by the poorest segments of society, **the payout is tied to (the development and approval of) a contingency plan.** Once the criteria are fulfilled (=approval of contingency plans), a country gets a payout. Contingency plans must be developed by the countries themselves and are approved through an international peer review process. The review committee consists of ARC member countries. Countries are, however, largely free regarding the policy they want to implement to limit the socioeconomic impact of a drought, yet, they must be targeted at the poorest segments of society (see Table 6.1).

Insurance and reinsurance structure

Insurance policies: terms and premiums

CCRIF, ARC and the PCRAFI Pilot are all instruments that offer countries flexible insurance contracts. Countries can buy and tailor the coverage according to their need and financial capability. The premium level varies according to the coverage countries wish to select and their 'risk profile' (basically determined by a country's financial exposure to hazards). Countries can select three parameters (trigger/exhaustion point and ceding percentage). The rule of thumb is the more insurance coverage a country wants to buy and the higher its financial exposure, the higher the premium.

It is notable that ARC offers insurance coverage for more frequent events than CCRIF and the PCRAFI Pilot. In ARC's case the trigger point—the modeled event frequency for which countries can buy an insurance coverage—is set at 1:5 years events; in the CCRIF and the PCRAFI Pilot the trigger point is set at 1:10 years events (the PCRAFI Pilot) and 1:15 years events, respectively (CCRIF). All payouts are capped at a certain level. In CCRIF's case this level is fixed at US$ 100 million, ARC's case at US$ 30 million.

The total value of the insurance policies also varies considerably. It summarizes to US$ 623 million in the Caribbean and US$ 30 million in Africa. The different values largely reflect diverging economic development of these regions. It is notable that CCRIF has (constantly) written more insurance policies than it can actually finance. In an extremely busy season—a season, for instance, in which a hurricane strikes the three biggest Caribbean economies—CCRIF may face insolvency. While CCRIF classified the probability of such a 'black swan' (Taleb, 2007) event as extremely low (it modeled such an event as a millennial event), it still remains a possibility. Moreover, is it questionable to what degree such predictions can hold in the face of climate change. Climate change might not only alter the frequency and severity of hurricanes but also their paths. Black swan events might become 'white swans' events. In any case, in the unlikely event of the case described previously, CCRIF has insolvency guidelines in place.

Reinsurance structure: retention vs non-retention

The insurance policies offered by CCRIF, ARC and the PCRAFI Pilot are structured as conventional 'excess-of-loss arrangements'. As mentioned previously,

Table 6.1 Generic goals and payout modalities of CCRIF, ARC and the PCRAFI Pilot

Unit of comparison	*CCRIF*	*ARC*	*PCRAFI Pilot*
Perils covered	Hurricanes; earthquakes excessive rainfall	Drought	Tropical cyclones earthquakes (incl. tsunamis)
Primary goal	Mitigate short-term cash-flow problems of public finances	Mitigate drought-related famine	Mitigate short-term cash-flow problems of public finances
Secondary goals	Ensure provision of public relief and emergency services and speed up the beginning of recovery and reconstruction efforts Smooth long-term financial implications of a disaster Supply tools and knowledge for disaster management	Prevent behavioral survival and coping strategies and limit the (long-term) socioeconomic consequences of a drought Accelerate the speed at which international aid and assistance is mobilized and distributed	Ensure provision of public emergency services and speed up the beginning of recovery and reconstruction efforts Smooth long-term financial implications of a disaster
Total payouts	US$ 138,8 million *(December 2018)*	US$ 38,8 million *(December 2018)*	US$ 6,6 million *(December 2018)*
Time frame until payout	Within 14 days after disaster	2–4 weeks *before* disaster	Within 14 days after disaster
Payout modality	Unconditional *(laissez-faire)*	Conditional (tied to approval of contingency plans)	Unconditional *(laissez-faire)*

CCRIF, ARC and the PCRAFI Pilot pay out once a certain loss level has been breached. CCRIF and ARC thereby retain losses up to a certain level: CCRIF retains up to US$ 25 million, ARC up to US$ 7 million (insurance policy year 2016/17). Losses beyond that level are paid by commercial re-insurers and/or the capital markets via cat bonds (CCRIF). By contrast, the PCRAFI Pilot has no retention. This means that in the case of the PCRAFI Pilot there is no 'risk layering' involved as in conventional excess-of-loss arrangements. All losses are passed on straightforward from the *de facto* insurer (the World Bank) onto the capital markets via a cat swap. The rule of thumb is the higher the self-retention, the lower the (re-)insurance prices (see Table 6.2).

Technical basis

Payout and catastrophe risk model

CCRIF and the PCRAFI Pilot offer insurance based on a modeled-cost trigger. This means that they use a model to establish the financial costs of a disaster and the associated payout. The level of the payout is calculated with the aid of a standardized catastrophe risk model. They comprise four modules: the hazard module, the exposure module, the vulnerability/damage and the insurance module. The model ties together information of (1) the hazard value of a peril at specific location (in the case of CCIRF hurricanes, earthquakes, excessive rainfall; in the case of the PCRAFI Pilot: tropical cyclones, earthquakes and tsunamis) (2) the distribution/value of assets at those specific locations (3) the damage of those assets associated with the hazard value at this specific location and the subsequent financial replacement costs associated with that damage and (4) specific insurance contract details (e.g., loss level at which a payout is due, the level at which a payout is capped and the maximum coverage). By the nature of their design, the models value in particular the concentration of wealth and, thus, key economic centers.

Also, ARC offers insurance policies based on a parametric index trigger. The software estimates the costs of a social interventionist program *ex ante*—that is before the peril—the drought—has actually materialized. It compromises three modules: the hazard module, the vulnerability module, the cost module. The model ties together information regarding (1) rainfall shortage and related reductions of agro-pastoral production in specific area (2) the number of *poor* people affected at those specific areas (in ARC's case: The definition of 'poor' can be adapted by countries) (3) the financial cost to support those people (in ARC's case standardized at US$ 50 per affected person, but this can be customized by countries). As in the case of CCRIF/ARC, the data is then fed into the insurance module which contains specific information on the insurance contract, such as the cost level when a payout is due, the level at which a payout is capped and the percentage of the 'response costs' that are covered. By nature of its design, the model values in particular the concentration of the poorest and, thus, the 'poverty centers' of a country (see Table 6.3).

Operational structure

CCRIF is a stand-alone insurance facility, ARC an international organization (ARC Agency) and a stand-alone insurance facility (ARC Ltd), the PCRAFI

Table 6.2 (Re-)insurance structure of CCRIF, ARC and the PCRAFI Pilot

Unit of comparison	*CCRIF*	*ARC*	*PCRAFI Pilot*
Insurance season	1 June–31 May	1 June–31 May	1 Nov–31 Oct
Total annual premium income	US$ 19,5–22 million	US$ 8,5–17,5 million	US$ 1,1–2,1 million
Premium per country	n/a	US$ 0,4–4,5 million*	US$ 0,1–0,4 million
Maximum coverage offered (per country)	US$ 100 million	US$ 30 million	US$ 11 million
Aggregated coverage limit	US$ 623 million	US$ 60 million	US$ 43 million
Retention (by facility)	US$ 25 million	US$ 7 million	n/A
Reinsured losses (via re-insurers/ capital markets)	25–US$ 163 million	7–US$ 60 million	0 –US$ 43 million
Value of insurance-linked securities issued	US$ 30 million (cat bond)	n/a	US$ 43 million (cat swap)

Note: data on CCRF and ARC refers to the insurance policy year 2016/17. Data on the PCRAFI Pilot refers to the insurance policy year 2014/15

*refers to insurance policy year 2013/14

Table 6.3 Technical basis of CCRIF, ARC and the PCRAFI Pilot

Unit of comparison	*CCRIF*	*ARC*	*PCRAFI Pilot*
Coverage (for modeled loss events in years)	≥ 1:10 (excessive rainfall) ≥ 1:15 (hurricanes) ≥ 1:20 (earthquakes)	≥ 1:5 (drought)	1:10, 1:15, 1:20 (all perils)
Trigger type	Modeled-loss	Modeled-loss	Modeled-loss
Key variable in the model	Concentration of valuable assets	Concentration of vulnerable population	Concentration of valuable assets

Pilot, a development program. Given the greater degree of comparability I will only outline the differences between CCRIF and ARC.

In the case of ARC there exists a division of labor between the ARC Agency and ARC Ltd. The ARC Agency is responsible for all strategic and critical 'political' decisions (such as the admission of new members and/or the approval of contingency plans etc.) and ARC Ltd for all insurance-related day-to-day decisions (such as collection of premiums, investment decisions, etc.). The relations between the ARC Agency and ARC Ltd are fixed in the Establishment Agreement, an international treaty and its respective obligations in an Operations Manual. CCRIF, by contrast unites both tasks (political and financial) in one facility.

CCRIF is registered in the Cayman Islands, the ARC Agency in Johannesburg and ARC Ltd, the insurance provider for the ARC Agency, in Bermuda. That means both insurance providers (CCRIF and ARC Ltd) are domiciled in offshore tax havens. The key drivers for the selection of these offshore islands were political reasons and/or the provision of a certain service infrastructure and not so much tax benefits (although CCRIF and ARC Ltd are exempted from taxation). Legally speaking, CCRIF is licensed as a *captive*, ARC Ltd, by contrast, as a *mutual*. While both may be understood as self-insurance vehicles, from a legal perspective there are substantial differences. A *captive* is generally defined as an insurance company that is wholly owned and controlled by its insurance policyholders.[1] A *mutual*, by contrast, may be owned and controlled by unrelated parties. The background is that, in practice, ARC Ltd is owned and controlled by the ARC Agency—and, thus, African governments, the insurance policyholders (see Table 6.4).

CCRIF and ARC are virtual organizations. Neither on the Caribbean Islands nor in South Africa exists physical buildings or staff *per se*. They become 'real' on the internet (www.ccrif.org; www.africanriskcapacity.org). CCRIF and ARC are based on contractual relationships. Service providers realize all legal obligations and operational tasks. Tasks and procedures for service providers are fixed in an Operations Manual. The same applies also for the PCRAFI Pilot.

Governance structure

CCRIF, ARC and the PCRAFI Pilot all are characterized by a different governance structure (see Table 6.5). CCRIF and ARC are formally governed by the countries themselves; the PCRAFI Pilot, by contrast, is governed by the World

Table 6.4 Operational structure of CCRIF, ARC and the PCRAFI Pilot

Unit of comparison	*CCRIF*	*ARC*	*PCRAFI Pilot*
Launch	2007	2012	2013
Institutional structure	Stand-alone insurance facility	International organization + stand-alone insurance facility (ARC Agency + ARC Ltd)	Development Program
Legal structure of insurance facility	Captive	Mutual	n/a
Domiciled	Cayman Islands	Bermuda	n/a
Member countries 2018/19 (*countries with coverage in 2016/17; ** countries with coverage in 2015/16)	Anguilla*, Antigua and Barbuda*, Bahamas*, Barbados*, Belize*, Bermuda, British Virgin Islands, Cayman Islands*, Dominica*, Grenada*, Haiti*, Jamaica*, Montserrat, Nicaragua*, Panama, St. Kitts and Nevis*, St. Lucia*, St. Vincent and the Grenadines*, St Maarten*, Trinidad and Tobago*, Turks and Caicos Islands*	Burkina Faso*, Burundi, Central African Republic, Chad, Comoros, Republic of Congo, Côte d'Ivoire, Djibouti, Gabon, The Gambia, Ghana, Guinea, Guinea Bissau Kenya, Liberia, Libya, Madagascar, Mali*, Malawi, Mauritania*, Mozambique, Niger*, Rwanda, Sahrawi Arab Democratic Republic, Sao Tome and Principe, Senegal*, Sierra Leone, Sudan, The Gambia, Togo, Zambia, Zimbabwe	Cook Islands**, Marshall Islands**, Samoa**, Solomon Islands, Tonga**, Vanuatu**,
Countries covered	17	6	5
Organizational structure	Lean (based on contracts)	Lean (based on contracts)	World Bank administered

Table 6.5 Governance structure of CCRIF, ARC and the PCRAFI Pilot

Unit of comparison	*CCRIF*	*ARC*	*PCRAFI Pilot*
Governance style	Public-private collaboration	Public-private collaboration	Public-private collaboration
Governance pole	Governing Board CCRIF	Governing Board AU Agency	World Bank
Steering relationships	Steers relations with governments (insurance policyholders) and private service suppliers Governing Board defines goals, strategy, provides technical oversight, takes key decisions, defines access	Steers relations with governments (insurance policyholders) and insurer (ARC Ltd) Governing Board defines goals, strategy, provides technical oversight, takes key decisions, defines access	Steers relations with governments (insurance policyholder) and private service suppliers
Composition Governing Board	Intergovernmental representatives (CARICOM) + international financial institution (CDB) + private service suppliers	AU Agency: Intergovernmental representatives (AU) + government representatives (ARC Members)	n/a

Bank, the *de facto* insurer. In the cases of CCRIF and ARC there exists a Governing Board that is responsible for all strategic directions and key decisions. In the case of the PCRAFI Pilot all relevant decisions were taken by the World Bank.

In the case of CCRIF, the Governing Board comprises representatives from multilateral institutions (CDB, CARICOM) and 'technical' experts—that is, financial experts. CDB and CARICOM (the Caribbean's finance ministers) appoint four out of five Board Members. This gives financial interests a great opportunity to shape the overall direction of the facility. This structural bias is remarkable insofar as since 2008–09—as CCRIF expanded its mandate and introduced the TA-program—CCRIF provides much more than insurance. It offers, among others, education and training risk management and small grants for risk minimization. Despite its expansion into the risk management field CCRIF did not integrate disaster management specialists (or bodies) into its Governing Body. This might be problematic given that it is typically "the least powerful—those lacking in power, resources, political voice and influence—who are the most severely impacted" (Handmer & Dovers, 2013, p. 73).

In the case of the ARC Agency, the eight-headed Governing Board is composed of six government representatives and two AU Commissioners. According to the Establishment Agreement, the representatives should have expertise in the area of insurance, food security, agriculture and/or disaster management. The government representatives are appointed by ARC members in accordance with a regional proportional representation system, the AU representative are appointed by the AU Chairperson in accordance with WFP. In the case of Pacific DRFIP, decisions are taken by a small unit within the World Bank.

Taken together, in the case of CCRIF and the PCRAFI Pilot international financial institutions play a greater, not to say, *the* main role in the governance of these instruments (CDB, CARICOM, World Bank). In the case of ARC there is a greater role for governments and/or (non-finance related) international entities (AU Commission; WFP).

CCRIF, ARC and the PCRAFI Pilot are all based on contractual relationships. That means, the Governing Board—respectively the World Bank—has to communicate and deal with a number of actors for various purposes; these are the governments, donors, catastrophe modelers, re-insurers and other financial market participants and other regional actors (e.g., CARICOM, CDM, AU). Moreover, CCRIF, ARC and the PCRAFI Pilot are constituted through a range of highly specialized service providers equipped with contracts. To grasp these public-private relations systematically and assess them is extremely difficult for an outsider.

Given its specific operational functioning, CCRIF is a hybrid between a public and private entity. It is responsible for political as well as for financial tasks. What makes it a public entity is that the Governing Board is composed of representatives from international financial institutions and its financial resources are (mostly) based on taxpayers' money. The premium, participation fees and its reserves have been sourced from governments and (multilateral) donors. Yet, while CCRIF is registered as a not-for-profit company that is formally owned by Caribbean governments, it functions very much like a for-profit company by investing premium incomes in various financial market products (among them, partly high risk bonds) to ensure its 'financial sustainability' and offer additional services to its 'clients' (the governments).

In the case of ARC the public-private boundaries are a clearly demarcated. ARC Agency is the political entity while ARC Ltd has a commercial role. Both are independent entities with their own Governing Boards and bodies. While ARC Ltd might still be sourced from public money—in particular with money from German and British taxpayers—it is responsible only towards its parent, the ARC Agency, to which it must report. In contrast to CCRIF, it does not need to find a balance between political demands and respective (insurance) policy implications. It 'simply' has to fulfill the guidelines set by the ARC Agency.

Instrument histories

The histories of the various instruments demonstrated that all three instruments have emerged from a fundamentally different packaging of events, actors, processes and dynamics (seeTable 6.6). The unique configurations of these events, actors, processes and dynamics have shaped the goal and design of the instruments and continue to do so: Since their first implementation, all instruments have gone through significant transformations over time and remain politically contested. The unique configurations can account for both the differences *between* and *within* the instruments (see Table 6.7).

The history of CCRIF

CCRIF, for instance, developed out of the interplay between previous *ideas*, an *exogenous event* (Hurricane Ivan), a *policy failure* (Grenada), *policy learning* (CARICOM), *external sponsorship* (World Bank/donors) and a *participatory design process* (World Bank). Grenada's efforts to fund early relief and recovery measures in the aftermath of the devastating impact of Hurricane Ivan did not work. In this 'crisis situation' and in an effort to 'learn lessons' from this experience, the CARICOM's Heads of State turned to the World Bank and mandated it to develop a regional catastrophe insurance scheme. The World Bank was 'mandated' because CARICOM knew the World Bank would have the concept readily available given its technical expertise due to its previous proposal for a regional scheme and its failed effort to set up a sub-regional insurance pilot. Once mandated, the World Bank learned its own lessons and modified its previous proposal by modeling it on the experiences of the policy failure in Grenada. Embracing a 'participatory design approach' it worked in close interaction with the broader policy community at the regional and local level, which inluded representatives from within and outside governments (e.g., CARICOM, CDB, finance ministries, disaster management agencies, NGOs), to secure the relevance of the proposal and political commitment. Once this policy window opened, the World Bank also made sure the instrument would be implemented by providing not only the technical leadership but also 'financial support'. Since its first implementation, CCRIF has made 36 payouts totaling to about US$ 130 million. Yet, a slight tendency of 'underinsurance' (negative basis risk) and criticism and demands led the Governing Board of CCRIF to review the instrument's purpose. As a consequence, CCRIF expanded gradually but substantially its activities and products. Today, it provides not only insurance coverage but also disaster-related services and capacity building.

Table 6.6 Distinctive features of CCRIF, ARC and the PCRAFI Pilot

Unit of comparison	*CCRIF*	*ARC*	*PCRAFI Pilot*
Unique features	First multi-country insurance pool in the world First World Bank-issued cat bond	Makes a payout before a drought has materialized *(forecast-based financing)*	World Bank acts as *de facto* insurer
Additional services	Capacity Building Program ('TA program'): includes scholarship program, workshops, conferences, small grants for risk reduction education and measures Access to 'risk maps' and satellite-based weather forecasting software Access to a dense network of scientific and disaster-related service agencies	Capacity Building Program: cross-sectoral trainings, workshops and conferences on contingency planning Access to 'risk maps' and satellite-based weather forecasting software	Capacity Building Program: regional knowledge sharing workshops Access to 'risk maps'
Knowledge transferred	Competencies specific for the arrangement through training and education in risk minimization and contingency planning	Competencies specific for the arrangement through training and education in contingency planning	Competencies specific for the arrangement

Table 6.7 Key characteristics of the instrument histories CCRIF, ARC and the PCRAFI Pilot

Unit of comparison	*CCRIF*	*ARC*	*PCRAFI Pilot*
Instrument constituency	World Bank	WFP-affiliated 'start-up' team	World Bank
Key financiers	Multi-Donor Trust Fund (World Bank, Japan, Canada, UK, France, Ireland, EU, Bermuda, CDB)	Rockefeller Foundation, Germany (KfW), UK (DFID)	World Bank, Japan, Asian Development Bank, Australia, EU
Regional vehicle	CARICOM	African Union	n/a
Design/implementation approach	Participatory	Participatory	Participatory
Nature of approach	World Bank engaged with the broader policy community actors at the regional and local level, including representatives from within and outside the government (e.g., CARICOM, CDB, finance ministries, disaster management agencies, NGOs)	WFP-team engaged with the broader policy community at the regional and local level, including representatives from within and outside the government and key stakeholders in CCRIF on how to best design the instrument and effectively influence the policy agenda and political process (e.g., AU, World Bank, FAO, finance ministries/ food security/ ministries, disaster management agencies, NGOs)	World Bank engaged with core members of relevant policy network, especially local government officials responsible for policy implementation (finance ministries)
Purpose of approach	Ensure relevance + political commitment	Gain mandate + ensure legitimacy	Ensure political commitment

The history of ARC

ARC, by contrast, developed largely out of a series of *policy experimentations*, a *financial sponsorship* (Rockefeller Foundation), *playfulness* (ARC Project Team), cross-regional *policy learning* (CCRIF) and *a participatory design approach*. Since the 1990s, various ideas have been discussed and pilot projects been implemented within the UN context in Africa to restructure the financing of development aid. The key trigger for the development of ARC was, however, the Rockefeller Foundation who wanted to support something 'off-the-shelves' and funded, based on a scientific article, a small project team to develop and realize some novel ideas concerning how to restructure disaster aid in selected countries. Informed by the scientific article of the team leader, the project team developed a software and played around with what it could do with it. At the same time, CCRIF was just implemented and the team reached out to key proponents to obtain knowledge and 'learn lessons'. It decided to transfer the arrangement as well as the political approach to the African context. Things, however, went differently as the AU insisted on local ownership. As a consequence, the team developed the instrument in close collaboration with the AU. While the project team took over the technical leadership and organized funding, the AU secured the political support. ARC came out of a very participatory process with strong local ownership. Since its implementation ARC has made five payouts totaling to about US$ 36 million.

The history of the PCRAFI Pilot

The PCRAFI Pilot, in turn, emerged through a different story. The history of the Pacific instrument resembles, in fact, a textbook case of a (cross-regional) *policy transfer*. The idea to set up a regional insurance arrangement was discussed within the region since the late 1990s, when the World Bank briefed the SIDS on a global conference on its pilot insurance project in the Eastern Caribbean. This triggered discussion and efforts within the Pacific Forum, the main regional political forum, to develop something similar, but due to the lack of adequate data the project stalled. In 2007 the World Bank increased its own efforts in the region and together with a number of other actors launched a massive, four-year long data collection program, which created new geo-scientific and economic knowledge on disasters. Once finished, this new body of knowledge represented the *conditio sine qua non* for the development of insurance instruments. Meanwhile the World Bank increasingly systematized its efforts by launching a more global orientated project—the *Disaster Risk Financing and Insurance Program*. With the aid of the Asian Development Bank it organized a large, regional, high-profile conference on the "Development Catastrophe Risk Insurance" in Tokyo to sensitize public and private representatives on the benefits of insurance and public-private collaboration. In 2011, inspired by the 'success story' of CCRIF, the World Bank offered the small Pacific states a fully sponsored pilot program on a regional insurance scheme. The World Bank largely copied the Caribbean model and transferred it to a different context to test demand and viability. Drawing on its experience in the Caribbean, it engaged with core members of relevant policy networks, especially local government officials responsible for policy implementation (finance ministries). Since its implementation, PCRAFI (Pilot and PCRIC) has made three payouts totaling to about US$ 6,6 million.

Part II: the puzzle and the answer

The comparison revealed that while CCRIF, ARC and the PCRAFI Pilot share a number of commonalties, more often than not they differ substantially in terms of their goal, form and history. All instruments have been shaped in a unique way to suit their specific purpose and have seen the light through a very different packaging of events, actors, processes and dynamics. CCRIF, for instance, is instrumental in providing 20 countries (including Nicaragua) with short-term liquidity in the aftermath of a major hurricane, earthquake or excessive rainfall. ARC, by contrast, is instrumental in providing five countries with limited funds to set up a social interventionist program in the advent of a drought. The PCRAFI Pilot, in turn, is instrumental in providing five countries with an insurance coverage against tropical cyclones, earthquakes and tsunamis. Accordingly, all three instruments are based on a *distinctive* (re-)insurance structure, a *distinctive* operational structure, a *distinctive* technical basis and a *distinctive* governance structure. A more abstract and *political* perspective, a perspective that is sensitive to power relations embedded in the arrangements and their history, reveals, however, that CCRIF, ARC and the PCRAFI Pilot have much more in common in terms their characteristics and histories than these empirical insights suggest on first sight:

First, all three instruments empower the executive. They provide governments with quick and diversified access to funding which reduces their dependency on international aid. This allows governments to skip erstwhile usual and 'nasty' budget reallocation processes and move on to planning and action. In addition, they provide governments with social capital: They give governments a wide-ranging access to technology and knowledge and, thus, to a disaster-related infrastructure and a number of disaster-related services they could possibly not afford otherwise.

Second, all three instruments can be employed in flexible ways. The insurance schemes are highly 'customizable' and 'adaptable' and thus compatible with various political agendas. Within certain parameters governments are largely free to decide how much coverage they want to obtain, and they pay accordingly. As the schemes are enabled through a number of (short-term) service provision agreements, they also can be quickly restructured to accommodate changing political demands and circumstances. This built-in flexibility (or openness) allows for the integration of various interests and actors.

Thirdly, all three instruments were sponsored by development actors. Development actors were strongly involved in devising the instruments and facilitating its adoption. They inserted the idea into local policy circles, provided the intellectual groundwork, steered and coordinated all activities related to instrument design and development, covered design costs, organized funding and/or provided their own financial support for implementation and provided financial and insurance literacy. Doing so, they took care to involve the local policy community early on in the design and implementation process to secure a political buy-in and provide the instruments with substantial input-legitimacy. In short, they facilitated the adoption.

I claim these three reasons of *empowerment*, *differentiated functionality* and *external facilitation* can explain why states from various regions across the Global South have come to adopt these regional catastrophe risk pools. In the next section I will elucidate my argument.

Empowerment

Regional catastrophe risk pools empower the executive in three ways:

First, all instruments **provide governments**—or more precisely speaking the finance ministries (!)—**with a quick and diversified access to funding.** CCRIF and the PCRAFI pilot pay out within 14 days, ARC even before the event has occurred. These characteristics give governments the capacity, that is, the power to act quickly after/just before a disaster. Hence, the instruments ensure agency and allow for accelerating executive processes. Governments can erstwhile skip typical and 'nasty' budget reallocation processes and/or circumvent the negotiation and design processes associated with alternative ex-post financing instruments (such as taxes, grants or loans) and proceed to the distribution of funds *immediately*. This implies that the instruments somewhat serve to depoliticize finance politics. Simply put, buying insurance means buying a 'right' to compensation. In this context, it is also important to note that instruments are not intended to crowd out existing disaster financing tools but to supplement them. They are intended to provide governments with *additional* funds and thus with a *diversified* access to funding. These additional funds are sourced from other governments from the region (pooling of premium payment), donors (initial capitalization), re-insurers and the capital markets.

Second, **the instruments free governments from political and economic constraints.** The insurance payouts come with no or comparatively little strings attached. In the case of CCRIF and the PCRAFI Pilot governments may use the payout for whatever purpose. This *laissez-faire* approach enables governments to act in a highly flexible manner. In contrast, for instance, to international aid, the funds are not tied to specific expenditures. Moreover, in contrast to loans or credits the insurance payouts do not have to be repaid. Countries do not have to service debts and thus may avoid the power relations that typically structure debt-creditor relationships. Countries are in full control of the money; this means that the insurance liberates them from conditionalities. The same argument applies also to ARC, albeit to a limited degree. Even though the insurance payout is tied to a contingency plan, countries enjoy a comparatively high degree of freedom in terms of how they want to use the payout. The key condition they need to meet is that the payout must serve the poorest segments of society. If this criteria is fulfilled is decided through a 'peer review' mechanism; the peers consist of the other insurance policyholders, that is, other ARC members. That means, the discussion of a fulfillment is *in practice* based on a political negotiation.

Third, all instruments **provide governments with access to new technology, knowledge and domains of knowledge exchange—and, thus with social capital.**[2] CCRFI, ARC and the PCRAFI Pilot are all based on a range of international, intra-regional and inter-sectoral collaborations. Through these collaborations countries gain more structured access to technologies, knowledge and disaster-related services. Countries are provided with hazard-related data and maps at the regional and country level and with opportunities to share and learn lessons of 'best practice'.

This is most notable with regard to CCRIF (which is also the 'oldest' instrument): It offers its members not only access to a satellite-based weather forecasting software but also access to a dense network of specialized scientific and disaster-related service agencies. Collaborations include programs with the Caribbean Disaster Emergency Response Agency, Caribbean Institute of Meteorology and Hydrology and University of the West Indies. The collaborations are funded, among others, with grants from the Global Facility for Disaster Risk Reduction (GFDRR), a World Bank-administered multi-donor trust fund and the International Climate Initiative (IKI), Germany's international climate policy vehicle, and involve, for instance, the development of 'risk maps' for natural hazards. A highlight is CCRIF's Technical Assistance (TA) program, which offers small grants for disaster-related prevention/adaption measures, workshops and training for government officials as well as scholarships for postgraduate studies in the US and UK. Within this program countries gain competencies specific for the arrangement through training and education in risk minimization and contingency planning. All these collaborations and services help to transfer, diffuse and build up disaster-related knowledge. Albeit these features are mostly by-products and fulfill mainly a secondary, legitimizing function (to keep up demand for policy renewal and, thus, secure the organizational survival) it is hard to deny that they make a membership in CCRIF ultimately more attractive. For a country to leave CCRIF, it would mean missing out on these opportunities.

ARC and the PCARFI Pilot offer these by-products too, albeit to a lesser extent than CCRIF. They provide its members with access to satellite-based 'risk maps' as well as capacity building measures. The data that underpins these maps and/or forecasting tools is based on satellite technology and software able to process big data. The data, technology and software have been out of reach for most countries that have bought the insurance—due to costs and/or capacity reasons. Hence, by buying into the insurances the countries buy into a range of services.

Differentiated functionality and high integration capacity

All instruments **can be employed in flexible ways, as they are compatible with various political agendas.** They are highly 'customizable' and 'adaptable'.

First, within certain limits countries **are largely free to decide how much coverage they want to obtain and pay accordingly**. Countries can basically choose their coverage level, attachment and trigger points. They can change their preferences every year, which gives them great flexibility to react to a changing fiscal space. Also, in the case of CCRIF, for instance, countries need not necessarily buy an insurance contract to enjoy other benefits such as the TA-program. The built-in flexibility manifests in the "differentiated functionality" (Capano & Lippi, 2013, p. 14) that characterizes the various instruments.[3] As highlighted previously, every instrument is unique in terms of its specific field of application and workings. All instruments have distinctive features that set them apart from instruments of the same family.

Second, **the instruments can be quickly restructured to accommodate feedback, changing demands of stakeholders or political and economic circumstances.** The reason is that they are enabled through a number of (short-term) service provision agreements. This characteristic reduces the barrier for the governing bodies embedded within the institutional setting of the instruments to restructure the goal and functioning of the instrument. It allows for a quick integration of various goals and interests. In other words, the instruments have a "high integration capacity" (Capano & Lippi, 2013, p. 16) and can fairly easily integrate an increasing number of actors, goals and interests.

This is best illustrated with reference to CCRF. The Governing Board of CCRIF, for instance, reacted quickly to political demands made by member states: In 2008, as a result of critics, CCRIF developed a new rainfall product and updated its payout model. Also, over the years and with the adoption of the TA-program, CCRIF has substantially expanded its activities and accommodated the interests of the local disaster management community. While initially conceptualized as a purely financial tool, over time CCRIF has become a regional hub for disaster-prevention. Also ARC was quick in reacting to demands from its member states as it has embarked on expanding its 'product line'.

External facilitation

Development actors created a 'supply push'. They made tremendous efforts to stimulate the adoption of regional catastrophe risk pools. In all cases development actors intervened at strategic leverage points to facilitate the adoption. I claim these activities made the adoption technically, financially and politically feasible in the first place. They prepared the grounds for a receptive environment.

First, in all three cases development actors **provided a design template and created an enabling environment**. They inserted a first concept note into regional policy circles and secured insurance literacy by providing high-level and upper-level policymakers with substantial knowledge about the workings of insurance in general and the instrument in particular.

In the case of CCRIF and the PCRAFI Pilot it was the World Bank, in the case of ARC it was a WFP-affiliated Project Team who came up with the idea to set up a regional insurance arrangement. They all disseminated the idea into regional policy circles (CARICOM, AU, Pacific Forum) using 'political champions', that is, top-level representatives. Once the World Bank and the ARC Project Team was mandated to develop the idea further, they worked closely and collaboratively with a number of local policy actors at various levels to assess needs, formulate goals and develop the instrument accordingly. In this context they created an enabling environment. Given the widespread lack of knowledge of insurance, as demonstrated in the case of ARC and the PCRAFI Pilot, the World Bank and the ARC Project Team organized workshops and trainings on the functioning of insurance for high-level and upper-level policymaker; they transferred competences specific for an insurance arrangement in the run-up of the formal adoption so that policymakers would make an 'informed' decision once they needed to decide if they really wanted to buy insurance. Through (in-)formal meetings, workshops and conferences, the instrument constituencies trained high-level policymakers on how insurance works in principle, how the regional insurance scheme would work in particular and how the software would work that underpins the arrangement.

I claim that by providing *insurance literacy* development actors created not only an enabling environment but had substantial—albeit hardly visible—influence to frame the decision and thus 'socialize' national decision-makers into the 'right' decision to adopt insurance. This influence derives from the unequal power that typically structures speaker-listener situations and the information and alternatives that were presented in these meetings, training, workshops and conferences. It is somehow notable that in all cases insurance development went rather smoothly into their adoption. In this context it is also worth pointing out that in no case did the development actors provide policymakers with alternatives to insurance.

Crudely simplified, development actors approached countries with 'solutions' and then tailored the 'solutions' to make sure they would match local needs and, while doing so, educated policymakers about their benefits. There was in fact no 'real' instrument selection process in the sense that regional policymakers sat down together and assessed different pros and cons of different options. Development actors prepared and developed the instruments carefully and created a receptive environment.

Second, in all three cases, development actors **sponsored the costs surrounding the instrument development process**. They covered not only the initial start-up costs (e.g., the development of data/software) but, once mandated, organized funding and/or provided financial support for implementation (e.g., premium support, contribution donor capital).

This central role that external sponsorship played is possibly best illustrated with reference to the history of the PCRAFI Pilot. In case of the PCRAFI Pilot, the World Bank together with regional partners (Asian Development Bank, Applied Geoscience & Technology Division) made the insurance instrument viable through its huge investments in the creation of a novel geo-scientific and economic database. As shown previously in Chapter 6, the Pacific Catastrophe Risk Assessment and Financing Initiative was a massive data collection program that, with the aid of state-of-the-art technologies and practices (analysis of satellite images; crowd-sourced mapping) took stock of the Pacific Island's exposure to hazards. This newly created data enabled vulnerability analyses and thus the creation of risk profiles. These profiles were necessary to develop an insurance instrument in the first place, as there is no insurance without exposure data and an inventory of assets.

Financial sponsorship was also of key relevance in the case of ARC and CCRIF: ARC emerged out of a project financed by the Rockefeller Foundation. Keen to fund something 'off-the-shelves' the foundation provided a grant to a project team to develop a software and some ideas for its application. It is hard to imagine the history of ARC without the seed funding provided by the Rockefeller Foundation—in particular given that WFP initially declined to support the project idea. Also, in the case of CCRIF, the World Bank invested substantial resources to make sure the idea was then also translated into practice: (a) It used its convening power to organize funding from various donors to cover initial start-up costs (b) it created a Multi-Donor Trust Fund to provide indirect premium support. The Multi-Donor Trust reimbursed some of CCRIF's expenditures. The possibility to reimburse some of its costs allowed CCRIF to invest more of its capital stock in bonds and thus accelerate the growth of capital reserves (through the gained investment incomes). Having greater capital reserves enabled CCRIF to buy less reinsurance coverage. This, in turn, translated into low(er) country premiums (c) It provided direct premium support. As some countries were still not able (or willing) to buy insurance, the World Bank along with the Caribbean Development Bank stepped in and provided direct premium support. They sponsored the premiums for some countries for the first three policy years. The premium of Haiti is still sponsored today.

Third, in all three cases development actors **worked closely and collaboratively in a participatory design process to tailor the instrument to the local context and secure a political buy-in.** They opted for an inclusive design approach and, in doing so, provided the instrument with substantial input legitimacy.

In contrast to a 'classic' instrument design process, both instrument constituencies (World Bank/WFP-affiliated start-up team), once mandated, opted for a collaborative, participatory instrument design process. They constantly involved policymakers at national and regional levels to learn about their needs and interests. The instrument constituencies closely engaged with representatives from within and outside a government (in particular representatives from the finance ministries but also representatives from disaster management agencies, NGOs and other relevant policy actors) to inform the design, tailor and revise the instrument to local preferences. This *adaptive learning* approach helped to foster a sense of local ownership and secured the political commitment over time at various political levels, as demonstrated in the various case studies. Most importantly, however, it provided the instrument with substantial input legitimacy. The input-legitimacy is most evident in the case of ARC: Here the AU played a key role in the instrument design process—paradoxically against the initial intention of the instrument constituency.

The observation that all instruments were developed in a participatory manner must, however, not veil the fact that the actual purpose of this participatory design process and degree to which it was realized varied tremendously across the cases: In the case of ARC the participatory design process embraced by the WFP-affiliated start-up team and was initially largely driven by a *political purpose*, namely to influence the AU agenda and receive the political legitimacy to develop the instrument on the AU's behalf (and not with them). The resistance or better insistence of the AU to develop the instrument in a collaborative manner resulted, however, in a lengthy, four-year long design processes (which included, among other things, the negotiation of an international treaty). While the WFP-affiliated start-up team was in charge of the technical design, the AU policy organs were involved in key decisions by rejecting and validating certain proposals. The design process was thus highly politicized and the project was sometimes close to failure. Yet, as the AU played a key role it shaped to a substantial degree ARC's final institutional design.

In the case of CCRIF, the participatory approach served to ensure local *relevance*. For years the instrument constituency, the World Bank, had been in search of a problem it could 'solve' with an insurance instrument. Once mandated by the CARICOM and endowed with financial resources, it engaged widely with the broader regional policy community to get the problem framing right and ensure the instrument's 'good fit'. It collected experiences and knowledge from a diversity of actors that came all with different frame, language and interests (e.g., government representatives, disaster management specialists, insurance representatives). It involved input and feedback from numerous policy and administrative actors at various levels through various circles. Yet, in contrast to ARC, no regional political bodies were involved in the key design questions.

In the case of the PCRAFI Pilot, the participatory approach served rather to *improve* the instrument and develop a regional version of CCRIF. The PCRAFI

Pilot was based on a policy transfer. The World Bank basically adapted the Caribbean model to the Pacific. Accordingly, the design process was initially 'only' informed by an internal and instrumental learning process. The World Bank team 'simply' transferred the concept of CCRIF, including the lessons it had learned from establishing and running CCRIF to the Pacific. The learning process was instrumental insofar as the World Bank did not start with a new problem investigation and an off-the-ground concept development—as it did in the Caribbean—instead, it considered what it could do better with reference to an instrument it had successfully designed and implemented elsewhere. It created the PCRAFI Pilot within a pre-defined problem description and thus with pre-defined policy goal: The starting point was the problem it identified in the Caribbean, a public 'liquidity gap' after a disaster and the need of a quick payout. The stakes of local actors from the Pacific in the design process were therefore much more limited. They were involved in reviewing the instrument during the implementation process, the customization of insurance policies and the tremendous capacity building.

Notes

1 The International Association of Insurance Commissioners (IAIS) defines a captive as "an insurance or reinsurance entity created and owned, directly or indirectly, by one or more industrial, commercial or financial entities, other than an insurance or reinsurance group entity, the purpose of which is to provide insurance or reinsurance cover for risks of the entity or entities to which it belongs, or for entities connected to those entities, and only a small part if any of its risk exposure is related to providing insurance or reinsurance to other parties" (Hall, 2006).

2 As probably any concept in the social sciences, 'social capital' is a highly contested one. I refer here to Bourdieu's definition as "the sum of the resources, actual or virtual, that accrue to an individual or a group by virtue of possessing a durable network of more or less institutionalized relationships of mutual acquaintance and recognition" (Bourdieu & Wacquant, 1992, p. 119) and apply it to the state level. For the applicability of Bourdieu's notions for International Relations see exemplarily Adler and Pouliot (2011) and the related 'practice turn' in IR.

3 With "differentiated functionality" I refer to the characteristics that make the instrument unique, that is, totally different from others. Differentiation means distinction (Capano & Lippi, 2013, p. 14).

References

Adler, E., & Pouliot, V. (2011). *International practices*. Cambridge: Cambridge University Press.

Bourdieu, P., & Wacquant, L. J. D. (1992). *An invitation to reflexive sociology*. Chicago: University of Chicago Press.

Capano, G., & Lippi, A. (2013). *Enlightening policy mix complexity: A typology of actors' choices*. Conference Paper, Grenoble. Retrieved from www.icpublicpolicy.org/IMG/pdf/panel_11_s3_capano.pdf

Hall, S. (2006). *Recent developments in the captive insurance industry*. Retrieved February 6, 2017, from www.naic.org/cipr_newsletter_archive/vol2_captive.htm#_edn1

Handmer, J., & Dovers, S. (2013). *Handbook of disaster policies and institutions*. Abingdon: Routledge.

Taleb, N. (2007). *Black swan—The impact of the highly improbable*. London: Allen Lane.

7 Conclusion and implications

Conclusion

The purpose of this book was to shed light on the increasing popularity of index insurance instruments to finance the impact of disasters. This is evidenced, among others, by the cross-regional diffusion of regional catastrophe risk pools across the Global South. Regional catastrophe risk pools are generically designed to provide states with an index-based insurance protection in the case of a major disaster. Despite there exists only limited experience with these tools and although instrument utility is usually highly context-dependent, regional catastrophe risk pools are increasingly used by various states in the Caribbean, Africa and the Pacific. To understand these disaster financing tools and explain their cross-regional uptake across the Global South has been the core of this book. So what have we learned?

The analysis showed first of all that regional catastrophe risk pools differ substantially in terms of their goal, form and history. CCRIF, for instance, has been designed to mitigate cash-flow problems and aims to provide Caribbean countries with short-term liquidity in the aftermath of a major hurricane, earthquake or excessive rainfall to kick off early relief measures. ARC, by contrast, aspires to provide African countries with the necessary funds in advance of a drought to facilitate the implementation of limited social interventionist program to prevent famine among the poorest segments of society. The PCRAFI Pilot, in turn, aims to provide the Pacific Island Countries with short-term liquidity to allow them to cover the emergency costs to address the immediate impacts of tropical cyclones, earthquakes and tsunamis. Depending on their specific purpose and the hazard context, they are all based on a *distinctive* (re-)insurance structure, a *distinctive* operational structure, a distinct technical basis and a *distinctive* governance structure. All instruments have been shaped in a unique way to suit their specific purpose.

Analogous to their unique characteristics, all three regional catastrophe risk pools have all seen the light through a different story. All instruments have emerged from a fundamentally different packaging of events, actors, processes and dynamics. The unique configurations of these events, actors, processes and dynamics have shaped the goal, design and ultimately also adoption of these instruments and continue to do so. CCRIF, for instance, developed out of the interplay between previous *ideas*, an *exogenous event* (Hurricane Ivan), a *policy failure* (Grenada), *policy learning* (CARICOM), *external sponsorship* (World Bank/donors) and a *participatory design process* (World Bank). ARC, by contrast, developed largely out of a series of *policy experimentations*, a *financial sponsorship* (Rockefeller Foundation), *playfulness* (ARC Project Team), cross-regional *policy learning* (CCRIF) and *a participatory design process*. The history of the Pacific instrument resembles, in fact, a textbook case of a (cross-regional) *policy transfer*. Inspired by the successful launch of CCRIF the World Bank offered the small Pacific states a fully sponsored regional insurance pilot program to test demand and viability in a different context.

This analysis suggests that regional catastrophe risk pools are highly generic tools with little commonalities but their diffusability. A more abstract, power-sensitive perspective showed, however, that CCRIF, ARC and the PCRAFI Pilot have much more in common in terms of their characteristic and genesis than at first sight.

First, **all three instruments empower the executive**. They provide governments with quick and diversified access to funding which reduces their dependency on international aid. This allows governments to skip erstwhile typical and 'nasty' budget reallocation processes and move on to action. In other words, they contribute to depoliticizing disaster aid and strengthen agency. In addition, the regional catastrophe risk pools provide governments with a wide-ranging access to technology and knowledge and, thus, to a disaster-related infrastructure and a number of disaster-related services they could possibly not afford otherwise.

Second, **all three instruments can be employed in flexible ways.** The insurance schemes offered by the pools are highly 'customizable' and 'adaptable' and thus compatible with various political agendas. Within certain parameters, governments are largely free to decide how much coverage they want to obtain and they pay accordingly. As the schemes are enabled through a number of (short-term) service provision agreements, they also can be quickly restructured to accommodate changing political demands and circumstances. This built-in flexibility (or openness) allows for the integration of various interests and actors.

Third, **all three instruments were sponsored through development actors.** They made tremendous efforts to stimulate the creation of regional catastrophe risk pools. They created a receptive environment and were strongly involved in devising and promoting the instruments. They inserted the idea to set up insurance mechanism into local policy circles, provided the intellectual groundwork, steered and coordinated all activities related to instrument design and development, made sure to involve the local policy community, covered design costs, organized funding and/or provided their own financial support for implementation and provided insurance literacy. Doing so, they were able to provide the instruments with substantial input-legitimacy and secure a political buy-in at the country level.

I claim the *empowering function* and the *differentiated functionality of these instruments* as well as the '*supply-push*' by development actors can explain why states from various regions across the Global South have increasingly come to adopt these regional catastrophe risk pools.

Implications

Implications and recommendations for future research

The book contributes foremost to debates in Public Policy and the policy instruments literature especially. Yet, it also contributes to International Relations and important debates in other sub-fields, among them, Financial History and Governance Studies.

Public Policy: Demonstrates that regional index insurance instruments are not just insurance tools

The study provided, to my best knowledge, **the first comprehensive comparative study on** regional catastrophe risk pools. Doing so, it demonstrated first and foremost that what appears (and, at times, is presented) as a monolithic bloc is, in fact, a very complex instrument that needs further specification. Regional catastrophe risk pools have been—and can be shaped—in various ways to serve different purposes and target groups. By elucidating these differentiated goals and functionalities, this study paves the grounds for future analyses to better categorize these instruments and speaks therefore directly to the functionalist policy instrument literature (Hood, 1983; Hood & Margetts, 2007; Howlett, 1991, 2011; Lascoumes & Le Gales, 2007; Salamon, 2002).

The study suggests, for instance, that **regional catastrophe risk pools are not 'classic' disaster financing tools**. By buying into a pool, countries buy not only a 'right' to access financial compensation but also a 'right' to access disaster-related knowledge and services. Countries, once a member of CCRIF or ARC, gain access to new technology, knowledge and knowledge exchange platforms. To some extent they allow countries to address the lack of public investment in disaster-related infrastructure and technology. From a development actor's perspective, the instruments serve also to build up institutions—rules and laws—and administrative capacities for disaster risk management. Doing so, they contribute to state- or governance-building. Given the numerous functions regional catastrophe risk pools serve I propose to categorize them more broadly as **'disaster management' or 'disaster governance' tools.**

Moreover, this study has demonstrated that the pools are comprehensive and generic tools that facilitate the bundling of various goals, actors and interests. They are characterized by a "differentiated functionality" and have a **high "integration capacity"** (Capano & Lippi, 2013, p. 16). In that sense they **might be superior to other similar (financial) instruments**. It is also precisely this differentiated functionality and the high integration capacity, which makes regional catastrophe risk pools 'diffusible' in the first place. These claims must, however, be subject to a further critical discussion. This study hopes to prompt a new generation of studies that seeks to categorize these instruments by putting them and their capacities in relation to other available tools.

By providing a comprehensive overview the study paved also the grounds for scholars interested in *explaining* the commonalities and differences between the instruments, rather than merely just highlighting and elucidating them as this study did. A root for such an explanation could be found in this book, namely in the unique configuration of f(a)ctors through which these instruments have come about. Historical comparativists or political sociologists may wish to dig deeper here and consider, for example, the larger ideational and or politico-economic context and/or various organizational cultures that accompanied instrument design, implementation and evolution.

Public policy: shows how instrument constituencies match solutions and problems

The study **provides empirical support for the accuracy and usefulness of the notion of 'instrument constituencies'** in order to understand the selection of a policy instrument. In this sense, this research should speak in particular to scholars interested in this emerging concept (Béland & Howlett, 2016; Jordan & Huitema, 2014; Mukherjee & Howlett, 2015; Voß & Simons, 2014) but also to scholars grasping to explain instrument choice and policy transfer/learning more in general (Boscarino, 2009; Bulmer & Padgett, 2005; Capano & Lippi, 2013; Dolowitz & Marsh, 2000; Hall, 1993; Howlett, 2011; Kingdon, 1995; Lascoumes & Le Gales, 2007; Linder & Peters, 1998; Wurzel, Zito, & Jordan, 2013).

As elaborated in Chapter 1, 'instrument constituencies' may be understood as policy actors that strongly dedicated to promoting and designing *one* specific policy instrument, independent of the problem context and can be distinguished from similar actors such as 'policy communities', 'advocacy coalitions' or 'epistemic communities' (Voß & Simons, 2014, p. 740). As this study has revealed, the World Bank and the WFP-affiliated project team put substantial efforts in developing a regional insurance 'solution' without seriously considering an alternative. The alternative they presented in available cost-benefit analyses was the case of not buying insurance. They did not create a choice of menu. Instead, they had an insurance 'solution' at hand that 'just' needed to be tailored. In doing so, they were successful. Albeit lacking a counterfactual, **it seems implausible that regional catastrophe risk pools would have been adopted without the tremendous support of these actors**. Against the background of their fixation and the tremendous efforts they have put into making the insurance arrangement work, the World Bank and the WFP-affiliated project team may be understood as 'instrument constituencies'.

This study should be of particular interest to the scholars working on the concept of 'instrument constituencies' because it **provides detailed empirical insights into *how* instrument constituencies defined and promoted a specific policy instrument**. Doing so, the study provides input for a theory on when instrument constituencies are successful and what resources they need to bring in.

This study indicates, for instance, that key success factors involve convincing 'political champions' (e.g., higher and upper-level policymakers), 'working strongly and collaboratively with local actors and processes' (e.g., regional organizations), comprehensive 'service provision' (e.g., design of templates, development of software, organization and provision of trainings and workshops, etc.), including 'financial sponsorship', and 'adaptive learning' (e.g., constant deliberation and revision).

This **success strategy could be described as a 'facilitation approach'**, as an approach that uses a 'light touch' but does not interfere too strongly. It intervenes at strategic leverage points (e.g., political top level) and provides certain incentives or opportunities (services, trainings, finance) so that countries 'self-select' the instrument. Instrument constituencies steered and coordinated all the instrument development processes—that means, they were still in control of them—and, yet, they provided substantial room for co-creation; this room for maneuver allowed countries to buy into the idea of the instrument. It also equipped the instrument with substantial input legitimacy.

These insights imply also that **instrument constituencies require** substantial resources and organizational, analytical and diplomatic skills to give life to a policy instrument. Put differently, it requires **a certain "delivery capacity"**—that is, a certain analytical, convening and financial power—"to make things happen".[1] In addition, it requires time. The instrument constituencies invested substantial time to build trust, learn more about local needs, how to meet them, build up local skills and improve local knowledge.

Successful instrument constituencies need therefore a team (or a network of people) with a diverse set of skills, a pragmatic attitude to overcome challenges (such as a model that barely fits the purpose, administrative turnovers, military coups, power games, as highlighted in this study) as well as flexible organizational structures. Overall, instrument constituencies must act as innovators, communicator, relationship-builders, coaches and trainers. Only densely networked actors should be capable of providing all these functions.

It is important to note, however, that the case studies also demonstrated that even best efforts and capacities do not guarantee success. The book showed that the instrument adoption processes were also driven by endogenous dynamics. Despite the substantial resources the instrument constituencies put on the table, the instrument adoption was surrounded by uncertainties and contingencies. Instrument constituencies act in a political environment that is shaped by competing and conflicting demands. They need to be able to deal with resistance or instrument competitors, mediate between potentially hostile actors (e.g., between various hostile states), find ways to engage with political crises or struggles (such as the illustrated rivalry within the AU Commission in Chapter 4) and navigate the tricky terrain between national and regional preferences as well as their own organizational ambition and capacities.[2] In other words, there is no automatism.

Scholars interested in advancing the notion of 'instrument constituencies' might use these insights for conceptual refinement—in particular to sharpen the notion of 'instrument constituencies' in relation to other concepts such as, for instance, 'advocacy coalitions'.

International relations: provides a new 'social constructivist' account

The study **provides an alternative, more agent-centered 'social constructivist' account of the history of index-based insurance.** By illuminating *how* macro-index insurance have come about and especially the role therein played by development actors, this book addresses a central gap left by Foucauldian-inspired accounts in International Relations (Grove, 2012; Lobo-Guerrero, 2011).

Taking the shortcomings of the Foucauldian accounts on index insurance as inspiration, the book made the political genesis of the macro index insurance schemes explicit. It scrutinized the risk assemblages of "shifting networks of insurers, risk management experts, and state and international agencies" (Grove, 2012, p. 145), identified key actors (e.g., World Bank/WFP-affiliated project team) and illuminated their activities in 'persuading' countries to adopt insurance instruments. In doing so, the book showed that the **adoption of regional catastrophe risk pools was circumscribed by contingencies, contestations and challenges and shaped by other, larger institutional developments.** Regulatory changes with regard to finance and insurance (see Chapter 2), the international institutionalization of disaster management (see Chapter 3), power struggles and practical challenges (see Chapter 4) and the specifics of the global reinsurance market (see Chapter 5) all played their role in the emergence, diffusion and design of these pools.

By locating the cross-regional adoption of index-based insurance instruments in an interplay of range of global and local processes and developments such as exogenous events, policy failures, learning processes, conceptual innovations or sponsoring activities of development actors the book provides arguably a more nuanced account of the rise of these tools. Pointing to complexities, contestations and contradictions the book calls into question the linear tale of a Western (neoliberal) project inscribed in the Foucauldian accounts.

A more direct issue this book raises is whether it is useful to understand the emergence of regional catastrophe risk pools as some kind of cultural preference by Western actors. Particular in the face of climate change and the increase in frequency and severity of disasters, highly exposed countries remain in urgent need of new forms of risk management. There is a growing problem that needs to be addressed.

Financial history: highlights the role of development actors in financialization processes

The study may also be of interest to students of financialization processes working in the fields of Financial History and Critical Global Political Economy (Frieden, 2006; James, Borscheid, Gugerli, & Straumann, 2013; Mader, 2015; Mirowski, 2013)

The study indicates that **there is a role for international financial institutions as well as for more societal agents (such as foundations) in the promotion and expansion of distinctive finance products of late twentieth century**

capitalism. As pointed out previously, in all three cases the adoption of the regional catastrophe risk pools was strongly promoted by development actors. By expanding the insurances sector and thus financial services/insurance-linked securities such as catastrophe bonds from the Northern OECD world to the Southern non-OECD world, one may find further evidence that these actors contribute to nurturing the ongoing *financialization of capitalism* and thus also contributing to fueling the power of the finance sector in the global political economy.

Why is this an important insight? The origin of the expansion of the finance sector is usually located within broader politico-economic changes, such as the demise of the Bretton Woods system, innovations by financial market actors and the rise of neoliberalism as economic thought in the 1970s/1980s (e.g., Frieden, 2006; Mirowski, 2013). Students of financialization processes tend to ascribe (neoliberal) state elites as well as private actors such as banks, hedge funds typically a leading role in promoting the expansion of financial markets. This study points out that international financial institutions (World Bank, Caribbean Development Bank, Asian Development Bank, etc.) as well as more societal or philanthropist agencies (such as the Rockefeller Foundation as financial sponsor of ARC) also play a role in expanding the frontier of finance. The study thus confirms the insights of Mader's study on the political economy of microfinance (Mader, 2015). In his critical politico-economic study Mader argued that NGOs, think tanks, philanthropists, 'social' businesses and (trans)national political bodies such as development banks play a key role promoting micro-finance products in the Global South.

By expanding insurance-linked securities such as catastrophe bonds—distinctive products of late twentieth century capitalism—**international financial institutions** (such as the World Bank) **as well as third sector agencies** (like the Rockefeller Foundation) **contribute to fueling the growth of what might be depicted as "green capitalism"** (Smith, 2011; Sullivan, 2009). From a social-democratic perspective the advent of this type of capitalism is not necessarily bad. In fact, the increasing investment into catastrophe bonds may also be understood as a way to give financial flows a more socio-ecologic purpose. From a Marxist perspective this is, of course, problematic. Whatever normative stance one takes, the promotion of these products does its share to change capitalism and demonstrates that financialization processes and thus the increasing accumulation of power of the finance sector is a *politically* desired outcome. It is not a process simply 'out there' enforced by arcane market actors. There exist *political* financializers.

Governance studies: raises critical questions for accountability students

Moreover, this study **indicates** what might be termed a '**creeping privatization of disaster management**' and calls governance scholars to critically investigate the shifting public-private boundaries and the underpinning accountability mechanism.

The study suggests a transfer of disaster management activities—primarily the organization of funds to cover emergency relief expenditures (but also the organization of disaster-related services such as the provision or risk maps, trainings or disaster plans)—from the national and public to regional and public-private entities; this could be interpreted as the *creeping privatization of disaster management*—creeping because thc regionalization makes it less visible.

Historically, disaster management activities have been a core domain of the modern state and developed largely over the twentieth century[3] (Moss, 2002; Schwartz, 2014). With the rise of regional catastrophe risk pools **part of these 'traditional' disaster management activities are now being outsourced from the public to the private sector via service provision agreements**. While the countries formally 'own' the facilities all insurance and disaster-related services are implemented by private and quasi-autonomous actors, hired by the Governing Board of these facilities on a contract-basis.

The problem is when numerous actors come together it becomes difficult for outsiders (e.g., citizens, civil society, media but possibly also the 'stakeholders', that is the countries themselves) to understand who is responsible for what when it comes to a decision by the Governing Board of the facility. The accountability chain of a regional, public-private insurance entity, an entity that is made up of numerous transnational service provision agreements, is very long and fuzzy. Also for the Governing Boards themselves it might be difficult to assess internal information correctly because of its specialized nature. There is a risk of abuse. Private actors might use CCRIF or ARC to fuel their own business interests by, for instance, successfully persuading the Governing Board to review some of its guidelines, structures and activities and/or invest in different financial market products. Private actors are not philanthropists; they want to make a profit. And they make their profit out of consultancy services in support of developing, installing and maintaining the insurance instrument. Such a risk exists, of course, in all public-private collaboration and various accountability mechanisms have been developed to address this what is commonly described as "regulatory capture".[4] How the accountability mechanisms of 'transnational governance networks' like CCRIF or ARC look like is *terra incognita* and is a subject of investigation students of accountability may wish to pay attention to. Such an investigation might raise interesting follow-up questions such as if the *New Public Management* ideology that apparently underpins regional insurance entities is really a "functionally superior way of organizing activities" (Lodge & Wegrich, 2014, p. 38). This short outline indicates that this might not necessarily be the case. It might simply lead to great legitimacy problems. But again, such claims must be subject to a critical empirical investigation.

Insurance history: explores diffusion across contexts

Finally, **the book speaks also to economic historians exploring** the development and diffusion of various forms of insurance from one context to another and the rise of alternative risk transfer mechanisms associated with natural hazards.

In contrast to most historical studies, the book focused rather on the agents of diffusion than on larger structural developments. More specifically, it dissected the activities by which key promoters—in this case, development actors—inserted cutting-edge insurance instruments into different contexts. It showed how they facilitated the adoption by financing the necessary research, involving (potential) policyholders in product design and by disseminating knowledge and skills on insurance. In doing so, the study hopes to add to a more fine-grained picture of the internationalization of the insurance business. Moreover, the book demonstrated that the expansion of catastrophe insurance from one context to another follows through different organizational vehicles, notwithstanding the fact that the regional catastrophe risk pools are registered in tax havens and insure, at least partly, the same peril (e.g., tropical storms). The factors that determined the organizational form can be found in the specific goals of the risk pool as well as in economic (e.g., operational costs), legal (e.g., inability to do cross-border deals) and political (e.g., to prevent conflicts between interested policyholders) considerations.

The study also indicates that the World Bank is playing a key role as market maker. Through its research activities on the macroeconomic impact of disasters in the late 1990s, the subsequent creation of the GFDRR in the 2000s and as facilitator and promoter of CCRIF and PCRAFI the Bank played a key role in bringing governments and insurance industry together to develop innovative insurance solutions in 'underinsured' regions and countries. With CCRIF, it even developed a new insurance model. By issuing insurance-linked securities, as in the case of CCRIF and PCRAFI, the World Bank directly transferred reinsurance capacity to the global capital markets, offsetting the limits of the traditional reinsurance market. One could argue the Bank has become a key player in shaping the development of catastrophe insurance market. This study indicates that the 'classic' re-insurer (e.g., Munich Re, Swiss Re) acted merely as capital providers rather than as innovators or comprehensive service providers. Yet, only history will tell us if it was part of a larger shift in the actor relationships of the market or a specific 'World Bank moment'. In any case, the study shows that the frontier of insurance is expanding.

Implications and recommendations for policymaking

What are the implications of this study for policy?

Is this a sustainable trend?

The study indicated that the emergence and diffusion of CCRIF, ARC and the PCRAFI Pilot was rather donor-driven than demand-driven. As pointed out previously, development actors created a receptive or enabling environment for insurance 'solutions'. Central to the creation of this enabling environment was the provision of finance (e.g., start-up capital; capital endowments to lower premiums), highly specialized expertise (e.g., studies, analyses) and insurance literacy (e.g., transfer of knowledge/skills/competences to service insurance schemes) at the senior policymaker level. **One implication thereof is that the trend of setting up regional catastrophe risk pools and buying insurance might be more fragile than it appears from the outset.** It might depend on the facilitating activities from development actors. In fact, it might **depend to a substantial degree on the ability and political willingness of the World Bank to assemble the necessary resources and skills.**

As this study demonstrated, the World Bank played a key role in the emergence of the catastrophe risk pools here, as the bank used its convening power to assemble the necessary resources and skills to facilitate the development of CCRIF and PCRAFI. In addition, through the experience gained in facilitating the establishment of two regional catastrophe risk pools, the World Bank positioned itself as a strategically indispensable actor or future 'partner' when it comes to setting up new regional catastrophe risk pools. This provides the World Bank with a unique power to shape the future of regional catastrophe risk pools. In this context it is important to keep in mind that the World Bank is not a neutral, 'interest-free' actor. The World Bank is not an impartial humanitarian actor acting needs-based but an actor with an economic mandate and its own agenda. Its core mandate is securing economic development and not saving lives. To its key activities belongs the financing of economic reconstruction. Accordingly, the Bank is interested in mitigating the impact of disasters on local economic activity. This economic approach is also evident in the goal and subsequent modeling that characterize the two catastrophe risk pools that have been facilitated by the World Bank. The underpinning catastrophe models weigh 'economically productive' areas (e.g., commercial centers; agricultural areas) rather than 'poor' and thus highly vulnerable areas (e.g., urban slums; smallholder farms). A focus on macroeconomic dimensions of disasters is, of course, legitimate and important but it has its own limitations when it does not account for the social stratification of the impact of disasters.

Are regional catastrophe risk pools good or bad?

The study indicates that regional catastrophe risk pools work—in principle, but are they 'good' tools? An answer is not straightforward and depends to a substantial degree on the perspective and the underpinning normative framework.

From a Marxist international relations perspective, for instance, one may reject regional catastrophe risk pools straight away. First, they contribute *to geographically expanding the frontier of finance capitalism* from North to South. By tapping new insurance markets in the Global South, they contribute to expanding the power of finance. Even more capital is poured into the finance world. Second, the regional catastrophe risk pools must be understood as investment houses that invest the annual premium payments—even they are little in volume—in sovereign and corporate bonds issued in Wall Street and the City of London. This means that the premium payments finance 'Western' political economies. Given that the premiums are paid by poor states, this represents an outflow of capital from South to North and reproduces structural global economies inequalities—at least, as long as it does not come to a payout. Third, the regional catastrophe risk pools suffer from accountability problems. It requires substantial expertise to understand who is responsible for what when it comes to key decisions. This makes the articulation of (productive) criticism difficult.

From a realist state-centric public policy perspective, one may take a more favorable stance instead. The insurance instruments provide states with a 'right' to compensation and thus with urgently needed funds when it comes to a disaster. The quick access to funds allows for a more timely public intervention. If spent properly, the funds may contribute to saving life and helping to restore 'normality'. A guaranteed payout is better than the predominant 'wait-and-see' strategy that, for various reasons, characterizes post-disaster financial planning of many countries from the Southern hemisphere. It is in this sense that the insurance offers to improve the *status quo*.

While I do sympathize with the more critical IR perspective, I have ultimately a rather pragmatic view on the insurance. In my opinion, **regional catastrophe risk pools are not an evil *per se* and should not be rejected dogmatically**. Overall, this research has demonstrated that **insurance instruments work *in principle*. Yet, I do see various problems with *how* they work**. There is room for improvement.

Policy recommendations: how to improve regional catastrophe risk pools

Taking some of the insights of this study as background, I will conclude by highlighting some of the key problems and outlining policy recommendations in response to them.

Address climate justice: provide full premium support for poor countries

My strongest criticism relates to premium financing. While G20 countries and (multilateral) development actors covered the start-up costs, local governments financed the annual premiums largely out of their own annual budget. This means that, in their current form, the regional catastrophe risk pools require those who are the least responsible for climate change and the least able to finance the premiums to pay the final bill. From a climate justice perspective this is unfair. The G20 produces over 80% of global GDP and is responsible for about three quarters of global Greenhouse Gas (GHG) emissions. It therefore has a special responsibility to deliver support to the most vulnerable. Rather than requiring the most vulnerable to pay, I argue that insurance schemes should be based a "combination of what is known as the 'polluter pays principle' and the 'ability to pay principle'" (Duus-Otterström & Jagers, 2011, p. 8).

Championing this equity concern translates into the concrete policy recommendation that the G20 and insurance promoters such as the InsuResilience Global Partnership should fully cover the insurance premiums or, at least, subsidize the premiums in a way that they address both affordability *and* equity concerns. Premium subsidies by G20 countries could be related, for instance, to a country's greenhouse gas footprint.

Any premium support should thereby be designed in a way to minimize potential distorting effects (see also MCII, 2016, p. 114). Premium support may lead to moral hazard and to non-action with regard to the implementation of risk reduction measures. Premium support may be granted, for example, on the condition that a government implements disaster risk reducing measures to improve its national disaster management (e.g., adoption of disaster management plan, implementation of a disaster education program, development and enforcement of building codes, tax waiver for those who move out of risky areas). The application of the conditionality principle would incentivize administrative action to save lives and assets. Premium support may also be designed in a more indirect way to prevent distortions: the G20 and the InsuResilience Global Partnership may endow the insurance facilities with additional capital reserves so that they can push up the self-retention level. A higher retention level would ultimately translate into cheaper reinsurance costs for the facility and, thus, also into lower premiums for the countries.

Encourage the disbursement plans and integrate them into an overarching disaster financing strategy

My second criticism refers to the payout modality. Experience from the disaster management field indicates that it is not so much finance *per se* but early, quick and coordinated action that saves lives and livelihoods (Clarke & Dercon, 2016). If the larger purpose of regional catastrophe risk pools is to reduce climate change related disaster impacts, I do think the pools should provide strong incentives for disaster planning and encourage the development of disbursement plans that clarify the usage of insurance payouts *ex ante*.

This is not the case with CCRIF and the PCRAFI Pilot, as highlighted in this book. In contrast to ARC, CCRIF and the PCRAFI Pilot do not tie the payouts to disbursement plans. Instead, the arrangements are based on a *laissez-faire* philosophy that put emphasis on speed and flexibility. This may reflect the US regulatory culture and, indeed, has certain advantages; yet, following Clarke and Dercon (2016), I claim that even a good payout cannot replace a good disbursement plan. For *effective* insurance arrangements—insurance arrangements that work to smooth the impacts of natural hazards—it requires a plan for how to use the payout.

As claimed by Clarke and Dercon, "Just because catastrophe reinsurance has been bought or a large contingency fund has been set up does not mean it is the right solution for the circumstances and the organization involved" (2016, p. 77). Insurance provides only the financial means. A quick payout is of limited use if the response is too late due to domestic coordination failures and/or if the payout is spent poorly. Without plans there are the risks of political fraud, ambiguity about who is in charge of what, delays in response, etc. "The post-disaster decision-making process is far too politicized, leading to delays, poor decisions, and bad coordination of effort" (ibid., p. 4). Hence, "The way forward is to act before disasters strike, preparing credible plans with rules-based decision-making and early action and held together with sound financial planning agreed beforehand" (ibid. p. 11).

Disbursement plans should comprise "pre-agreed objectives, decision processes and implementation modalities" (ibid., p. 78). They should clarify who does what, when, how. In other words, they clarify aims and responsibilities to allow for quick and concerted action. The development of such a plan, including finding the most appropriate activities, might not be easy. ARC, however, provides a 'good practice' example for the development of disaster disbursement plans that gives necessary room for maneuver.

In that sense, I claim that CCRIF and the PCRAFI Pilot could be strengthened by taking inspiration from ARC. The development of disbursement plans should also be easier in the Caribbean context than African context, where some countries such as Jamaica have already a fairly sophisticated disaster risk management system in place. Here, the political challenge lies rather in setting incentives for countries to integrating the insurance payout in an overarching national disaster financing strategy. The payouts of CCRIF and PCRAFI are fairly small. As such, there is only a minor role to play for insurance and, yet, there is a role and that role has to be made explicit to improve effectiveness.

Insurance promoters such as Global InsuResilience Partnership could act as facilitator and develop knowledge platforms to facilitate cross-regional and cross-national learning on how to best use insurance payouts and where to place insurance in an overarching disaster financing strategy. Humanitarian actors such as the Red Cross, Welthungerhilfe or Oxfam have tremendous expertise and experience in providing aid.

CCRIF and the PCRAFI Pilot could also extend their services to assist local governments in identifying which critical infrastructures (e.g., key roads, bridges, hospitals) could be rebuilt with a payout (cf. Clarke & Dercon, 2016, p. 57); they could assist governments to identifying the agency who should take over the rebuilding. If the regional catastrophe risk pools get involved in post-disaster planning they would need to make sure, however, that the voices of those who are most affected by disasters, the poorest segments of society, are heard – that their needs are incorporated in the disaster plan. Finally, to secure a successful implementation it requires functioning distribution systems. The payouts must reach those they should reach.

Integrate disaster management specialist in the governance structure

My third point of criticism relates to the governance structure. As has come out of this study, CCRIF, ARC and the PCRAFI Pilot provide much more than insurance. They provide countries with access to disaster-related technology and services. CCRIF has even become a hub for regional disaster management. While this development is certainly positive given the rising number of natural catastrophes, it is notable that disaster management specialists are (still) sidelined within the governance structure of these facilities. CCRIF, ARC and the PCRAFI Pilot are largely run by finance and insurance specialists. In other words, financial interests and considerations tend to be more dominant in shaping the overall strategic direction of the facilities and designing the insurance policies. I think the Governing Board of CCRIF and ARC should give other concerns a greater voice and integrate disaster management specialists.

An over-reliance on financial experts limits the potential of these insurances to strengthen disaster risk reduction. Finance or insurance experts have their place but they are simply not disaster experts. If the broader goal of the insurances is to progressively limit the impact of disasters and enhance the resilience of countries, they should ensure that those who have the appropriate expertise are given a greater say. To argue for a revision of the governance structure is strongly related to my previous recommendation to link insurances more systematically to disbursement plans and national disaster risk governance. Such efforts require the necessary expertise.

The proposed revision in the governance structure of the catastrophe risk pools should also be mirrored at the country level. As highlighted by this study, currently the finance ministries are at the forefront of funding coordination when it comes to an insurance payout. This is somehow at odds with the proclaimed goal of the pools. To accelerate the disaster response, those coordinating the efforts—typically national disaster risk management agencies—should receive the payouts. If this, for political reasons, is not feasible specific then efforts should be undertaken to integrate the work of disaster risk management authority and the finance ministry. A possibility could be the formation of standing committees that involve representatives from the ministry of finance and the disaster management agency. The committee could coordinate disaster-related activities. A real world example for such an inter-agency committee would be the National Disaster Risk Reduction and Management Council in the Philippines. The committee pools various expertise, including expertise from civil society and the private sector, and guides disaster-related activities.

Establish a monitoring and evaluation system

My fourth and last point of criticism relates to the evidence gap and the largely uncritical endorsement of the macro-index insurance schemes. Little is known about the actual performance of the insurance. Setting up a systematic monitoring and evaluation framework component has not been a priority for the instrument constituencies so far. This is somehow remarkable given the tremendous efforts and resources they put into setting up these instrument. Whatever the precise reason might have been for not setting up a systematic monitoring and evaluation framework, the lack of evidence regarding the *effectiveness* of the insurances (including with regard to their secondary goals) and their *impact* is highly problematic.

From this study we know that insurance schemes work *in principle*: The regional catastrophe risk pools provide cash-stripped countries quickly with limited funds in the aftermath of a disaster. As of December 2018, they have made 44 payouts to 19 countries totaling to about US$ 173 million. This is certainly an achievement. But are these payouts adequate and effective? The payouts must not only be timely but also at the right scale and reach those they want to reach to be beneficial. In short: they must be adequate and effective.

This study indicates that there is the risk that the insurances are not adequate – that they do not provide sufficient cover. The following table compares some payouts vs the countries' humanitarian appeals launched via the UN system:

Table 7.1 Payouts vs humanitarian flash appeals

Country	*Peril/Year*	*Funding requirement*	*Payout*	*Share of funding requirement*
CCRIF				
Haiti	Earthquake 2010	US$ 1502 million	US$ 7,7 million	0,5%
Haiti	TP Matthew 2016	US$ 159,4 million	US$ 20,4 million	12,7%
Dominica	TP Maria 2017	US$ 50,3 million	US$ 19,2 million	38%
ARC				
Niger	Drought	US$ 264 million	US$ 3,5 million	1,3%
Senegal	Drought 2015	US$ 75,9 million	US$ 16,5 million	21%
PCRAFI				
Vanuatu	TP Pam 2015	US$ 37,2*million	US$ 1,9 million	5,1%
Tonga	TP Gita 2018	US$ 13,8* million	US$ 3,5 million	5,1%

Source: data taken from OCHA, 2019

* Given that humanitarian appeals have been formulated *after* the notification of a payout, I added the payout to the UN appeal to capture the total immediate funding requirements. I did not add payout in the case of Vanuatu and Tonga since available information captures received funding. I assumed received funding included the payout.

Although the insurance schemes are by design limited in scope, the table makes visible that payouts accounts for a relatively small share of the immediate financing requirements after a disaster; moreover, it shows that this share varies tremendously (0,5%–38%) across the various countries. It is not clear if payouts were really beneficial to those in need and if their quick disbursement (within 14 days)—the substantial advantage vis-à-vis appeals—really offsets this limitation. Unfortunately, there exists also no systematic information if the insurance payouts actually reached their target beneficiaries (e.g., disaster-affected households); that is, if they were spent on what they were supposed to be spent on to smooth disaster impacts. So far, the CCRIF and the PCRAFI Pilot have no tracking system in place; instead they trust the governments. ARC, has a limited monitoring and evaluation system in place but did not collect systematically data to assess ARC's effectiveness. Also here ARC seems to trust government reports.

It is also not clear if the models that underpin CCRIF, ARC and the PCRAFI Pilot are really accurate. This study indicates that the models used have a tendency to underestimate disaster-related costs. In at least five cases (Jamaica 2007, Belize 2013, Solomon Islands 2013 and 2014, Malawi 2016), no (immediate) payouts were triggered despite there actually being substantial damage. The most prominent case is probably the (initial) non-payout to Malawi (see Chapter 4). Only in one case (Dominica 2007) did a country benefit from a payout despite there being very limited damage. This basis risk is inherent to index-based insurances, as highlighted in Chapter 2. Index-based insurance indemnifies insurance policyholders based on modeled losses and not real losses. This means that payouts are only as accurate as the models and, by implication, the assumptions that underpin them and the data that is fed into them. In this context it is also important to keep in mind that the models are not apolitical. Definitions of who counts as 'poor' and assumptions of the reference crop or replacement are far from technical but deeply political questions: They have very real consequences for the people on the ground. They affect the payout level—if it comes at all to a payout. These assumptions must be made explicit, regularly updated to reflect changes and clearly communicated.

Adequacy and accuracy issues aside, the insurance schemes are also likely to have (negative) unintended consequences. They are instruments that distribute resources. Whenever resources are distributed, they affect political dynamics. This is especially true for fragile and conflict-affected contexts. A lot of them are to be found in Africa. When insurance payouts are distributed without greater attention to existing divisions and tensions, they might reinforce them and, in doing so, contribute to social conflicts, even violence. A distribution according to a beneficiary list defined by a tribal or political leader or a security force, rather than transparent and neutral selection criteria, might create new dependencies and facilitate corruption. So far, these impacts have not been captured by the insurers. To acquire knowledge of them is, however, of great political relevance.

There is also risk that the insurance schemes crowd out other instruments. On the international level, the availability of 'disaster risk insurance' might prevent an agreement on a legally binding 'loss and damage' compensation mechanism in the climate negotiations. As the industrialized countries support 'disaster risk insurance', they have a good excuse to continue to further block the debate in the context of the Warsaw Mechanism on compensation payments to the non-industrialized world for climate change-related losses. They can claim they already help these countries to cope with climate risks, without committing to legally binding compensation payments. While beneficial to the non-industrialized world, this would be, from a global climate justice perspective, a bad outcome for the non-industrialized world.

On a national level some studies on micro index insurance field already indicate that they crowd out traditional coping mechanisms (Müller, Johnson, & Kreuer, 2017). Others show also that they incentivize the cultivation of (insured) cash crops which leads to monocultures (Cole, Giné, & Vickery, 2017). 'Disaster risk insurance' might also reinforce inequalities. The payouts are not gender-sensitive despite the fact that women are typically much more affected from the impacts of disasters than men. Undifferentiated payouts might exacerbate these problems. 'Disaster risk insurance' might also contribute to worsen conflicts, fraud and corruption. No safeguards have been developed to make sure that payouts do not end up with warlords, corrupt politicians or administrators and, in doing so, trigger or exacerbate political tensions and conflicts on the ground. Moreover, the availability of an insurance solution might also create a false sense of security. If countries are not aware of the inherent limitations of the insurance provided by the regional catastrophe risk pools, they might reduce or even give up on their emergency budget. As a result they might end up in a situation where the funds coming from the payout do not match their original emergency budget. In short: There is also a risk that the insurance schemes make people and governments worse-off.

All these potential adverse consequences serve to make a case for setting up a systematic and comprehensive monitoring system and to build up an evidence base to systematically evaluate the usefulness of regional catastrophe risk pools. The importance of setting up a comprehensive monitoring system becomes even more evident if one considers that only in November 2017 the Global InsuResilience Partnership has been provided with US$ 550 million to expand climate risk insurance.[5] According to a recent study by Oxfam, the sum is as much as international donors spend on disaster risk reduction measures every year (Oxfam, 2018, p. 6). It is not clear if this money is well spent given other instruments.

A final reminder: there are always alternatives

Against the backdrop of the modest reform proposals being made, we must also not forget that there are always alternatives to insurance: Insurance is not a panacea. The fact that some actors strongly promoted insurance instruments in the first place without seriously considering alternatives, especially non-financial instruments, should worry us. There is a risk that insurance is neither cost-effective nor really support those they actually aim to serve: the poorest.

I believe insurance can help to address the impacts of disasters but it should be coupled with other measures. Admittedly, leading actors in the disaster finance field—among them the World Bank, InsuResilience Global Partnership and the Munich Climate Insurance Initiative—have increasingly pointed out that insurance 'is not a silver bullet' and can only be a 'part of the solution' to deal with disasters. Yet, this is of little value when, at the same time, the very same actors put so much effort into expanding these insurances.

I do understand that the policy world cannot (and should not) wait until everything has been weighted and checked for by science. Yet, I do think it is also important to look beyond financial solutions and consider, for example, a proper international human 'right' to aid under the Humanitarian Law or a GHG-emission based tax to fund climate change adaptation (especially nature-based 'no-regret' solutions, such as the planting of mangroves or afforestation in risk zones) and disaster risk reduction measures. Developing or scaling-up social safety programs in a coordinated manner could be another alternative. Only in 2017, a meta-evaluation study investigating 27 safety net programs in 14 African countries found that these programs have significant positive impacts on total food consumption and show promising results on asset accumulation, such as livestock ownership (Ralston, Andrews, & Hsiao, 2017). Expanding existing safety programs horizontally (by increasing the benefit value/duration of an existing social protection program) or vertically (by adding new beneficiaries) could be viable alternatives to support people through disasters (Oxford Policy Management, 2017). The goal of all of these measures must be to address the *structural* drivers of vulnerability and they vary from country to country. Some of these proposed measures might not be politically viable options at this point in history. Yet, if they are the better solutions vis-à-vis insurance instruments, they should be seriously explored, discussed and developed.

A critical evaluation of the *effectiveness* and the *impact of* current insurance solutions, that is, an evaluation that goes beyond a mere cost-benefit analysis and assesses them in relation to other *non-financial* instruments would be a small but first step in the right direction. It would allow for a more balanced and evidence-based discussion. From an international development perspective, the current macro-index insurances must be beneficial for the poor; if they are not, there is little point in sustaining them. I claim their benefit has yet to be demonstrated.

Notes

1 By "delivery capacity" I refer to the capacity "to make things happen" (Lodge & Wegrich, 2014, p. 36).
2 In the Governance Report 2014, Wegrich and Lodge describe key challenges of a "delivery capacity" of an administration, among them "the capability to navigate the tricky terrain between local demands, national priorities and organisational ambitions" (Lodge & Wegrich, 2014, p. 38).
3 Moss speaks of "risk management". Risk Management comprises "any government activity designed either to reduce or to reallocate risk" (Moss, 2002, p. 1). Disaster Management is but one area of risk management.
4 Simply put, regulatory capture refers to a failure of a public entity to act in the public interest as private actors charged with regulatory tasks advance their own interest at the expenses of public actor interests. For an overview see (Moss & Carpenter, 2014).
5 The US$ 715 million includes the start-up capital provided for ARC as well as capital to expand micro-insurances. The mandate of the InsuResilience Secretariat is to expand climate risk insurance coverage for 400 million people by 2020: www.insuresilience.org/.

References

Béland, D., & Howlett, M. (2016). How solutions chase problems: Instrument constituencies in the policy process. *Governance*, 29(3), 393–409. http://doi.org/10.1111/gove.12179

Boscarino, J. E. (2009). Surfing for problems: Advocacy group strategy in U.S. forestry policy. *Policy Studies Journal*, *37*(3), 415–434. http://doi.org/10.1111/j.1541-0072.2009.00321.x

Bulmer, S., & Padgett, S. (2005). Policy transfer in the European Union: An institutionalist perspective. *British Journal of Political Science*, *35*(1), 103–126. http://doi.org/10.1017/S0007123405000050

Capano, G., & Lippi, A. (2013). *Enlightening policy mix complexity: A typology of actors' choices*. Conference Paper, Grenoble. Retrieved from www.icpublicpolicy.org/IMG/pdf/panel_11_s3_capano.pdf

Clarke, D., & Dercon, S. (2016). *Dull disasters? How planning ahead will make a difference*. Oxford: Oxford University Press.

Cole, S., Giné, X., & Vickery, J. (2017). How does risk management influence production decisions? Evidence from a field experiment. *The Review of Financial Studies*, *30*(6), 1935–1970. http://doi.org/10.1093/rfs/hhw080

Dolowitz, D. P., & Marsh, D. (2000). Learning from abroad: The role of policy transfer in contemporary policy-making. *Governance*, *13*(1), 5–23. http://doi.org/10.1111/0952-1895.00121

Duus-Otterström, G., & Jagers, S. C. (2011). Why (most) climate insurance schemes are a bad idea. *Environmental Politics*, *20*(3), 322–339. http://doi.org/10.1080/09644016.2011.573354

Frieden, J. (2006). *Global capitalism—Its fall and rise in the twentieth century*. London and New York: W. W. Norton.

Grove, K. (2012). Preempting the next disaster: Catastrophe insurance and the financialization of disaster management. *Security Dialogue*, *43*(2), 139–155. http://doi.org/10.1177/0967010612438434

Hall, P. (1993). Policy paradigms, social learning, and the state: The case of economic policymaking in Britain. *Comparative Politics*, *25*(3), 275–296. http://doi.org/10.2307/422246

Hood, C. (1983). *The tools of government*. London: Palgrave Macmillan.

Hood, C., & Margetts, H. Z. (2007). *The tools of government in the digital age*. Basingstoke: Palgrave Macmillan.

Howlett, M. (1991). Policy instruments, policy styles, and policy implementation: National approaches to theories of instrument choice. *Policy Studies Journal*, *19*(2), 1–21. http://doi.org/10.1111/j.1541-0072.1991.tb01878.x

Howlett, M. (2011). *Designing public policies: Principles and instruments*. London: Routledge.

James, H., Borscheid, P., Gugerli, D., & Straumann, T. (2013). *The value of risk*. Oxford: Oxford University Press.

Jordan, A., & Huitema, D. (2014, November). Innovations in climate policy: The politics of invention, diffusion, and evaluation. *Environmental Politics*, 1–20. http://doi.org/10.1080/09644016.2014.923614

Kingdon, J. W. (1995). *Agendas, alternatives and public policies*. New York: Longman.

Lascoumes, P., & Le Gales, P. (2007). Introduction: Understanding public policy through its instruments? From the nature of instruments to the sociology of public policy instrumentation. *Governance*, *20*(1), 1–21. http://doi.org/10.1111/j.1468-0491.2007.00342.x

Linder, S., & Peters, G. (1998). The study of policy instruments: Four schools of thought. In G. Peters & F. K. M. van Nispen (Eds.), *Public policy instruments: Evaluating the tools of public administration*. Cheltenham: Edward Elgar Publishing.

Lobo-Guerrero, L. (2011). *Insuring security: Biopolitics, security and risk*. Abingdon and New York: Routledge.

Lodge, M., & Wegrich, K. (2014). Administrative capacities. In *The governance report 2014* (pp. 27–48). Oxford: Oxford University Press.

Mader, P. (2015). *The political economy of microfinance*. New York: Palgrave Macmillan. http://doi.org/10.1057/9781137364210

MCII. (2016). *Climate risk insurance for the poor and vulnerable*. Bonn. Retrieved from www.climate-insurance.org/fileadmin/mcii/pdf/COP-21/G7_InsuResilience_MCII_Factsheet_SuccessFactors___EnablingConditions.pdf

Mirowski, P. (2013). *Never let a serious crisis go to waste*. New York: Verso.

Moss, D. (2002). *When all else fails*. Cambridge, MA: Harvard University Press.

Moss, D., & Carpenter, D. (2014). Introduction. In *Preventing regulatory capture: Specal interest influence and how to limit it* (pp. 1–23). New York: Cambridge University Press.

Mukherjee, I., & Howlett, M. (2015). Who is a stream? Epistemic communities, instrument constituencies and advocacy coalitions in public policy-making. *Politics and Governance*, *3*(2), 65–75. http://doi.org/10.17645/pag.v3i2.290

Müller, B., Johnson, L., & Kreuer, D. (2017, June). Maladaptive outcomes of climate insurance in agriculture. *Global Environmental Change*, *46*, 23–33. http://doi.org/10.1016/j.gloenvcha.2017.06.010

OCHA. (2019). *Financial tracking service*. Retrieved September 1, 2019, from https://fts.unocha.org

Oxfam. (2018). *Facing risk: Options and challenges in ensuring that climate/Disaster risk finance and insurance deliver for poor people*. Oxford. Retrieved from www.oxfam.org/en/research/facing-risk-options-and-challenges-ensuring-climatedisaster-risk-finance-and-insurance

Oxford Policy Management. (2017). *Shock-responsive social protection systems research: Literature review*. Oxford. Retrieved from www.opml.co.uk/sites/default/files/Shock_responsive_social_protection_Literaturereview_EN.pdf

Ralston, L., Andrews, C., & Hsiao, A. (2017). *The impacts of safety nets in Africa: What are we learning?* (Research Working Paper No. 8255). Washington, DC.

Salamon, L. M. (2002). *The tools of government—A guide to the new governance*. Oxford: Oxford University Press.

Schwartz, H. (2014, November 11). *Whistling up a storm: Toward a history of emergency*. Presentation held at the Amercian Academy Berlin, Berlin.

Shorten, G., Goosby S., Granger, K., Linday, K., Naidu, P., Oliver S., Walker G. (2003). *Catastrophe insurance pilot study, Port Vila, Vanuatu: Developing risk management options for disasters in the Pacific region – Volume 1* (SOPAC Joint Contribution Paper No. 147 prepared for World Bank Office Sydney, and AusAid, Canberra, Australia). Retrieved from https://www.gfdrr.org/en/publication/catastrophe-insurance-pilot-study-developing-risk-management-options-disasters-pacific

Smith, R. (2011, January). Green capitalism: The god that failed. *Ecology*, 112–144. Retrieved from www.climateandcapitalism.com/wp-content/uploads/2011/05/Green-capitalism-the-god-that-failed.pdf

Sullivan, S. (2009). Green capitalism, and the cultural poverty of constructing nature as service provider. *Radical Anthropology*, *3*, 18–27. http://doi.org/10.1170/T908

Voß, J. P., & Simons, A. (2014). Instrument constituencies and the supply side of policy innovation: The social life of emissions trading. *Environmental Politics*, *23*(5), 735–754. http://doi.org/10.1080/09644016.2014.923625

Wurzel, R., Zito, A., & Jordan, A. (2013). *Environmental governance in Europe*. Cheltenham and Northampton, MA: Edward Elgar Publishing. http://doi.org/10.4337/9781849804721

Index

Note: Page numbers in *italics* indicate figures; page numbers in **bold** indicate tables.

For Product Safety Concerns and Information please contact our EU representative GPSR@taylorandfrancis.com
Taylor & Francis Verlag GmbH, Kaufingerstraße 24, 80331 München, Germany

www.ingramcontent.com/pod-product-compliance
Lightning Source LLC
LaVergne TN
LVHW011119300926
849160LV00008B/441
* 9 7 8 1 0 3 2 1 7 3 5 7 3 *